Current Topics in Microbiology 141 and Immunology

Mechanisms in B-Cell Neoplasia 1988

Workshop
at the National Cancer Institute,
National Institutes of Health,
Bethesda, MD, USA, March 23–25, 1988

Organized and Edited by
M. Potter and F. Melchers

With 122 Figures

Springer-Verlag
Berlin Heidelberg New York
London Paris Tokyo

Michael Potter, M.D.
Laboratory of Genetics
National Institutes of Health
National Cancer Institute
Bethesda, MD 20892
USA

Prof. Dr. Fritz Melchers
Basel Institute for Immunology
Grenzacherstr. 487
CH-4005 Basel
Switzerland

ISBN-13: 978-3-642-74008-4 e-ISBN-13: 978-3-642-74006-0
DOI: 10.1007/978-3-642-74006-0

Library of Congress Catalog Card Number 15-12910
Softcover reprint of the hardcover 1st edition 1988

2123/3130-543210 – Printed on acid free paper

Preface

The papers in this book were presented at the 6th Workshop on Mechanisms in B-Cell Neoplasia, held in Bethesda, March 23-25, 1988. On alternate years this meeting is sponsored by the Basel Institute of Immunology in Basel, Switzerland and by the National Cancer Institute in Bethesda, and is attended by 100 to 150 participants. This 6th workshop, like the preceding five, was characterized by intense and enthusiastic discussion which reflects, we think, the exciting growth and development of this field. It is quite clear, however, that despite many general advances an understanding of the precise underlying mechanisms in B-cell tumor development is not yet defined.

Probably, there is no single mechanism for all the various forms of B-cell neoplastic development. Many different forms of B-cell neoplasms are known, and these are distinguished by several characteristics: 1) the stage of development attained by the tumor stem cells; 2) mode of growth (slow or fast); 3) association with natural or inductive etiologic agents and 4) specific and consistent mutational mechanisms such as retroviral insertion, chromosomal rearrangement. Those characteristic forms which arise naturally in relatively high frequency or those tumors with hallmark properties which can be induced consistently are the models most frequently studied, e.g., endemic Burkitt's lymphoma, follicular lymphoma, acute and chronic lymphocytic leukemia and multiple myeloma in man; bursal lymphoma in chickens; Abelson virus induced pre B cell lymphomas and plasmacytomas in mice and immunocytomas in rats. Each model system, has special problems and advantages.

The processes of retroviral insertion and transduction and chromosomal rearrangement have facilitated the identification of host genes that are consistently mutated in B-cell tumor formation. C-myc, c-abl, bcl-2, c-raf, and c-myb are examples of genes that are consistent targets of mutagenesis in one or more forms of B-cell neoplasia. Understanding the normal physiological function of these genes presents a major area of research as the biochemical mode of action of the respective gene products are at present poorly understood. The mutant forms of these genes affect functions related to the regulation of cellular proliferation and differentiation.

It is widely accepted that neoplastic development in general results from multiple genetic changes and B-cell tumors are no exception. There is very little information available on the identity of multiple mutations in a single naturally occurring neoplastic form. There are inductive methods in which two oncogenes accelerate the development of B cell tumors. In pristane injected mice, plasmacytomas can be induced rapidly when transforming retroviruses are introduced as with Abelson virus which introduces v-abl. J3 (virus) that contains an avian myc and a defective v-raf gene or RIM virus that contains a myc gene under the control of the heavy chain enhancer and promoters and a V-Ha ras gene, are potent inducers of plasmacytomas in pristane conditioned mice.

The E μ-myc transgenic mice have provided a valuable system for studying the cooperative action of other oncogenes. In this model system the plasmacytomas all possess chromosomal translocations that are associated with regulatory disorders of c-myc. In other inductive methods, two oncogenes are introduced simultaneously in the same retrovector. Cooperative effects have been described utilizing a variety of ways of introducing the second oncogene: by infection with a transforming retrovirus or by crossing two different transgenic mice carrying different oncogenes. C-abl, c-raf and ras emerge as important cooperating oncogenes. Products of these oncogenes are known to be components of growth factor signal induction pathways. The recent finding of a special mode of c-abl activation in B-ALL in man by t9;22 chromosomal translocations further emphasizes the importance of the abl gene in a natural form of B-cell neoplasia (B-ALL). Rather surprisingly a new transgenic C57BL mouse carrying N-myc alone develops B-cell tumors with plasma cell characteristics.

Many defined growth factors affect normal B cell development and new factors continue to be identified. Different growth factors act on different stages in B cell development. The most recent addition to this list of factors is IL-7. Suspicion of this kind of factor was strongly suggested by the studies of Osmond et al., who have shown various peripheral stimuli (e.g., infections, injection of mineral oil) send stimuli to the bone marrow which stimulate expansion of the pre B population.

IL-7 is normally produced by stromal cells in interactions with cells at early stages of B cell development. The recombinant protein has been found to expand precursor B cells. Since IL-3, in combination with IL-4, also appears to stimulate pre B cell growth, the possibility exists that pre B cells could alternatively use either of the growth factors, one produced endogenously by stromal cells (IL-7), the others under exogenous stimulation of helper T cells (IL-3, IL-4). Studies now in progress with the recently described IL-6 have shown that it is a growth factor for B-cell proliferation.

Many of the papers in this book focus on the oncogenes that are important in B cell tumor formation. Of these c-myc stands out as the central problem. Four current sets of experiments on c-myc described in this book deal with aspects of the problem: 1) the mechanism of c-myc activation by cytoplasmic signal pathways; 2) the nuclear regulators of c-myc transcription; 3) the mode of action of the c-myc gene product; 4) the pathological mechanism of myc gene activation. The pathological mechanism of c-myc activation is still not clearly defined. There are many ways by which c-myc is pathologically activated.

The c-myc gene is normally translated into two proteins, p64 and p67, which are made by transcription from the initiation sites, followed by the appropriate translation of the two resulting mRNAs. One common pathological event is the mutation of the CTG start codon which abolishes the expression of p67, resulting in unbalanced p64 expression. Ways to deregulated expression of c-myc are the many point mutations in the coding regions of the c-myc gene which have so far been analyzed. Higher levels of p64 and p67 expression can also deregulate cellular growth, and this, again, can happen by point mutation in regulatory sequences of the gene.

A major problem is presented by the chromosomal translocations that occur near but not within the domain of the c-myc gene (e.g., the Pvt-1/IgL translocations t(2;8), t(8;22) in man, rcpt(6;15) in the mouse) and the forms of t(8;14) in endemic Burkitt's Lymphoma that occur 5' of the known regulatory sequences in c-myc. Most of these later types remain to be mapped. The above mentioned translocations appear to activate c-myc indirectly through base substitution mutations. The underlying mechanism for this is not yet known.

The regulation of c-myc transcription by nuclear factors is currently being worked on in several laboratories. A number of potential factor binding sites have been identified extending from intron-1, 5' to base -2000. Future studies will be required to evaluate how factors interact with these sites and with each other to regulate transcription.

The cytoplasmic signal pathway leading to c-myc activation has been studied in many cell types including B-cells. Here again, regulation of c-myc transcription may vary in different cell types as well as with the stage of development of the B-cells.

With the availability of data from a variety of sources, some clarity is emerging in all of these areas. New concepts of c-myc function are still being described. For example, Sullivan, et al., have described the activation of c-myc in peripheral blood lymphocytes following radiation damage.

Many of our current models and hypotheses on oncogene action come from transformation studies using cultured cell lines. The process of in vitro adaptation probably depends in part upon some of the same kinds of mutational or adaptive changes that occur in neoplastic development. For example, many mouse cell lines spontaneously become neoplastic in vitro. Neoplastic transformation in vivo is a more complex and difficult problem to analyze but it is becoming increasingly more clear that mutations that activate proto-oncogenes produce biological effects that do not directly transform cells from a normal to a neoplastic state. Thus many problems involve interpreting the biological effects of pathological oncogene activation.

We thank the National Cancer Institute for sponsoring this meeting and Prof. Dietrich Goetze for his willingness to publish these papers. We are very grateful to Ms. Victoria Rogers for her help in organizing this meeting and getting the documents ready for publication.

Michael Potter

Fritz Melchers

Table of Contents

Part I: Pathogenetic Mechanisms

Early B Cell Tumors

B Cell and Plasma Cell Tumors

Part II: Studies of B Cell Relevant Oncogenes

c-*myc*

List of Contributors

You will find the adresses at the beginning of the respective contribution

Part I:
Pathogenetic Mechanisms

Early B Cell Tumors

B Cell Precursors in Bone Marrow: In Vivo Proliferation, Localization, Stimulation by Activated Macrophages and Implications for Oncogenesis

D. G. Osmond, Y.-H. Park[1], and K. Jacobsen

INTRODUCTION

Precursor cells in mammalian bone marrow continuously give rise to virgin B lymphocytes by undergoing a series of genetic, mitotic and selection events (Osmond 1985, 1986, Osmond & Park 1987). During the differentiation of individual precursor cells the B cell antigen-binding specificities are determined by the formation of functional Ig variable (V) region genes from a combination of gene segments (Tonegawa 1983). The enzyme terminal deoxynucleotidyl transferase (TdT) can insert additional nucleotides during rearrangement of the heavy (H) chain gene, contributing to B cell diversity (Yancopoulos et al 1984, Kunkel et al 1986). Thereafter, however, the mitotic activity of the precursor B cells largely dictates the ultimate output of B cell clones, especially with respect to clone size (the number of B cells having a given antigen-binding specificity) and the number of different clones produced per unit time. This process may be modified by cell loss. The combined genetic and mitotic activities of early precursor B cells make them susceptible to genetic errors which may result in cell death. Other non-lethal aberrations could predispose to gene dysregulation and oncogenesis (Greaves 1986, Lenoir & Bornkum 1987, Harris et al 1988). The population dynamics and regulatory mechanisms of precursor B cells in the bone marrow thus underly considerations of both normal immune competence and neoplasias of the B cell lineage.

The present article briefly considers our recent studies of B lineage precursor cells in mouse bone marrow in vivo, aiming to define successive differentiation stages through which individual precursor cells pass with time, to quantitate the proliferation of cells at each stage, to examine their organization in the intact bone marrow and to assess their proliferative stimulation by exogenous agents causing macrophage activation. Finally, the possible relevance of these findings to mechanisms of B cell neoplasia will be proposed.

Department of Anatomy, McGill University,
3640 University Street, Montreal, Quebec, Canada H3A 2B2.
[1]Present address: Kitano Hospital, Osaka, Japan.

Current Topics in Microbiology and Immunology, Vol. 141

PROLIFERATION OF B LYMPHOCYTE PRECURSOR CELLS IN BONE MARROW

As differentiation markers of B lineage precursor cells in mouse bone marrow, we have examined the expression of 1) free cytoplasmic μ chains (cμ) 2) the B lineage-associated B220 glycoprotein, detected at the cell surface by monoclonal antibody (mAb) 14.8 (Kincade et al 1981) and 3) nuclear TdT, detected by a rabbit anti-TdT antibody, providing a marker for a precursor cell stage at which H chain genes are being rearranged. By double immunofluorescence labeling, we have defined 6 phenotypically distinct populations of presumptive B lymphocyte precursor cells, summarized in Table 1 as a working model of their sequence of development.

Table 1. Phenotype of B lineage cells in mouse bone marrow[a]

Cell population	Phenotype TdT	14.8	cμ	sIgM	Mitoses[b]	Cell[bc] Size (μm)	Incidence[bd] (%)
1. Early B precursors	+				Yes	9.0	0.8
2. Early B precursors	+	+			Yes	10.0	0.9
3. Early B precursors		+			Yes	11.5	4.3
4. Pre-B cells: large		+	+		Yes	11.5	3
5. Pre-B cells: small		+	+		No	8.0	8
6. B lymphocytes		+		+	No	8.0	8

[a]In 8-10 wk C3H/HeJ mice
[b]See Opstelten & Osmond 1983, Osmond & Park 1987, Park & Osmond 1987
[c]Median cell diameter in cytocentrifuged preparations
[d]Frequency relative to all nucleated bone marrow cells

Three cell populations comprise <u>Early B Precursors</u> before the detectable expression of μ chains 1) TdT^+, 2) $TdT^+14.8^+$ 3) $TdT^-14.8^+$. The subsequent 3 populations, all 14.8^+, include <u>Pre-B Cells</u> ($c\mu^+s\mu^-$), both large dividing and small post-mitotic forms, as well as small, non-dividing <u>B lymphocytes</u> ($sIgM^+$). Precursor-product relationships between Pre-B and B cells are well established (Osmond 1986, Osmond & Park 1987, Kincade 1987); those among early B precursors, which include the stage of IgH chain gene rearrangements, are presumptive but consistent with all our available data, including current studies on the sequence of regeneration after sublethal X-irradiation (Park & Osmond, unpublished observations).

A combination of immunofluorescence and stathmokinetic techniques has enabled us to analyse precursor cell proliferation of cells under normal conditions <u>in vivo</u> (Opstelten & Osmond 1983, Park & Osmond 1987). After ip injection of vincristine sulfate, dividing precursor cells become blocked in metaphase. The rate at which cells of a given phenotype accumulate in metaphase indicates their rate of entry into mitosis, and thus the turnover time of the population which, when combined with the total population size gives the actual number of cells passing through cell cycle within each population per unit time in the entire bone marrow (Table 2).

Table 2. Turnover of dividing B cell precursors in mouse bone marrow

	Phenotype			Turnover rate (%/hr)	Total bone marrow turnover (cells X 10^6/day)
	TdT	14.8	cμ		
1.	+			5.1[bc]	2.5
2.	+	+		9.0[bc]	5
3.		+		13.5[b]	36
4.		+	+	15.3[d]	35

[a]Osmond and Park 1987
[b]Park and Osmond 1986
[c]Park and Osmond [manuscript in preparation]
[d]Opstelten and Osmond 1983

The 3 populations of early B precursors before μ chain expression show a progressive increase in frequency, average cell size (Table 1) and production rate (Table 2). They show substantial proliferative expansion of cell numbers from stage to stage particularly at the $14.8^+\mu^-$ stage, resulting in a large cell flow through this compartment (3.6×10^7 cells/day). Assuming the V_H specificity to be established by gene rearrangements within the first TdT^+ cell population, a total of at least 4 mitotic cell cycles would appear to take place in the course of developing a clone of mature B cells. Each of the phenotypically distinct populations of early B precursors and large pre-B cells accumulate in mitosis during vincristine-induced metaphase arrest. Thus, each of these 4 stages of differentiation must be associated with at least one mitotic event. Possibly, more than one mitosis could occur within a given phenotypic compartment. Only tentative correlations between cell cycle and phenotypic boundaries can be drawn from current data, however. If a phenotypic compartment coincides with a single cell cycle and ends at mitosis, the "outflow" of cells from the compartment should be twice the rate at which cells flow through cell cycle in the compartment itself. This number of cells becomes the "inflow" into the next compartment. On the other hand, if cells of a given phenotype undergo a series of mitotic cell cycles before differentiating to the next phenotype, only approximately half the cells in the compartment will be in the last generation, leaving after the last mitosis: the outflow will then approximate the total number of cells flowing through cell cycle within the compartment itself. From these considerations, the data in Table 2 would be consistent with single mitoses in each of the 2 TdT^+ subpopulations of early B precursors and a direct flow of cells from one to the next. The further flow of cells would match the observed compartment turnover rates if cells are postulated to pass through 3 successive mitotic cycles within the $14.8^+\mu^-$ stage, and one mitosis at the large pre-B cell stage. Thus, a total of approximately 6 mitoses would occur between the initiation of V specificity and the terminal post-mitotic maturation phase in the resulting B cell clone. The extent to which the successive mitotic cycles may be associated with the "fixation" of successive genetic events, eg., $D\text{-}J_H$ and $V\text{-}DJ_H$ joining, remains to be verified.

A reduction of cell flow has been observed at the transition from dividing to post-mitotic pre-B cells, suggestive of substantial cell loss, possibly resulting from ineffectual or unacceptable gene rearrangements (Osmond, 1986, Osmond & Park, 1987). It cannot be excluded that additional loss of precursor B cells could also occur at earlier stages, in which case the observed population turnover values (Table 2) would represent the sum of cellular proliferation less cell loss. The number of mitoses among the surviving cell fraction could then be correspondingly greater than that deduced above.

In the light of present data concerning the number of successive mitoses in B cell clonal development in vivo, the clone size of virgin B cells produced in the bone marrow would appear to be small (~50 cells or less) and the bone marrow could be generating much of the available repertoire of virgin B lymphocyte specificities on a daily basis.

IN VIVO LOCALIZATION AND STROMAL CELL ASSOCIATIONS OF PRECURSOR B CELLS

Bone marrow cells in vivo lie between prominent venous sinusoids within a complex network of reticular stromal cells. The in situ organization and microenvironment of the B lineage cells, however, has been almost completely unknown.

Developing methods for the in vivo administration of ^{125}I-labeled antibodies in young mice we have previously analysed in detail the localisation of sIgM$^+$ B cells by light and electron microscope radioautography (Osmond & Batten 1984, Batten & Osmond 1984), while current work has examined 14.8$^+$ B lineage cells (Fig. 1).

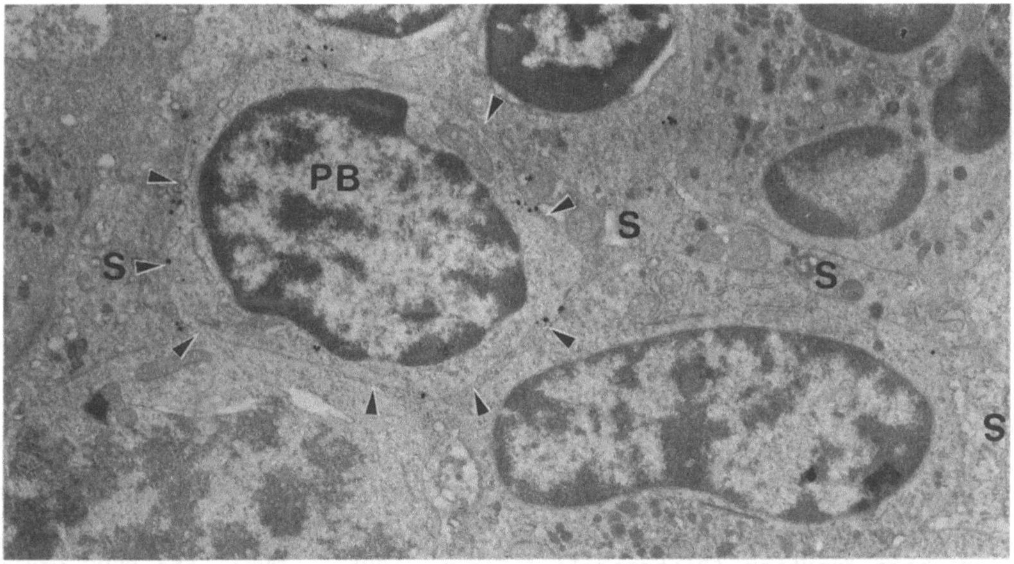

Fig. 1. Electron microscope radioautograph of a labeled precursor B cell (PB) binding ^{125}I-mAb 14.8 and adjacent to cytoplasmic processes of a stromal cell (S) in mouse bone marrow. Note 1) the localization over the cytoplasmic membrane of the precursor B cell of 20 radioautographic grains revealed by a fine grain development method (Kopriwa 1975) 2) extensive areas of contact (arrowheads) between the precursor B cell and stromal cell processes. X 8,100

Whereas mAb 14.8-binding B lineage cells as a whole are distributed throughout the bone marrow of the femoral shaft, the large dividing 14.8^+ precursor B cells tend to predominate peripherally towards the surrounding bone while mature $sIgM^+$ B cells are concentrated more centrally around the central venous sinusoid. In mice treated with anti-IgM antibodies from birth to delete mature $sIgM^+$ B cells, the central area of the bone marrow lacks 14.8^+ cells. These findings are consistent with the concept of a radial organisation of B cell genesis in bone marrow, the early precursor cells tending to proliferate in a peripherally located microenvironment from which they then may migrate centrally as they develop into mature B cells, finally traversing the wall of the sinusoids to enter the blood stream.

Large 14.8^+ precursor B cells occur either singly or in clusters, commonly of 5-10 cells, in each case being closely associated with the extensive processes of stromal cells (Fig. 1). Often, the precursor B cells are associated with stromal cells together with erythroid and granulocytic cells (Fig. 1) but occasionally, several 14.8^+ cells are observed to be aligned along the surface membrane of a single stromal cell.

Long term cultures of bone marrow cells can sustain B cell genesis, dependent upon a supportive layer of stromal cells, the B lineage cells clustering over certain stromal cells (Whitlock, Robertson & Witte 1984, Dorshkind 1986, Kincade 1987). Cloned stromal cell lines produce factors capable of promoting B cell genesis <u>in vitro</u> (Collins & Dorshkind 1987, Landreth & Dorshkind 1988, Namen et al 1988). However, the <u>in vivo</u> counterparts of these interactions so far have been unclear.

Our current findings demonstrate that close associations between many 14.8^+ precursor B cells and stromal cells normally exist in vivo, suggesting that stromal cell interactions may well be important for the regulation of B cell genesis in the normal intact bone marrow, as in bone marrow cultures. The precursor B cells could be stimulated by specific short-range growth factors produced by the associated stromal cells, whose activity in turn could be susceptible to a variety of general systemic influences.

STIMULATED PROLIFERATION OF EARLY B PRECURSORS BY ACTIVATED MACROPHAGES

Systemic influences on primary B cell genesis are becoming apparent. We have previously examined the effect of administering sheep red blood cells (SRBC) and other foreign agents to conventionally reared mice. A single ip injection of SRBC produces a well-marked wave of increased pre-B cell proliferation and B lymphocyte production (Fulop & Osmond 1983, Opstelten & Osmond 1985). The stimulating effect still occurs both in anti-IgM suppressed mice and in congenitally athymic (nude) mice suggesting that it does not require the mediation of B cells, T cells or an immune response. On the other hand, the effect is prevented by pretreating mice with silica particles to depress macrophage function (Pietrangeli & Osmond 1985). Thus, the SRBC would appear to be acting in a non-specific polyclonal way, mediated by macrophages. This is consistent with our observation that ip mineral oil, a non-specific "irritant", also stimulates bone marrow B cell genesis. Because prior splenectomy completely abolishes all the SRBC effects on B cell genesis, the responsible macrophages appear to be located in the spleen (Pietrangeli & Osmond 1985). The effects of SRBC can be adoptively transferred by spleen cells. Organ fragments and cell suspensions from the spleens of donor mice 4h after an ip SRBC injection transferred to the peritoneal cavity of normal mice stimulate pre-B cell proliferation in the recipients, thus confirming the role of spleen cells in mediating the stimulatory effect of SRBC on marrow B cell genesis (Pietrangeli & Osmond 1987). The nature of the relevant splenic macrophage-derived factors and whether they may act directly on precursor B cells or indirectly via an effect on bone marrow stromal cells remain to be established.

To simulate exposure to ongoing environmental stimuli we have examined the effects of protracted stimulation by SRBC (Fulop, Pietrangeli & Osmond 1986). Repeated SRBC injections for 4 wk result in considerable expansion in the population sizes of marrow B cell precursors and B cells, as well as a marked absolute increase in the production of small lymphocytes and the proliferation of pre-B cells. Such sustained stimuli thus raise the whole level of continuous B lymphocyte production to a new elevated kinetic steady state.

Current studies demonstrate that the foregoing stimulatory effects also operate on early B precursor cells, before μ expression. After a single ip injection of SRBC the proliferation of $14.8^{+}\mu^{-}$ cells increased 2-3 fold by 4 days, compared with saline-injected controls, as indicated by the total number of $14.8^{+}\mu^{-}$ cells in metaphase per femur at a constant interval after inducing metaphase arrest (Table 3). This effect was silica-sensitive and dependent upon the spleen (Table 2). Repeated SRBC injections produced a 60% expansion in population size of $14.8^{+}\mu^{-}$ cells (data not shown).

Table 3. Effect of SRBC injection, silica treatment and splenectomy on proliferation of $14.8^{+}\mu^{-}$ precursor B cells in bone marrow[a]

Injections	$14.8^{+}\mu^{-}$ cells in metaphase[b] (per femur X 10^{-5})
Saline	1.33 ± 0.23
SRBC[c]	4.27 ± 0.36
Saline + Silica[d]	0.84 ± 0.20
SRBC + Silica[d]	1.16 ± 0.23
SRBC[c] + sham splenectomy[e]	3.80[f]
SRBC[c] + splenectomy[e]	0.86[f]

[a] 8-9 wk old C3H/HeJ mice
[b] 2h 40m after ip vincristine sulfate, 1 mg/kg body wt.
[c] $4x10^{8}$ SRBC in 0.5 ml saline by ip injection
[d] ip silica particles (0.012 μm; 3 mg in 0.3 ml saline) 2 hr after saline or SRBC
[e] splenectomy or sham splenectomy 2 hr after SRBC
[f] cells pooled from groups of 3 mice

These results raise the concept that marrow B lymphocyte genesis may normally represent the summation of two regulatory steps (1) a basal level, regulated mainly by local microenvironmental stromal cell-associated mechanisms 2) a polyclonal amplification step resulting from environmental stimuli. This model has implications for both normal regulation and derangements of B cell genesis:

a) Under normal circumstances, the findings suggest that external stimuli may play a significant role in determining the final number and clone sizes of primary B lymphocytes generated in the bone marrow under conventional environmental conditions.

b) Under pathological circumstances, the findings raise the possibility of an etiological link between processes associated with prolonged macrophage activation and oncogenesis in the B cell lineage.

POLYCLONAL STIMULATION OF PRIMARY B CELL GENESIS AND B CELL NEOPLASIA

Oncogenesis in general is currently viewed as a multistep process, consisting of successive genetic events including the activation of 2 or more oncogenes, sometimes virally associated. The initial genetic event in the sequence could thus occur at a considerably earlier stage in cellular differentiation than that at which the tumor cells are eventually "fixed" by the final clonal step in oncogenesis. In the case of the B cell lineage, the early precursor B cells may be particularly prone to genetic error because of the combinational, junctional and hypermutational events naturally occurring in their DNA during the formation of functional Ig genes, coupled with their intense mitotic activity. It has been proposed that spontaneous mutations in precursor B cells may constitute the major, if not sole, initiating step in human childhood acute lymphoblastic leukemia (Greaves 1986). Any factor which increases the Ig combinatorial activity in precursor B cells in vivo, particularly by increasing the proliferative activity and population size of the vulnerable cells, could increase the risk of genetic aberrations, including the chromosomal translocations and myc/Ig H chain locus juxtaposition which characterise many B cell neoplasias (Potter et al 1985, Klein & Klein 1986, Lenoir & Bornkum 1987). The importance of this consideration is emphasized by work with mice transgenic for a c-myc gene driven by the Ig H enhancer (Eμ-myc) (Langdon et al 1986, Harris et al 1988). Such mice initially show a persistent benign polyclonal over-proliferation of pre-B cells in the bone marrow before ultimately developing clonal B lymphoid neoplasias.

Our recent findings, summarized above, suggest that similar conditions can be produced by the macrophage-mediated stimulation of early precursor B cell proliferation in the bone marrow, which could in turn predispose to the initiation of B cell neoplasias at the early precursor B cell stage. A primary phase of macrophage activation characterizes 2 widely studied B cell neoplasias, pristane-induced murine plasmacytoma (PCT) and malaria-associated Burkitt lymphoma (BL). Susceptible BALB/c mice given ip pristane develop chronic ip granulomata rich in macrophages which show evidence of activation before the eventual development of PCT (Potter et al 1985, 1987). In malaria, gross splenomegally is associated with a marked increase in the macrophage population, phagocytosis of parasitized RBC and macrophage activation (Wyler 1982, Weiss, Geduldig & Weidanz 1986, Klein & Klein 1986, Lenoir & Bornkum 1987). Our continuing work aims to test the hypothesis raised by our findings that such conditions of non-specific sustained macrophage activation could predispose to B cell neoplasias by operating at the level of the central genesis of the B cell lineage, viz. by elevating the total proliferative activity in the bone marrow of early precursor B cells and thus increasing the odds of genetic errors which could constitute the first of a series of steps towards neoplasia.

Acknowledgement. Supported by the Medical Research Council of Canada and the National Cancer Institute of Canada.

REFERENCES

Batten DG & Osmond DG The localization of lymphocytres in mouse bone marrow: Radioautographic studies after in vivo perfusion of radio-labeled anti-IgM antibody. J Immunol Meth 72:381-399.

Collins L, Dorshkind K (1987) A stromal cell line from myeloid long-term bone marrow cultures can support myelopoiesis and B lymphopoiesis. J Immunol 138:1082-1087

Dorshkind K ((1986) In vitro differentiation of B lymphocytes from primitive hemopoietic precursors present in long-term bone marrow cultures. J Immunol 136:422-429

Fulop GM & Osmond DG (1983) Regulation of bone marrow lymphocyte production. III Increased production of B and non-B lymphocytes after administering systemic antigens. Cell Immunol 75:80-90

Fulop GM, Pietrangeli CE & Osmond DG (1986) Regulation of bone marrow production. IV. Altered kinetic steady state of lymphocyte production after chronic changes in exogenous stimuli. Exp Hemat 14: 27-34.

Greaves MF (1986) Differentiation-linked leukemogenesis in lymphocytes. Science 234:697-704

Harris AW, Pinkert CA, Crawford M, Langdon WY, Brinster, RL, Adams JM (1988) The Eμ-myc transgenic mouse: A model for high-incidence spontaneous lymphoma and leukemia of early B cells. J Exp Med 167: 353-371

Kincade PW (1987) Experimental models for understanding B lymphocyte formation. Adv. Immunol 41:181-267

Kincade PW, Lee G, Watanabe T, Sun L and Scheid MP (1981) Antigens displayed on murine B lymphocyte precursors. J Immunol 127:2262-2268

Klein G, Klein E (1986) Conditioned tumorigenicity of activated oncogenes. Cancer Res 46:3211-3224.

Kopriwa B (1975) A comparison of various procedures for fine grain development in electron microscope radioautography. Histochemie 44: 201-244.

Kunkel TA, Gopinathan KP, Dube KP, Snow ET and Loeb LA (1986) Rearrangements of DNA mediated by terminal transferase. Proc Nat Acad Sci USA 83:1867-1871

Landreth K, Dorshkind K (1988) Pre-B cell generation potentiated by soluble factors from a bone marrow stromal cell line. J Immunol 140: 845-852

Langdon WY, Harris AW, Cory S, Adams JM (1986) The c-myc oncogene perturbs B lymphocyte development in Eμ-myc transgenic mice. Cell 47:11-18

Lenoir GM, Bornkum GW (1987) Burkitt's lymphoma, a human cancer model for the study of the multistep model of cancer: proposal for a new scenario. Adv in Viral Oncology 7:173-206

Namen AE, Schmierer AW, March CJ, Overell RW, Park LS, Urdal DL and Mochizuki DY (1988) B cell precursor growth-promoting activity. Purification and characterization of a growth factor active on lymphocyte precursors. J Exp Med 167:988-1002

Opstelten D & Osmond DG (1983) Pre-B cells in mouse bone marrow: immunofluorescence stathmokinetic studies of the proliferation of cytoplasmic μ chain-bearing cells in normal mice. J Immunol :2635-2640

Opstelten D & Osmond DG (1985) Regulation of pre-B cell proliferation in bone marrow: immunofluorescence stathmokinetic studies of cytoplasmic μ chain-bearing cells in anti-IgM-treated mice, hematologically deficient mutant mice and mice given sheep red blood cells. Eur J Immunol 15:599-605

Osmond DG (1985) The ontogeny and organization of the lymphoid system. J Invest Dermatol 85:2s-9s

Osmond DG (1986) Population dynamics of bone marrow B lymphocytes. Immunol Rev 93:103-124

Osmond DG & Batten SJ (1984) Genesis of B lymphocytes in the bone marrow: Extravascular and intravascular localization of surface IgM-bearing cells in mouse bone marrow detected by electron microscope radioautography after in vivo perfusion of ^{125}I anti-IgM antibody. Am J Anat 170:349-365

Osmond DG & Park YH (1987) B lymphocyte progenitors in mouse bone marrow. Int Rev Immunol 2:241-261

Park YH & Osmond DG (1987) Phenotype and proliferation of early precursor cells in mouse bone marrow. J Exp Med. 165:444-458.

Pietrangeli CE & Osmond DG (1985) Regulation of B-lymphocyte production in the bone marrow: Role of macrophages and the spleen in mediating responses to exogenous agents. Cell Immunol 94:147-158

Pietrangeli CE & Osmond DG (1987) Regulation of B lymphocyte production in the bone marrow: Mediation of the effects of exogenous stimulants by adoptively transferred spleen cells. Cell Immunol 107: 348-357

Potter M, Mushinski JF, Mushinski EV, Brust S, Wax JS, Wiener F, Babonits M, Rapp UR, Morse HC (1987) Avian v-myc replaces chromosomal translocation in murine plasmacytomagenesis. Science 235:787-789

Potter M, Wax JS, Anderson AO, Nordan RP (1985) Inhibition of plasmacytoma development in BALB/c mice by indomethacin. J Exp Med 161:996-1012

Tonegawa S (1983) Somatic generation of antibody diversity. Nature 302:575-581

Weiss L, Geduldig U, Weidanz W (1986) Mechanisms of splenic control of murine malaria: Reticular cell activation and the development of a blood-spleen barrier. Am J Anat 176:251-285

Whitlock CA, Robertson D, Witte ON (1984) Murine B cell lymphopoiesis in long term culture. J Immunol Meth 67:353-369.

Wyler DJ (1982) Malaria: Host-pathogen biology. Rev Infect Dis 4:785-797

Yancopoulos GD, Desiderio SV, Paskind M, Thomas E, Boss M, Landau N, Alt FW, Baltimore D (1984) Preferential utilization of the most J - proximal V gene segments in pre-B cell lines. Nature 311:727-733

Lymphoid-Restricted Stem Cells

R. A. PHILLIPS, D.-D. WU, and G. M. FULOP

INTRODUCTION

Work from many laboratories has demonstrated the existence in bone marrow of a multipotent stem cell able to give rise to progeny in all of the hematopoietic lineages (Abramson et al 1977, Dick et al 1985, Keller et al 1985). Studies on the differentiation of myeloid cells have demonstrated that among the progeny of multipotent stem cells are stem cells with restricted differentiation potential (Abramson et al 1977). The spleen colony-forming stem cell (CFU-S)(Till and McCulloch 1961) produces myeloid and erythroid progeny but lacks the ability to produce lymphocytes (Paige et al 1979). The existence of a similar lymphoid-restricted stem cell remains controversial. In studies on the reconstitution of irradiated animals or mutant mice with normal bone marrow cells, most groups detect the pluripotent stem cells, S_P (see Fig. 1) (Abramson et al 1977, Keller et al 1985, Lemischka et al 1986), myeloid-restricted stem cells, S_M (Abramson et al 1977) and CFU-S (Wu et al 1968), but no one has yet convincingly identified a lymphoid-restricted stem cell, S_L, in normal bone marrow. Fig. 1 shows a differentiation scheme assuming that restricted stem cells exist.

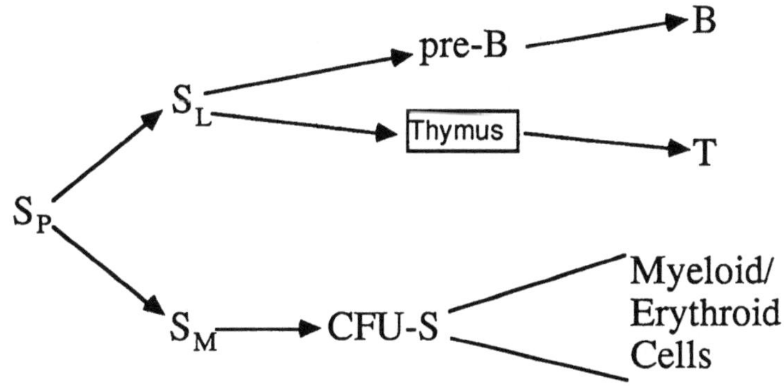

Fig. 1. Proposed lineage diagram for differentiation in the hematopoietic system. S_P, pluripotent stem cell; S_M, myeloid-restricted stem cell; S_L, lymphoid-restricted stem cell; CFU-S, spleen colony-forming stem cell

Bone marrow stem cells can be propagated for long periods of time, up to six months, in a culture system described by Dexter and colleagues (Dexter et al 1977). Predominant cell types in these cultures are granulocytes and macrophages with a few erythroid cells; lymphoid cells are not detectable (Phillips et al 1984). The cultures, nevertheless, contain stem cells able to reconstitute the lymphoid system (Jones-Villeneuve and Phillips 1980). In fact, we have demonstrated in several different systems the ability of stem cells from such long-term bone marrow cultures (LTBMC) to reconstitute lymphoid function (Jones-Villeneuve and Phillips 1980, Dorshkind and Phillips 1982, Fulop and Phillips 1986). The following sections summarize the evidence for such reconstituting ability. The conclusion from these experiments is that LTBMC contain a lymphoid restricted stem cell.

Department of Medical Genetics, University of Toronto and Research Institute, Hospital for Sick Children, Toronto, Canada

Current Topics in Microbiology and Immunology, Vol. 141

LONG-TERM BONE MARROW CULTURES

The LTBMC system established by Dexter requires a complex stromal layer for the support of hematopoietic stem cells (Dexter et al 1977). In general, bone marrow cells are placed in a culture flask with medium, 20% horse serum and hydrocortisone (Greenberger 1978). During the next two weeks, most of the hematopoietic cells including the stem cells die, but the stromal cells proliferate and form a layer of adherent cells. When the stromal layer is nearly confluent, additional bone marrow is added to the cultures; the stem cells in this innoculum attach to the stromal layer and begin to self-renew, proliferate and differentiate (Dexter et al 1977). With this culture system, it is possible to maintain murine hematopoietic stem cells for a long period of time. At any time during the culture period, it is easy to recognize cells of the granulocytic, monocytic and erythroid pathways. Similarly, the progenitors of these cells can be detected either in vitro as colony-forming cells or in vivo as part of the progeny within a spleen colony (Keller and Phillips 1984, Magli et al 1987).

Not surprisingly, lymphoid cells find the environment in LTBMC hostile due to the high concentrations of horse serum and hydrocortisone. Consequently, all lymphoid cells disappear rapidly from such cultures with no T, B or pre-B cells detectable, morphologically or functionally, in LTBMC after the first week of incubation (Phillips et al 1984). Despite this absence of detectable lymphoid cells, it is clear from the ability of these cultures to reconstitute irradiated recipients that they contain stem cells with lymphoid potential. The failure of the stem cells to express their lymphoid potential in LTBMC depends on the growth medium and the microenvironment. If LTBMC are switched to a growth medium known to support B cell differentiation (Whitlock et al 1984), B cells appear in the cultures after a lag of several weeks (Dorshkind 1987). Unfortunately, stem cells die under conditions that support lymphopoiesis (Dorshkind 1987).

RECONSTITUTION OF LETHALLY IRRADIATED RECIPIENTS WITH LTBMC

We first detected a possible lymphoid restricted stem cell when we reconstituted lethally irradiated mice with cells from LTBMC (Jones-Villeneuve and Phillips 1980). To distinguish between donor and recipient cells, we established LTBMC from CBA mice carrying the T6 chromosome marker. Cells from such cultures were used to reconstitute coisogenic normal CBA recipients which had been exposed to 950 cGy of whole body radiation prior to reconstitution. Two weeks after transplantation, the majority of the spleen colonies and the proliferating cells contain the T6 chromosome marker identifying them as progeny of stem cells and early progenitors in LTBMC. At this time after transplantion, almost all of the dividing cells are myeloid and erythroid. Lymphoid reconstitution does not begin until approximately two weeks after transplantation and is not complete until four to six weeks post transplantation. Two months after transplantation, the recipients were analyzed for the presence of the T6 chromosome marker in LPS blasts (B cells), Con-A blasts (T cells) and bone marrow cells (predominantly myeloid/erythroid cells).

In contrast to the results obtained at two weeks after transplantation, at eight weeks post transplant, few of the myeloid cells but many LPS and Con-A blasts contained the T6 chromosome marker. In addition, there was a very high correlation between the percent of Con-A blasts with the T6 chromosome marker and the percent of LPS blasts with the T6 chromosome marker (r^2=0.90). There was no correlation between the percent of lymphoid blasts with the T6 marker and the percent of myeloid cells with the T6 chromosome marker. A likely explanation of these results is that LTBMC contain lymphoid-restricted stem cells capable of producing B and T cells but unable to differentiation into myeloid cells. The absence of donor-derived myeloid cells in the irradiated recipients suggest that LTBMC lack myeloid stem cells with extensive self-renewal potential. It is also likely that LTBMC lack significant numbers of multipotent stem cells, S_P in figure 1.

Harrison et al. have also observed reduced erythropoietic reconstituting ability in LTBMC (Harrison et al 1987). Using a competitive repopulation assay to evaluate the proliferative and differentiative potential of stem cells in LTBMC, these investigators observed that stem cells maintained for greater than 7 weeks in LTBMC lost up to 90% of their reconstituting ability compared to fresh bone marrow.

RECONSTITUTION OF CBA/N MICE WITH LTBMC

CBA/N mice carry an X-linked mutation resulting in a decreased response to several antigens and mitogens (Sher et al 1975). One consequence of this mutation is an inability of the B cells in CBA/N mice to form B lymphocyte colonies (CFU-B) in agar. Paige et al (1979) demonstrated that transplantation of normal CBA bone marrow into irradiated CBA/N recipients reconstituted CFU-B activity. To test the ability of normal CBA stem cells in LTBMC to reconstitute CFU-B in CBA/N mice, we injected 2 x 10^6 cells into irradiated recipients. At various times after transplantation, the number of CFU-B in bone marrow and spleen were measured (Dorshkind and Phillips 1982). Several different tissues were examined for reconstituting ability, normal bone marrow, non-adherent cells in LTBMC and adherent cells in LTBMC. As expected, normal bone marrow gave rapid and full reconstitution of CFU-B activity. From LTBMC, the adherent population, but not the non-adherent population, gave full reconstitution of CFU-B (Dorshkind and Phillips 1982). It should be noted that both the adherent and non-adherent populations contained equal numbers of CFU-S despite a greater than one hundred-fold difference in CFU-B reconstituting ability. This result was not particularly surprising since other investigators had demonstrated that CFU-S from the adherent layer in LTBMC had greater proliferative potential than the non-adherent population of CFU-S (March et al 1980). Our results confirm that the stem cells in the adherent layer have properties different from those in the non-adherent population; these adherent stem cells have more proliferative and reconstituting potential than the non-adherent population.

RECONSTITUTION OF scid MICE WITH LTBMC

Scid mice lack functional B and T lymphocytes due to a recessive mutation on chromosome 16 (Bosma et al 1983, Dorshkind et al 1984). Mice homozygous for this mutant allele appear unable to correctly rearrange either the immunoglobulin genes or the T cell receptor genes, leading to a marked immune deficiency with an absence of normal lymphocytes in all tissues (Schuler et al 1986, Malynn et al, in preparation). We have demonstrated previously that normal bone marrow stem cells will reconstitute B and T cell function in scid mice, but that full reconstitution requires sublethal irradiation of the scid recipients (Fulop and Phillips 1986).

Table 1. Stem cell frequency in normal bone marrow and in LTBMC

Cells	Curing units/10^6 cells[a]
Normal Bone Marrow	120
LTBMC	670

[a] A cure was defined as a scid mouse having $\geq 10^3$ CFU-B/femur (Fulop and Phillips, in preparation). A curing unit was defined by limiting dilution, and the frequency was calculated as described by Porter and Berry (1964). The value shown for normal bone marrow is an average of 2 experiments, and for LTBMC, 3 experiments.

The major advantage of scid mice as recipients is that their survival does not depend on their receipt of a graft of normal bone marrow cells. Thus, as opposed to lethally irradiated recipients, one can test very small numbers of restricted stem cells for their repopulating ability in scid recipients. Analysis of the ability of LTBMC to reconstitute scid recipients demonstrates that in contrast to the reduced ability of stem cells in LTBMC to reconstitute myeloid function such cultures are highly efficient at reconstituting B and T cell function in scid mice. In fact, full lymphoid reconstitution in scid recipients occurs with many fewer LTBMC cells than with normal bone marrow cells. In order to quantitate the number of lymphoid stem cells in LTBMC, we have applied a limiting dilution analysis by determining the minimum number of cells required for reconstitution. This procedure is analogous to the one used by Boggs et al in their determination of the number of cells able to reconstitute congenitally anemic, W/W^v, mice (Boggs et al 1982). The details of this limiting dilution assay in scid mice will be described in detail elsewhere (in preparation). A brief summary of the number of stem cells in several experiments is shown in Table 1. It is clear from the data that LTBMC contain between 5- and 10-fold more reconstituting stem cells than normal bone marrow.

To determine whether or not the LTBMC cells able to reconstitute scid mice were multipotent stem cells capable of reconstituting all hematopoietic lineages or stem cells restricted to lymphoid differentiation, we reconstituted scid mice with LTBMC infected with a retrovirus containing the bacterial gene for neomycin resistance (Dick et al 1986). All of the scid mice reconstituted with such LTBMC cells were reconstituted for CFU-B and for T cell function. In many recipients, we detected B and T lymphocytes carrying the neomycin gene (Wu, unpublished observations). However, we were unable to detect the neomycin resistance gene in granulocyte-macrophage precursors, either by direct Southern blotting or by functional tests, i.e. no granulocyte-macrophage precursors were detected which were resistant to the neomycin analogue, G418. On the basis of these results, we conclude that the stem cells in LTBMC which can reconstitute scid mice are restricted to lymphoid differentiation and are significantly reduced in myeloid reconstituting ability.

SUMMARY

LTBMC definitely contain stem cells able to reconstitute lymphoid function in deficient animals. The stem cells with greatest reconstituting ability are found in the adherent layer of such cultures; the non-adherent population of cells contain little if any lymphoid reconstituting ability. Our results on the ability of LTBMC to reconstitute lethally irradiated CBA or CBA/N mice and sublethally irradiated scid mice all indicate that lymphoid reconstitution occurs from a lymphoid restricted stem cell in LTBMC. Although such a lymphoid restricted stem cell has not been unequivocally demonstrated in normal bone marrow cells, it is likely from our data that such a precursor exists. The characterization of the lymphoid reconstituting stem cell in LTBMC should provide a description which will allow the identification of this stem cell in normal bone marrow.

ACKNOWLEDGEMENTS

Excellent technical assistance was provided by S. Chance. The research was supported by grants from the Medical Research Council and by The National Cancer Institute of Canada.

REFERENCES

Abramson S, Miller RG, Phillips RA (1977) Identification of pluripotent and restricted stem cells of the myeloid and lymphoid systems. J.Exp.Med. 145: 1567-1579

Boggs DR, Boggs SS, Saxe DF, Gress LA, Canfield DR (1982) Hemopoietic stem cells with high proliferative potential. Assay of their concentration in marrow by the frequency and duration of cure of W/W^v mice. J Clin Invest 70: 242-53

Bosma GC, Custer RP, Bosma MJ (1983) A severe combined immunodeficiency mutation in the mouse. Nature 301: 527-530

Dexter TM, Allen TD, Lajtha LG (1977) Stimulation of differentiation and proliferation of hemopoietic cells in vitro. J Cell Physiol 82: 461-473

Dick JE, Magli M, Huzzar, D, Phillips RA, Bernstein A (1985) Introduction of a selectable gene into primitive stem cells capable of long-term reconstitution of the hematopoietic system of W/W^v mice. Cell 42: 71-79

Dorshkind K (1987) In vitro differentiation of B lymphocytes from primitive hemopoietic precursors present in long-term bone marrow cultures. J Immunol 136: 422-429

Dorshkind K, Keller GM, Phillips RA, Miller RG, Bosma GC, O'Toole M, Bosma MJ (1984) Functional status of cells from lymphomyeloid tissues in mice with severe combined immunodeficiency disease. J Immunol 132: 1804-1808

Dorshkind KA, Phillips RA (1982) Maturational state of lymphoid cells in long-term bone marrow cultures. J Immunol 129: 2444-2450
Fulop GM, Phillips RA (1986) Full reconstitution of the immune deficiency in scid mice with normal stem cells requires low dose irradiation of the recipient.J Immunol 136: 4438-4443
Fulop GM, Phillips RA (in preparation) Identification and quantitation of lymphoid-restricted stem cells in long-term bone marrow cultures.
Greenberger JS (1978) Sensitivity of corticosteroid-dependent insulin-resistant lipogenesis in marrow preadipocytes of obese (db/db) mice. Nature 275: 752-54
Harrison DE, Lerner CP and Spooncer E (1987) Erythropoietic repopulating ability of stem cells from long-term marrow culture. Blood 69: 1021-1025
Jones-Villeneuve E, Phillips A. (1980) Potentials for lymphoid differentiation by cells from long-term cultures of bone marrow. Exp Hematol 8: 65-76
Keller GM, Phillips RA (1984) Maintenance of hemopoiesis in long-term bone marrow cultures from Sl/Sl^d and W/W^v mice. Exptl Hematol 12: 822-824
Keller G, Paige C, Gilboa E, Wagner E (1985) Expression of a foreign gene in myeloid and lymphoid cells derived from multipotent hematopoietic precursors. Nature 318: 149-154
Lemischka IR, Raulet DH, Mulligan RC (1986) Developmental potential and dynamic behavior of hematopoietic stem cells. Cell 45: 917-927
Magli MC, Dick JE, Huszar D, Bernstein A, Phillips RA (1987) Modulation of gene expression in multiple hematopoietic cell lineages following retroviral vector gene transfer. Proc Natl Acad Sci USA 84: 789-793
Malynn BA, Blackwell TK, Fulop GM, Ferrier P, Rathbun GA, Furley AJW, Morrow MA, Oltz EM, Kaplan KB, Heinke LB, Phillips RA, Yancopoulos GD, Alt FW (in preparation) The scid defect affects the final step of the immunoglobulin VDJ recombinase mechanism.
March P, Greenberger JS, Botnick L, Hannon E, Hellman S (1980) Evidence for structural variation in self-renewal capacity within long-term bone marrow cultures. Proc Natl Acad Sci USA 77: 2927-2930
Paige C, Kincade PW, Moore MAS, Lee G (1979) The fate of fetal and adult B-cell progenitors grafted into immunodeficient CBA/N mice. J Exp Med 150: 548-563
Phillips RA, Bosma M, Dorshkind K (1984) Reconstitution of immune-deficient mice with cells from long-term bone marrow cultures. In: Long-Term Bone Marrow Cultures. Liss, A.R. New York p 309
Porter EH, Berry RJ (1964) The efficient design of transplantable tumor assays. Brit J Cancer 17: 583-595.
Scher I, Steinberg AD, Berning AK, Paul WE (1975) X-linked B-lymphocyte immune defect in CBA/N mice. II. Studies of the mechanisms underlying the immune defect. J Exp Med 142: 637-650
Schuler W, Weiler IJ, Schuler A, Phillips RA, Rosenberg N, Mak TW, Kearney JF, Perry RP, Bosma MJ (1986) Rearrangement of antigen receptor genes is defective in mice with severe combined immune deficiency. Cell 46: 963-972
Till JE, McCulloch EA (1961) A direct measurement of the radiation sensitivity of normal mouse bone marrow. Radiat Res 14: 213-222
Whitlock CA, Robertson D, Witte ON (1984) Murine B cell lymphopoiesis in long term culture. J Immunol Meth 67: 353-369
Wu AM, Till JE, Siminovitch L, McCulloch EA (1968) Cytological evidence for a relationship between normal hemopoietic colony-forming cells and cells of the lymphoid system. J Exp Med 127: 455-63

Recombinant Interleukin-7 Supports the Growth of Normal B Lymphocyte Precursors

G. LEE[1], A. E. NAMEN[2], S. GILLIS[2], and P. W. KINCADE[1]

INTRODUCTION

Hemopoiesis occurs under the influence of regulatory elements in bone marrow. Long term culture of bone marrow (LTBMC) cells under conditions described by Dexter and associates (Dexter et al. 1977) demonstrated that association with adherent stromal cells is critical in the maintenance of hemopoietic stem cells and myelopoiesis (reviewed in Allen and Dexter 1984). Similarly, lymphopoiesis in Whitlock-Witte type LTBMC (Whitlock and Witte 1982) is dependent upon adherent cells (Whitlock et al. 1985). Detection of macrophage, granulocyte or granulocyte/macrophage colony stimulating activity in culture supernatants of stromal cell lines from Dexter cultures (Shadduck et al. 1983; Song et al. 1985; Rennick et al. 1987) suggests that stroma derived soluble factors may account in part for myelopoiesis in these cultures. Furthermore, stromal cell lines able to support lymphopoiesis have also been isolated from LTBMC (Hunt et al. 1987; Whitlock et al. 1987; Collins and Dorshkind 1987; Namen et al. 1988; Pietrangeli et al. 1988) and some of these have been shown to secrete factors that independently maintain the growth of B cell precursors from LTBMC (Hunt et al. 1987; Whitlock et al. 1987; Namen et al. 1988). A 25,000 da protein, designated interleukin-7 (IL-7), with this activity has recently been purified from a bone marrow derived stromal cell line (Namen et al. 1988) and the cDNA for it cloned (Namen et al. submitted). We have now examined the activity of recombinant IL-7 on subpopulations of normal B lineage cells.

RESULTS

IL-7 induced extensive cell proliferation, measured by ^{3}H-thymidine uptake, when cultured with adult Balb/c bone marrow, adherent cell depleted bone marrow or isolated B cell precursors bearing Ly5/220, a cell surface antigen preferentially expressed on B lineage cells (reviewed in Kincade 1987). The latter cells were prepared by positive selection of B cell depleted bone marrow on anti-Ly5/220 antibody coated plates. In contrast, no proliferation was observed in cultures when bone marrow depleted of Ly5/220 bearing cells were cultured with IL-7 until 4-8 days after initiation of the culture period. Examination of the proliferating foci at this point revealed large, sIg$^-$,Ly5/220$^+$,BP-1$^+$ lymphoid cells, many of which also had detectable cytoplasmic μ (cμ) heavy chain. These foci did not appear in control cultures over the same culture period. Spleen or lymph node cells and B cells isolated from bone marrow did not proliferate in response to IL-7. In addition, IL-7 did not support the formation of B cell colonies (CFU-B) in semi-solid agar medium. These results indicate that IL-7 preferentially promotes the growth of sIg$^-$,Ly5/220$^+$ B cell precursors, but not earlier progenitors which do not yet display this marker, or more mature lymphocytes in bone marrow or peripheral tissues. Furthermore, separation of sIg$^-$,Ly5/220$^+$ bone marrow cells into large and small size fractions by flow cytometry, followed by culture with IL-7, showed that the proliferative activity of IL-7 is directed at the large rather than small cells.

B cell precursors isolated from bone marrow have been maintained in liquid culture with IL-7 for >40 days in the apparent absence of stromal or non-lymphoid cells. At the initiation of culture, these cells were sIg$^-$,Ly5/220$^+$, Mac1$^-$, and did not display N-CAM, an adhesion molecule recently reported to be expressed by stromal cells (Thomas et al. 1988). Most (50-65%) were cμ^+ and small fractions were BP1$^+$, Thy1$^+$ or Ia$^+$. During the first two weeks of culture, total numbers of cells increased greater than 50 fold. Cells arising during this period were Ly5/220$^+$,BP1$^+$ and the majority also expressed Thy1, Ia and/or cμ. Cells cultured in

[1]Oklahoma Medical Research Foundation, Oklahoma City, OK 73104
[2]Immunex Corporation, Seattle, WA 98101

Current Topics in Microbiology and Immunology, Vol. 141

control medium died within 3-4 days. Surface IgM$^+$ cells appeared transiently in both control and IL-7 cultures and were not detectable after 2 weeks. A heterogeneous range of surface densities of BP1 and Ly5/220 was observed on cells bearing these antigens at the initiation of culture. By 40 days *in vitro*, cells with a high surface BP1 and low surface Ly5/220 expression were predominant in IL-7 cultures. Less than 5% had detectable cμ and numbers of Thy1$^+$ and Ia$^+$cells also declined. Cultured cells remained negative for a panel of T cell, macrophage and granulocyte cell surface markers. The results indicate that pre-B cells rapidly expand during the first growth phase, but cells with a relatively immature lymphoid phenotype are preferentially maintained during long term culture with IL-7. Expression of the BP1 (Cooper et al. 1986) and Ly5/220 antigens by these cells suggests that they are of the B lineage. Studies are in progress to determine their Ig gene configuration. Display of class II antigens on cultured cells is also interesting because this is a feature of mature cells in mice. It is noteworthy that Ia is inducible by IL-4 on some pre-B cell lines (Polla et al. 1986, Lee et al. 1987).

When cultured in semi-solid agar medium containing IL-7, isolated large sIg$^-$,Ly5/220$^+$ B cell precursors as well as unfractionated bone marrow cells formed discrete colonies within 4-5 days in numbers proportional to numbers of cells plated. *In situ* double labeling of cells within the colonies for surface and cytoplasmic Ig showed that all colonies consisted of pre-B cells. In contrast to LPS stimulated B cell colonies cultured under similar conditions, IL-7 induced pre-B colonies were not inhibitable by anti-Ig.

Among the diverse activities of transforming growth factor β (TGF-β) are the abilities to inhibit factor induced proliferation of B cells and several transitional stages during pre-B and B cell differentiation (Kehrl et al. 1986; Lee et al. 1987). Co-culture studies indicate that TGF-β is also inhibitory to IL-7 induced proliferation of B cell precursors. Furthermore, it prevented the development of cells able to proliferate in response to IL-7 during culture of bone marrow from which Ly5/220 bearing cells had been initially eliminated.

SUMMARY

Transfection of adherent bone marrow cells from LTBMC with genes encoding the transforming sequences of SV40 led to the isolation of a stromal cell clone from which the cDNA encoding IL-7 was derived. This 25,000 da protein was originally defined by its ability to support the growth of pre-B cells arising from LTBMC. Our current results show that recombinant IL7 sustains the survival and growth of B cell precursors freshly isolated from normal bone marrow. Within the B lineage, the proliferative activity of IL-7 is targeted primarily at large B cell precursors. In addition to expansion of large numbers of these cells in liquid culture, IL-7 allowed the quantitative cloning of pre-B cells in semi-solid agar directly from whole bone marrow. Furthermore, this activity was inhibitable by TGF-β, which has previously been reported to inhibit B cell differentiation and function. IL-7 did not elicit proliferation of mature B cells, or precursors which do not yet express the Ly5/220 cell surface antigen. No evidence of differentiation towards Ig heavy or light chain expression was observed in several early murine B lineage cell lines treated with IL-7. Thus, the prinicipal activity of IL-7 seems to be a replication stimulus for cells at a discrete stage of differentiation. However, this factor may also directly or indirectly promote the survival and differentiation of early progenitors in bone marrow.

REFERENCES

Allen TD, Dexter TM (1984) The essential cells of the hemopoietic microenvironment. Exp Hematol 12:517-521.

Collins LS, Dorshkind K (1987) A stromal cell line from myeloid long-term bone marrow cultures can support myelopoiesis and B lymphopoiesis. J Immunol 138:1082-1087

Cooper MD, Mulvaney D, Coutinho A, Cazenave P-A (1986) A novel cell surface molecule on early B-lineage cells. Nature 321:616-618

Dorshkind K, Witte ON (1987) Long-term murine hemopoietic cultures as model systems for analysis of B lymphocyte differentiation. Curr Top Microbiol Immunol 135:23-41

Hunt P, Robertson D, Weiss D, Rennick D, Lee F, Witte ON (1987) A single bone marrow-derived stromal cell type supports the *in vitro* growth of early lymphoid and myeloid cells. Cell 48:997-1007

Kehrl JH, Roberts AB, Wakefield LM, Jakowlew S, Sporn MB, Fauci AS (1986) Transforming growth factor β is an important immunomodulatory protein for human B lymphocytes. J Immunol 137:3855-3860
Kincade PW (1987) Experimental models for understanding B lymphocyte formation. Adv Immunol 41:181-267
Lee G, Ellingsworth LR, Gillis S, Wall R, Kincade PW (1987) β transforming growth factors are potential regulators of B lymphopoiesis. J Exp Med 166:1290-1299
Namen AE, Schmierer AE, March CJ, Overell RW, Park LS, Urdal DL, Mochizuki DY (1988) B cell precursor growth-promoting activity. Purification and characterization of a growth factor active on lymphocyte precursors. J Exp Med 167:988-1002
Pietrangeli CE, Hayashi S-I, Kincade PW (1988) Stromal cell lines which support lymphocyte growth: Characterization, sensitivity to radiation and responsiveness to growth factors. Eur J Immunol (In Press)
Polla BS, Poljak A, Ohara J, Paul WE, Glimcher LH (1986) Regulation of class II gene expression: Analysis in B cell stimulatory factor 1-inducible murine pre-B cell lines. J Immunol 137:3332-3337
Rennick D, Yang G, Gemmell L, Lee F (1987) Control of hemopoiesis by a bone marrow stromal cell clone: Lipopolysaccharide-and interleukin-1-inducible production of colony-stimulating factors. Blood 69:682-691
Shadduck RK, Waheed A, Greenberger JS, Dexter TM (1983) Production of colony-stimulating factor in long-term bone marrow cultures. J Cell Physiol 114:88-92
Song ZX, Shadduck RK, Innes DJ, Waheed A, Quesenberry PJ (1985) Hematopoietic factor production by a cell line (TC-1) derived from adherent murine marrow cells. Blood 66:273-281
Thomas PS, Pietrangeli CE, Hayashi S-I, Schachner M, Goridis C, Low M, Kincade PW (1988) Demonstration of neural cell adhesion molecules on stromal cells which support lymphopoiesis. Leukemia 2:171-175
Whitlock C, Denis K, Robertson D, Witte O (1985) *In vitro* analysis of murine B-cell development. Ann Rev Immunol 3:213-235
Whitlock CA, Tidmarsh GF, Muller-Sieburg C, Weissman IL (1987) Bone marrow stromal cell lines with lymphopoietic activity express high levels of a pre-B neoplasia-associated molecule. Cell 48:1009-1021
Whitlock NA, Witte ON (1982) Long term culture of B lymphocytes and their precursors from murine bone marrow. Proc Natl Acad Sci USA 79:3608-3612

Mac-1^{+} Bone Marrow Cells Include Precursors of B Cells and T Cells

K. L. HOLMES[1,2], J. S. LEE[2], and H. C. MORSE III[2]

ABSTRACT

Previous studies of lymphomas and normal cells suggested that the putatively macrophage-specific antigen, Mac-1 may be expressed by precursors of the B cell lineage. This possibility was tested by analyzing the potential of sorted Mac-1^{+} cells from adult bone marrow (BM) to reconstitute CB.17-scid mice or sublethally irradiated B6-Ly-5.2 congenic mice. The results show that Mac-1^{+} BM cells include precursors to both T and B lymphocytes as determined by FACS analysis of spleen, bone marrow and thymus from mice inoculated with Mac-1^{+} cells. Three color FACS analysis showed that thymocyte subpopulations from B6-congenic mice inoculated with Mac-1^{+} cells were composed largely of donor cells, indicating that the donor-derived peripheral T cells had undergone normal thymic processing.

INTRODUCTION

The bone marrow (BM) is the primary site of hematopoiesis in adult animals, containing pluripotent stem cells capable of differentiating into all lymphoid, myeloid and erythroid elements as well as precursors with more restricted potentials for differentiation. Analyses utilizing mAb directed against cell surface differentiation-associated antigens have provided a powerful approach to evaluating the relationships among these precursors, particularly within the B cell lineage. Early B cell differentiation has been divided into stages based upon the sequential expression of Ly-5(B220), ThB (Coffman 1982) surface Ig and Ia (Dasch and Jones 1986). However, the phenotypic characteristics of the immediate precursor to the Ly-5(B220)$^{+}$, ThB^{-}, sIg^{-} large pre-B cell are not well defined. Studies of normal animals have shown that Thy-1^{+}, Ly-5(B220)$^{-}$, Mac-1^{-}, 8C5^{-}, Ly-2^{-}, Ly-4^{-} BM cells have the capacity to reconstitute all cell lineages in vivo (Muller-Sieberg et al. 1986). In addition, these cells may initiate long term B lineage cultures, presumably via a precursor defined by the absence of these antigens (Muller-Sieberg et al. 1986). Additional studies of normal cells and a large number of lymphomas (Davidson et al. 1984; Holmes and Morse 1984; Holmes, Pierce, et al. 1986) suggested the existence of the pro-B and pro-GMB cell stages of differentiation characterized by the expression of Lyb-2, Ly-17 (the Fc gamma receptor; Holmes et al. 1985) and Mac-1 (the C3bi receptor; Springer et al. 1984) a putatively macrophage specific antigen.

[1]Flow Cytometry Section, Biological Resources Branch, NIAID, National Institutes of Health, Bethesda, MD 20892
[2]Laboratory of Immunopathology, NIAID, National Institutes of Health, Bethesda, MD 20892

Current Topics in Microbiology and Immunology, Vol. 141

In order to examine the postulated relationship between Mac-1+ expression and lymphopoiesis, we evaluated the capacity of sorted Mac-1+ cells to reconstitute immune deficient or irradiated mice. Our results show that Mac-1+ BM cells from adult mice contain precursors of both the B and T cell lineages.

MATERIALS AND METHODS

Mice

CB.17-scid mice (Bosma et al. 1983) were from our colony established with breeding stock provided by Dr. D. Mosier (Medical Biology Instítiute, LaJolla, CA). C57BL/6-Ly-5.1 congenic mice, BALB/cAnN and C57BL/6N mice were obtained from the colonies of the National Institutes of Health. The nomenclature for the Ly-5 antigen used in this report follows guidelines established for Ly nomenclature (Morse et al. 1987) in which B6 mice carry the $Ly\text{-}5^b$ allele and the Ly-5.2 specificity.

Reagents and FACS Analyses

mAb to Mac-1 (clone M1/70), Ly-5(B220) (RA3-6B2), Ly-4 (GK-1.5) were used labeled directly with FITC. FITC-labeled mAb to Thy-1.2 (30-H12) and Ly-2 (53-6.7) were purchased from Becton Dickinson (Mountain View, CA). FITC-labeled mAb to mouse kappa was purchased from Serotec (Indianapolis, IN). FITC-labeled mAb to IgM of the b allotype (AF6-78.25), detecting BALB/c but not CB.17-scid IgM was a gift of Dr. P. Lalor (Stanford University, Palo Alto, CA). mAb to Mac-1, labeled with amino-hexanoyl-biotin-N-hydroxysuccinimide ester (Zymed, San Francisco, CA) were detected using either Texas Red (TR) avidin (Zymed) or allophycocyanin (PC) avidin (Biomeda, Foster City, CA). Biotin labeled mAb to Ly-5.2 (104-2) was a gift of Dr. B.J. Fowlkes (NIH). To prevent non-specific, Fc-mediated binding of antibodies, unlabeled anti-Ly-17 (2.4G2) was added prior to staining in all two and three color experiments and in experiments using the anti-Thy-1.2 mAb.

Single and two color FACS was conducted on a Becton Dickinson FACS 440 flow cytometer equipped with argon and argon/dye lasers. Three color analysis was performed on a Becton Dickinson FACStar Plus flow cytometer equipped with argon and argon/dye lasers. Non-viable cells were excluded from analysis by narrow forward angle light scatter and uptake of propidium iodide (6) All sorting was done on a Coulter Electronics (Hialeah, FL) Epics 753 equipped with argon and argon/dye lasers.

Experimental protocol:

Reconstitution of Scid Mice

In initial experiments, Mac-1+ cells detected with FITC-labeled anti-Mac-1 were sorted from whole BM of BALB/c mice and injected iv into CB.17-scid mice irradiated with 400R from a Cs radiation source (Gammacell 40, Type B; Atomic Energy of Canada, Ltd, Ottowa, Canada) according to Fulop and Phillips (1986). In experiments 4 and 5, BALB/c BM cells were mixed 3:1 with spleen cells from the same animal. From this BM/Spleen mix, Mac-1+ cells, detected with biotin-labeled anti-Mac-1 and TR or PC-avidin, or Ly-2+/Ly-4+ cells, detected with FITC-labeled mAb, were sorted and injected iv into separate scid recipients. This procedure was done to discount the possibility that the 1-4% contamination of Mac-1+ sorted cells with Mac-1- cells was responsible for any reconstitution observed. Mice were analyzed for reconstitution by FACS analyses of spleen and BM 3 to $5^1/_2$ weeks after inoculation.

Construction of B6 Radiation Chimeras

Mac-1$^+$ or Ly-2$^+$/Ly-4$^+$ cells from BM/spleen cell mixtures of C57BL/6 (Ly-5.2) mice were inoculated into 750R-irradiated C57BL/6-Ly-5.1 mice. Recipient mice were maintained on laminar flow racks in sterilized cages with sterilized food and bedding and received water containing 2g/l antibiotics (Biosol 325; Upjohn Co., Kalamazoo, MI). Mice were analyzed for reconstitution by FACS analyses of spleen, BM and thymus at 5 weeks after injection of sorted cells.

RESULTS AND DISCUSSION

Sublethally irradiated *scid* mice were injected with 5 X 10^5 to 1 X 10^6 sorted Mac-1$^+$ or Ly-2$^+$/Ly-4$^+$ cells from BM or BM/spleen mixtures of BALB/c donor mice aged 2-4 weeks. Mac-1+ sorted cells were always 96% or more pure upon reanalysis. Reconstitution was assessed 3 to 5 1/2 wk later by FACS analyses of spleens. The results of these experiments, summarized in Table I, show that the spleens of mice injected with Mac-1$^+$ BM cells contained between 20% and 60% B cells as indicated by expression of Ly-5(B220), kappa or IgM.

Table I. Reconstitution of *scid* Mice with Mac-1$^+$ Bone Marrow Cells[1]

Donor Cells			Analysis[2]		
Exp.	Age (weeks)	Cells injected	Weeks after injection	Splenic B cells (%)	Splenic T cells (%)
1	6	Mac-1$^+$	5.5	47	20
2	9	Mac-1$^+$	5	63	13
3	2	Mac-1$^+$	5.5	60	23
4	3	Ly-2$^+$/4$^+$	3	<1	3
5		Mac-1$^+$	5.5	20	19
		Ly-2$^+$/4$^+$		<1	2

[1] CB.17-*scid* mice were irradiated with 400R and injected iv with 5 x10^5 to 1 x10^6 sorted Mac-1$^+$ BM cells from Balb/c mice in experiments 1 to 3 or with Mac-1$^+$ or Ly-2$^+$/Ly-4$^+$ cells from BM/spleen mixtures of Balb/c mice in experiments 4 and 5.

[2] Spleen cells were analyzed by FACS at times indicated. B cells were detected by reactivity with mAb to kappa, Ly-5(B220) or IgM; T cells were detected by reactivity with Thy-1.2, Ly-2 or Ly-4.

The splenic kappa$^+$ B cells of the mice inoculated with Mac-1$^+$ cells failed to react with the anti-IgM mAb specific for the host (b) allotype demonstrating that they were of donor origin (Figure 1). In contrast, spleens from uninoculated, irradiated <u>scid</u> mice or mice inoculated with Ly-2+/Ly-4+ cells had no cells reactive with mAb to kappa, IgM or Ly-5(B220).

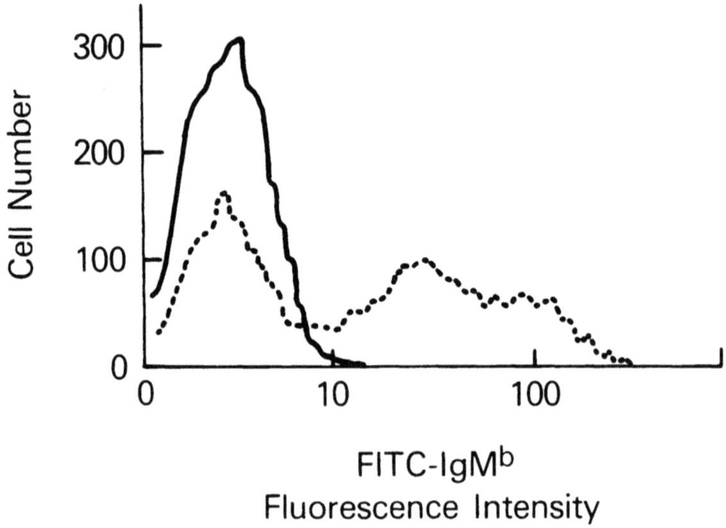

Figure 1:
FACS analysis of spleen cells from a C57BL/6 mouse (IgM^b allotype; dashed line) or from a scid mouse inoculated with Mac-1$^+$ cells from BALB/c (IgM^a allotype;solid line) BM. Cells were reacted with FITC-labeled anti-IgM^b mAb

Surprisingly, when spleen cells from mice inoculated with Mac-1$^+$ cells were analyzed for expression Thy-1, Ly-2 and Ly-4, all had between 13% and 23% T cells (Table 1). Spleens of mice receiving Ly-2$^+$/Ly-4$^+$ cells contained 2-3% T cells (Table I) but no T or B cells were detected in scid mice receiving 400R alone.

To determine if both the T and B cells in mice reconstituted with Mac-1+ cells were of donor origin, radiation chimeras were constructed using 5 X 10^5 sorted Mac-1$^+$ or Ly-2$^+$/Ly-4$^+$ cells from B6(Ly-5.2) mice for transfer into B6-Ly-5.1 congenic hosts. Since Ly-5 is a pan-leukocyte antigen, donor cells can be distinguished from host for lymphoid and myeloid cells by their reactivity with mAb specific for Ly-5.2. Two color FACS analysis of Ly-5.2 vs kappa, Ly-2, Ly-4 and Mac-1 showed that B cells and T cells in spleen and Mac-1$^+$ cells in BM of mice inoculated with Mac-1$^+$ BM cells were comprised of both donor (Ly-5.2$^+$) and host (Ly-5.2$^-$) cells (Figure 2;Ly-2 profile not shown). Spleens of mice inoculated with Ly-2$^+$/Ly-4$^+$ cells were also comprised of donor and host Thy-1+ cells, but contained only host kappa+ cells, and only host Mac-1+ cells were detected in BM of these animals (Figure 2).

To evaluate the composition of T lineage cells in the thymuses of chimeric mice, thymocytes from B6-Ly-5.1 mice inoculated either with Mac-1+ or Ly-2+/Ly-4+ cells were examined by three-color FACS analysis for the expression of Ly-2, Ly-4 and Ly-5.2. The results show that in the thymus of a mouse inoculated with Mac-1+ cells, all four major thymocyte subpopulations, ie. Ly-2+, Ly-4+, Ly-2+,4+ and Ly-2-,4- consisted of donor and host cells (Figure 3). Each subpopulation was comprised of >75% donor cells, except for the Ly-2-,4- population which consisted of 46% donor cells. In contrast, no donor cells were detected in any of the thymocyte

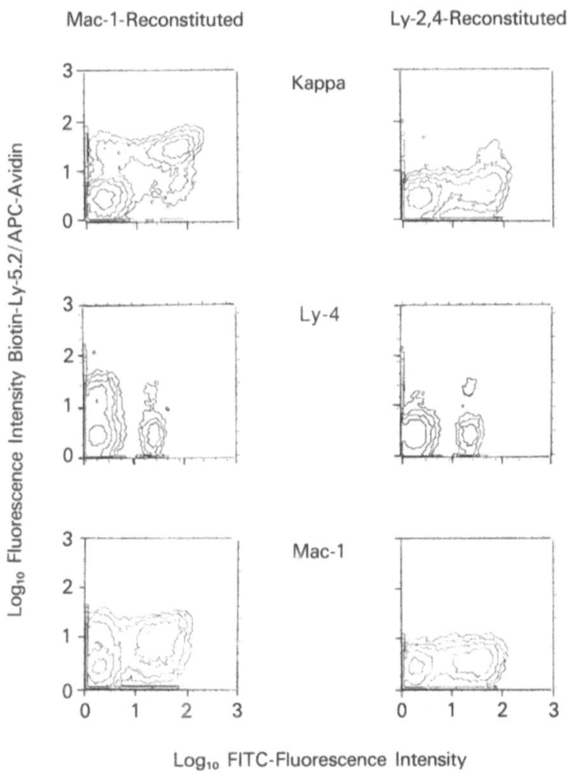

Figure 2: Two color FACS analysis of spleen cells (top four panels) and BM cells (bottom two panels) from C57BL/6-Ly-5.1 congenic mice inoculated either with Mac-1+ cells (left panels) or Ly-2+/4+ cells (right panels) from a C57BL/6 (Ly-5.2) donor. Cells were reacted with biotin anti-Ly-5.2 mAb and PC-avidin and with FITC-labeled anti-kappa, anti-Ly-4 or anti-Mac-1 mAb as shown

subpopulations of the mouse inoculated with Ly-2+/Ly-4+ cells (Fig. 4).

The results of these studies show that Mac-1+ cells in BM include cells capable of differentiation within the B and T cell lineages. The expression of Mac-1 by B lineage precursors was predicted by the observed coexpression of Mac-1 and B lineage antigens Lyb-2 and Ly-5(B220) on normal cells (Holmes and Morse 1984) and on several pre-B cell lymphomas (Davidson et al. 1984; Holmes, Pierce et al. 1986) The reconstitution of T cells with Mac-1+ cells may also have been predicted from recent observations that 9 of 52 T lineage lymphomas, induced with a unique MCF murine leukemia virus, Cas-NS-6 (Holmes, Langdon et al. 1986), were Thy-1+ Ly-4+ Mac-1+ (Holmes, manuscript in preparation).

Mac-1-Reconstituted

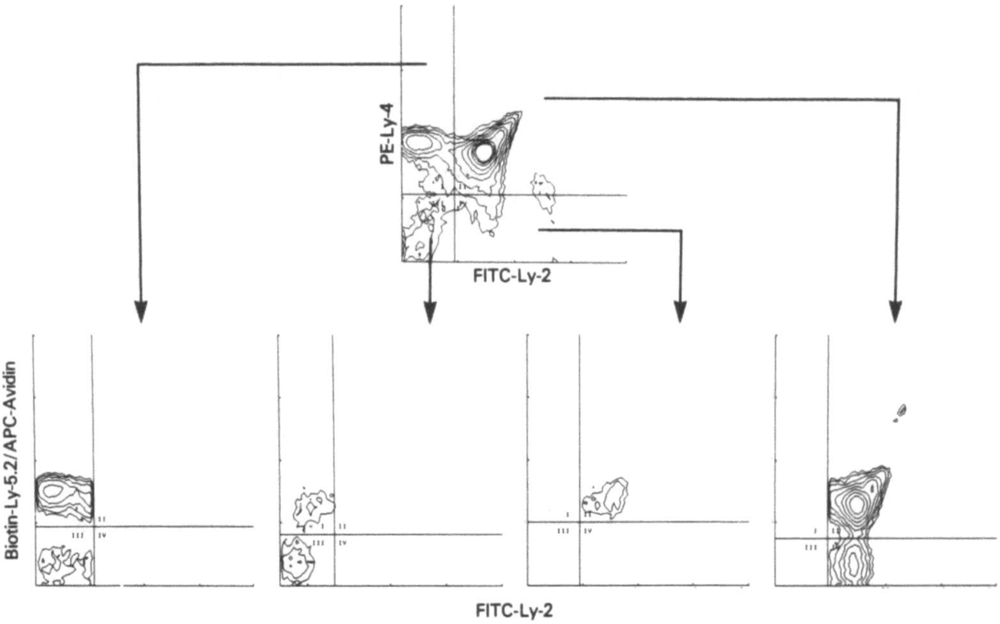

Ly-2,4-Reconstituted

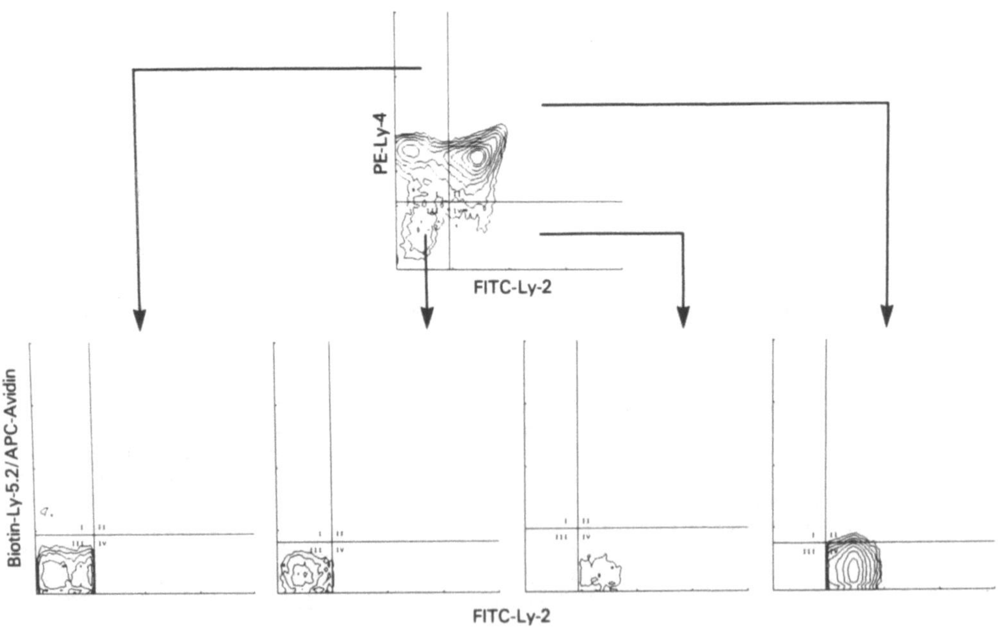

Figure 3:
Three color FACS analysis of thymocytes from a C57BL/6-Ly-5.1 congenic mouse inoculated with Mac-1$^+$ cells from a C57BL/6 (Ly-5.2) donor. Cells were reacted with FITC-labeled anti-Ly-2, phycoerythrin-labeled anti-Ly-4 and with biotin labeled anti-Ly-5.2 mAb and PC-avidin. Each of the four thymocyte subpopulations, identified in the Ly-2 vs. Ly-4 profile, are shown for their expression of Ly-5.2 vs. Ly-2 (four bottom panels). Note that the thymus contains cells of both donor (Ly-5.2$^+$) and host (Ly-5.2$^-$) origin

Figure 4:
Three color FACS analysis, as above, of thymocytes from a C57BL/6-Ly-5.1 congenic mice inoculated with Ly-2$^+$/4$^+$ cells from a C57BL/6 donor. Note that the thymus contains only cells of host (Ly-5.2$^-$) origin

The expression of Mac-1 in lymphoid differentiation appears most often to be transient as Mac-1$^+$ Ly-5(B220)$^+$ or Mac-1$^+$ Thy-1$^+$ cells were not detected by two color FACS analysis of BM, spleen or thymus of the scid mouse inoculated with Mac-1$^+$ cells in experiment #2 (data not shown). However, the observation that the Ly-1$^+$ sIg$^+$ B cell lymphoma, BCL_1 is Mac-1$^+$ (Davidson et al. 1984) and that Mac-1 is expressed by Ly-1$^+$ and Ly-1$^-$ B cells in the peritoneum of normal mice (Herzenberg et al. 1987) suggests that the Mac-1 antigen may persist in some B cell populations. The observed lack of Mac-1 expression on splenic B cells of Mac-1-inoculated scid mice is consistent with transfer experiments showing that adult BM is a poor source of precursors for Mac-1$^+$ mature B cells (Hayakawa et al. 1985).

It remains to be determined whether these injected Mac-1+ BM cells can mediate long term reconstitution of lymphoid cells. Although it is clear that Mac-1$^+$ BM cells include thymocyte precursors, experiments in which mice are analyzed for lymphoid reconstitution at time points longer than 5 weeks are necessary to assess their self-renewal capacity.

REFERENCES

Bosma GC, Custer RP, Bosma MJ (1983) A severe combined immunodeficiency mutation in the mouse. Nature (Lond.) 301:527-530.

Coffman RL (1982) Surface antigen expression and immunoglobulin gene rearrangement during mouse pre-B cell development. Immunological Rev 69:5-23.

Dasch JR, Jones PP (1986) Independent regulation of IgM, IgD, and Ia antigen expression in cultured immature B lymphocytes. J Exp Med 163:938-951.

Davidson WF, Fredrickson TN, Rudikoff EK, Coffman RL, Hartley JW, Morse HC III (1984) A unique series of lymphomas related to the Ly-1 lineage of B lymphocyte differentiation. J Immunol 133:744-753.

Fulop G, Phillips RA (1986) Full reconstitution of the immune deficiency in scid mice with normal stem cells requires low-dose irradiation of the recipients. J Immunol 136:4438-4443.

Hayakawa K, Hardy RR, Herzenberg LA, Herzenberg LA (1985) Progenitors for Ly-1 B cells are distinct from progenitors for other B cells. J Exp Med 161:1554-1568.

Herzenberg LA, Stall AM, Braun J, Weaver D, Baltimore D, Herzenberg LA, Grosschedl R (1987) Depletion of the predominant B-cell population in immunoglobulin u heavy-chain transgenic mice. Nature (Lond.) 329:71-73.

Holmes KL, Langdon WY, Fredrickson TN, Coffman RL, Hoffman PM, Hartley JW, Morse HC III (1986) Analysis of neoplasms induced by Cas-Br-M MuLV tumor extracts. J Immunol 137:679-688.

Holmes KL, Morse HC III (1984) Coexpression of Lyb-2 and Mac-1 by murine lymphomas and subpopulations of normal spleen and bone marrow cells. Curr Top Microbiol Immunol 113:224-230.

Holmes KL, Palfree RGE, Hammerling U, Morse HC III (1985) Alleles of the Ly-17 alloantigen define polymorphisms of the murine IgG Fc receptor. Proc Natl Acad Sci (USA) 82:7706-7710.

Holmes KL, Pierce JH, Davidson WF, Morse HC III (1986) Murine hematopoietic cells with pre-B or pre-B/myeloid characteristics are generated by in vitro transformation with retroviruses containing fes, ras, abl, and src oncogenes. J Exp Med 164:443-457.

Morse HC III, Shen F-, Hammerling U (1987) Genetic nomenclature for loci controlling mouse lymphocyte antigens. Immunogenetics 25:71-78.

Muller-Sieburg CE, Whitlock CA, Weissman IL (1986) Isolation of two early B lymphocyte progenitors from mouse marrow: A committed pre-pre-B cell and a clonogenic Thy-1 lo hematopoietic stem cell. Cell 44:653-662.

Springer TA, Thompson WS, Miller LJ, Schmalstieg FC, Anderson DC (1984) Inherited deficiency of the Mac-1, LFA-1, p150,95 glycoprotein family and its molecular basis. J Exp Med 160:1901-1918.

Molecular Characterization of a Transforming Retrovirus Involved in Pre-B Cell Lymphomas

W. Y. LANGDON[1], J. W. HARTLEY[2], S. P. KLINKEN[1], S. K. RUSCETTI[3], and H. C. MORSE III[2]

INTRODUCTION

Cas-Br-M is a biologically cloned ecotropic murine leukemia virus [MuLV] originally isolated from wild mice of the Lake Casitas region of California. Inoculation of Cas-Br-M into newborn NFS/N mice induces hematopoietic neoplasms. The tumors are evident at 18-30 weeks and include T- and B-cell lymphomas, myeloid leukemias and erythroleukemias [Fredrickson et al, 1984]. It is unclear how Cas-Br-M is able to génerate such a wide spectrum of hematopoietic neoplasms, although envelope-recombinant mink cell focus-forming [MCF] MuLVs are involved in some of these tumors [Holmes et al, 1986]. This was determined by preparing cell-free extracts from a range of tumors and injecting the extracts into newborn NFS/N mice. It was reasoned that if Cas-Br-M was the etiological agent then the mice should develop a range of tumor phenotypes with long latencies. If, however, MCF viruses were involved, the mice should develop tumors with the same phenotype as the extracted tumor and with reduced latency. Of 12 extracts examined, three appeared to support the latter hypothesis [Holmes et al, 1986]. On further examination one of these extracts, called Cas NS-1, was found to contain a virus capable of transforming mouse fibroblasts. This extract had been prepared from the spleen of a mouse that developed a pre-B cell lymphoma following Cas-Br-M inoculation. When the extract was inoculated into newborn NFS/N mice, 8 of 10 tumors were found to be pre-B lymphomas. This suggested that a transforming virus, and not an MCF virus, induced these tumors. Here we describe the molecular cloning of the transforming virus and preliminary characterization of its oncogene.

RESULTS AND DISCUSSION

Molecular cloning of the Cas NS-1 transforming virus.
The Cas NS-1 extract contains a mixture of ecotropic, MCF and transforming viruses [Langdon et al, 1984]. To enable the cloning of the transforming virus we attempted to isolate a transformed nonproducer cell line using normal rat kidney [NRK] cells. NRK cells were chosen in preference to a mouse cell line since the endogenous viral sequences present in the rat genome do not hybridize to MuLV probes at high stringency. Initial screenings to isolate a nonproducer cell line were unsuccessful. However, a transformed line free of MCF virus was obtained. DNA from these cells was used for preliminary genomic restriction mapping and subsequent molecular cloning. As shown in Fig 1, DNA from the transformed NRK cells showed more viral insertions than would be expected from the ecotropic Cas-Br-M virus and the transforming virus. Using the viral probes shown in Fig 1 to hybridize Eco RI digested NRK DNA it was determined that there were 4 viral integrations. These were 2 integrations of an envelope negative virus [in the 11 and 17 kb Eco RI fragments], one

[1]Walter and Eliza Hall Institute of Medical Research, PO Royal Melbourne Hospital, Victoria 3050, Australia
[2]Laboratory of Immunopathology and [3]Laboratory of Genetics, NIH, Bethesda, MD 20892

Current Topics in Microbiology and Immunology, Vol. 141

integration of Cas-Br-M [in the 13 kb Eco RI fragment] and one integration of a virus with a recombinant xenotropic/MCF envelope [spanning three Eco RI fragments of 4.0, 3.6 and 2.3 kb]. Further mapping showed this virus to be smaller than a replication-competent MCF virus. The presence of two internal Eco RI sites also distinguished this virus from any known MCF viruses.

Since either the envelope-negative virus or the "defective MCF" could be candidates for the transforming virus, attempts were made to molecularly clone both. Two genomic libraries were prepared; one from Eco RI digested DNA into bacteriophage Charon 4A and one from partial MboI digested DNA into λBF101. The libraries were screened with an LTR probe and the Charon 4A library yielded one of the envelope-negative virus integrations [11 kb Eco RI fragment], the ecotropic virus [13 kb Eco RI fragment] and the envelope and 3′ LTR of the "defective MCF" [3.6 kb Eco RI fragment]. The λBF101 library yielded many partial clones, one of which contained most of the "defective MCF". This clone overlapped with the Charon 4A "defective MCF" envelope clone at a BclI site common to MCF viruses. This site was used to join the clones and the complete "defective MCF" was subcloned into a pUC plasmid vector. Similarly, the envelope-negative virus was subcloned into a pUC plasmid.

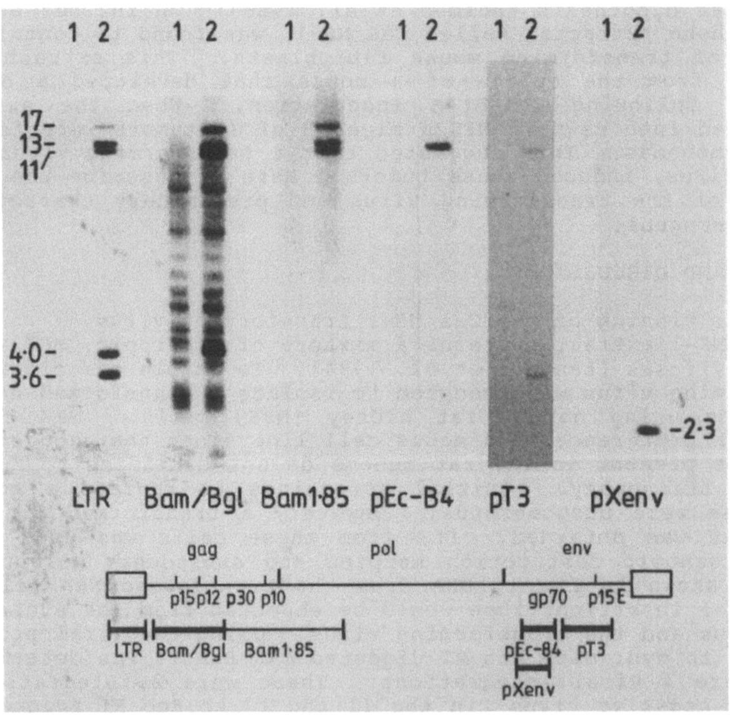

Figure 1
MuLV insertions in normal rat kidney [NRK] cells transformed with Cas NS-1 virus. The DNA was digested with Eco RI and hybridized to the MuLV probes shown above. 1. NRK control DNA, 2. NRK DNA from cells transformed with Cas-NS-1. The pEc-B4 probe is an ecotropic-specific envelope probe, and pXenv is a xenotropic/MCF-specific envelope probe. The Bam/Bgl probe detected endogenous rat gag sequences since the washes (55°C, 0.1XSSC) were not as stringent as for the other probes (65°C, 0.1XSSC)

Transforming properties and restriction mapping of the "defective MCF" virus.
DNA transfection of the viral clones into mouse fibroblasts conclusively proved the "defective MCF" to be the transforming virus in the Cas NS-1 extract. In contrast, the envelope negative viral DNA had no effect on mouse fibroblasts. The morphologies of the transformed and normal fibroblasts are shown in Fig 2.

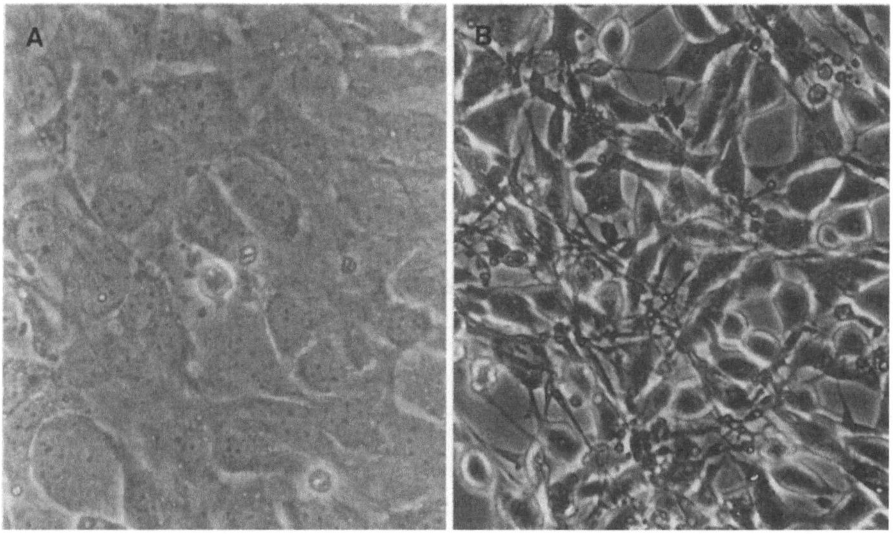

Figure 2
Morphology of A. normal Ψ2 mouse fibroblasts and B. Ψ2 fibroblasts transfected with Cas NS-1 viral DNA.

The molecular cloning of the transforming Cas NS-1 virus allowed for more extensive restriction mapping. A detailed restriction map is shown in Fig 3. By comparing the Cas NS-1 virus with Cas-Br-M, and an MCF virus [NS-6] derived from Cas-Br-M, it was possible to determine the sites where recombination had occurred. As shown in Fig 3 the Cas NS-1 oncogenic virus contains the gag p30 Xba I site [at 2.1 kb] common to both Cas-Br-M and NS-6 viruses. However, restriction sites 3′ of the XbaI site differed in Cas-NS-1. The BclI site of Cas-Br-M and NS-6, present in the 5′ end of pol at 3.1 kb, had been lost and a Sac I site at 2.6 kb gained. This suggested the virus-oncogene junction was in gag p10, between p30 and pol. This is consistent with the inability of anti-gag p10 antibodies to immunoprecipitate the 100,000 dalton gag-onc fusion protein [S.R. unpublished observations]. Further mapping showed the Cas NS-1 virus to have a typical recombinant MCF envelope and the coding region to be intact. This indicated the oncogene terminates at the 3′ end of pol.

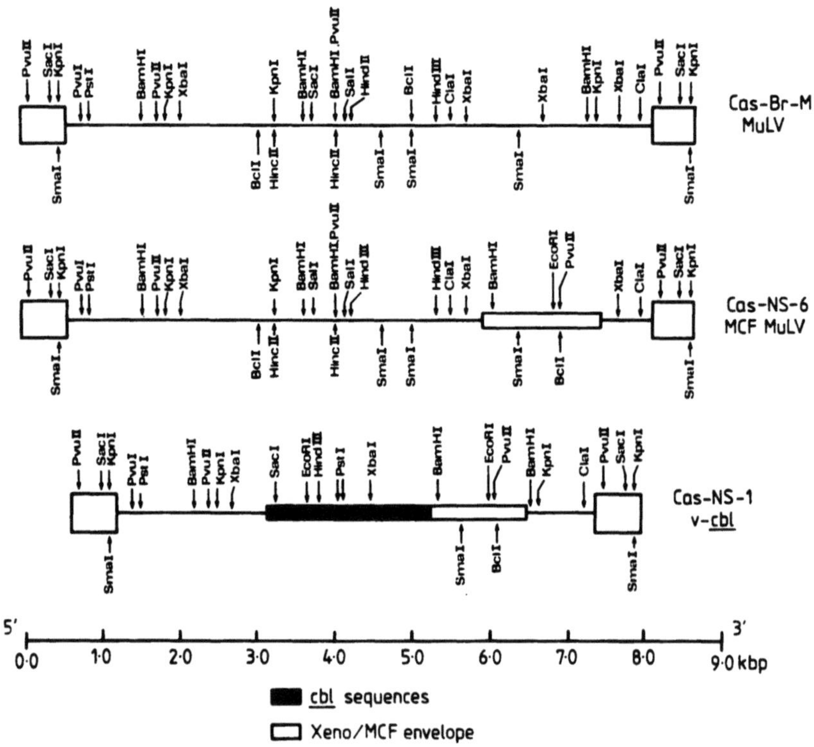

Figure 3
Restriction enzyme maps of Cas Br-M ecotropic MuLV, Cas NS-6 MCF MuLV and Cas NS-1 transforming MuLV. The Cas NS-1 virus has provisionally been called v-cbl (pronounced Sybil) for Casitas B-lineage Lymphoma.

In summary, the restriction mapping showed Cas NS-1 to be a dual recombinant virus, 7.4 kb long, of which 2.2 kb is non-viral. Since the parental Cas-Br-M virus is 8.8 kb it can be deduced that Cas NS-1 has lost 3.6 kb of viral sequence - equivalent to the size of the pol gene. Sequencing of the oncogene in Cas NS-1 is in progress. Preliminary results, coupled with searches of sequence data bases suggest that this virus might contain a new oncogene.

References

Fredrickson TN, Langdon WY, Hoffman PM, Hartley JW and Morse H C III (1984) Histologic and cell surface antigen expression of hematopoietic tumors induced by Cas-Br-M murine leukemia virus. JNCI 72:447-454

Holmes KL, Langdon WY, Fredrickson TN, Coffman RL, Hoffman PM, Hartley JW and Morse HC III (1986) Analysis of neoplasms induced by Cas-Br-M tumor extracts. J Immunol 137:679-688

Langdon WY, Hartley JW, Holmes KH, Fredrickson TN and Morse HC III (1984) Identification of a transforming virus from a lymphoma of a mouse infected with a wild mouse retrovirus. Curr Topics Microbiol Immunol 113:241-249

In Vitro Transformation of Murine Bone Marrow Cells with a v-*raf*/v-*myc* Retrovirus Yields Clonally Related Mature B Cells and Macrophages

M. PRINCIPATO[1], S.P. KLINKEN[1], J.L. CLEVELAND[3], U.R. RAPP[3], K.L. HOLMES[1], J.H. PIERCE[2], and H.C. MORSE III[1]

INTRODUCTION

Observations by various laboratories have indicated that lymphoid and myeloid cells derive from distinct precursors (Abramson, et al. 1977; Williams, et al. 1984) while other analyses have suggested that the differentiative programs of B cells and macrophages may be closely related. Some tumors (Davidson, et al. 1984; Neame et al. 1985; Strong, et al. 1985) and perinatal spleen and bone marrow cells were shown to coexpress putative B and myeloid lineage specific antigens (Holmes and Morse III, 1984), and tumors of B cell precursors have differentiated into cells with characteristics of mature macrophages (Boyd and Schrader, 1982; Holmes et al. 1986; Davidson et al. 1988). The demonstration of shared immunoglobulin heavy and light chain gene rearrangements in myeloid and B tumor cell lines have also suggested a close developmental relationship between these lineages (Holmes et al. 1986; Davidson et al. 1988; Bauer et al. 1986).

Morse and coworkers have previously reported a striking synergism between raf and myc oncogenes for transformation of hematopoietic cells in vivo utilizing a series of retroviral constructs containing v-myc, v-raf, or both oncogenes (Rapp et al. 1985; Cleveland et al. 1986; Morse et al. 1986). Thus, while v-raf induced erythroid hyperplasia and fibrosarcomas in muscle and spleen, deregulation of v-myc expression following inoculation of J3 produced lymphomas of T, pre-B, and B cell lineages. Animals inoculated with the v-raf/v-myc containing J-2 virus uniformly exhibited an expansion of the target population and a summation of the diseases induced with either v-raf or v-myc alone. Flow cytometry analysis of the J2 induced tumors demonstrated mixtures of T cell and B lineage blasts.

The in vivo induction of multiple hematopoietic tumors, and subsequent in vitro derivation of myeloid transformants following infection of fetal liver cells by J-2 (Blasi et al. 1985; Blasi et al. 1987) presents the question as to whether the target for v-raf/v-myc transformation represents the independent targeting of cells committed to each of these hematopoietic lineages, or whether the target is a common precursor, which can give rise to clonally related hematopoietic tumors. To evaluate this possibility, we compared the cell types obtained following in vitro infection of NFS/N murine bone marrow cells with v-raf (3611 MSV), v-myc (J-3), or v-raf/v-myc (J-2) oncogenes in a transformation assay system which permits growth of B and myeloid lineage cells (Pierce and Aaronson 1982; Holmes, Pierce et al. 1986).

Laboratory of Immunopathology, NIAID[1]; Laboratory of Cellular and Molecular Biology[2], and Laboratory of Viral Carcinogenesis[3], NCI, National Institutes of Health, Bethesda, MD

Current Topics in Microbiology and Immunology, Vol. 141

In Vitro Transformation of Bone Marrow Cells

J-3 and 3611-MSV stimulated moderate growth of lymphoid cells but were ineffective in transforming myeloid precursors. In contrast, the J-2 raf/myc construct induced the rapid outgrowth of lymphocytes and myeloid cells in the presence of 2-ME. Single lines of 3611-MSV or J-3 transformed cells were selected for further study along with three lines, designated J2C3, J2C5 and J2C9, transformed by J-2. The 3611-MSV and J3 lines retained a homogeneous lymphoid morphology in long-term culture. In contrast, each of the J-2 lines rapidly developed into heterogeneous mixtures of cells with nonadherant lymphoid features and adherant cells with characteristics of macrophages. FACS analyses of these lines (Table 1) showed that the 3611-MSV and J-3 lines had characteristics of pre-B cells; they were uniformly Lyb-2^+, Ly-5(B220)$^+$ and heterogeneous for ThB expression but did not express surface Ig (Davidson, et al. 1984; Holmes, et al. 1986). In addition, they lacked the T cell marker, Thy-1, and were negative for expression of the putatively myeloid specific marker, Mac-1. The demonstration by radioimmunoassays (RIA) that both lines contained cytoplasmic mu chains confirmed their designations as pre-B cells. By comparison, each of the J-2 lines was uniformly Ly-5(B220)$^+$ but contained subpopulations of Mac-1^+, IgM^+, and $kappa^+$ cells, consistent with the heterogeneous appearances of the cultures. These results suggested that the J-2 lines had differentiated along two pathways, becoming either mature B cells or macrophages, or that some cells coexpressed myeloid and B lineage markers.

Table 1. FMF analysis of cell lines derived from *in vitro* transformation with 3611-MSV J-3 and J-2

	Cell surface antigens						
Cell line	Thy-1	Mac-1	Lyb-2.1	Ly-5(B220)	Kappa	IgM	Ly-1
3611-1	-	-	+++	+	-	-	-
J3	-	-	+++	++	-	-	-
J2C3	-	-/+	-	++	-/+	-/+	+
J2C5	-	-/+	-	++	-/+	-/+	+
J2C9	-	-/+	-	++	-/+	-/+	+

Single cell suspensions were prepared and stained for FMF assays as described. The origins and specificities of the monoclonal antibodies are detailed elsewhere (Holmes et al. 1986).

Reactivity of the cell lines with the indicated antibodies in relation to staining on normal NFS/N spleen cells. -, non reactive; +, reactive but with lower intensity than normal cells; ++, reactive with intensity equal to normal cells; +++, reactive with intensity greater than normal cells; +/-, positive and negative subpopulations of cells with more than 50% of cells positive; -/+, positive and negative subpopulations with less than 50% of cells positive.

Characterization of J-2 Subclones

We therefore cloned the J-2 lines by limiting dilution and analyzed the subclones for expression of lineage-specific determinants. All of the subclones resulted from seedings at 0.3 cell/well or 1.0 cell/well which resulted in growth frequencies of less than 25%

(data not shown). Approximately 30 subclones were analyzed and of these, six J-2C3 clones (clones 1 through 6), two J-2C5 clones (clones A and E), one J-2C9-lymphoid clone (C9-1), and one J-2C9-myeloid subclone (C9-2), were characterized extensively by flow microflourometry (Figure 1). Four phenotypic patterns were defined by this analysis.

In the first category were four clones; (J2C3 clone 2, J2C5 clones A and E, and J2C9-2) which differed from any other normal or transformed cells previously examined in our laboratory in expressing Ly-24 but no other antigens associated with lymphoid or myeloid lineages. Despite their different origins, they were morphologically quite similar, growing as clustered adherant and nonadherant cells with ruffled cytoplasmic membranes and prominent cytoplasmic vacuoles. In spite of their resemblance to macrophages, these cells were not phagocytic and they did not produce nonspecific esterase or lysozyme. In addition, they were negative for cytoplasmic mu chains by RIA. This overall lack of distinguishing characteristics made it impossible to assign these clones to any hematopoietic lineage.

The second category, representative of pre-B cells, included only the J2C3 subclone 1. These cells were uniformly Ly-17(FcR)$^+$, sIg$^-$, and contained a subpopulation of Ly-5(B220)$^+$ cells. In culture, these cells resembled lymphoblasts with large nuclei and sparse cytoplasm. By RIA these cells contained cytoplasmic mu chains but not kappa light chains, confirming their designation as pre-B cells.

The third category included clones with many features of mature B cells. J2C3 clones 4 and 5, and J2C9 lymphoid clone C9-1 were uniformly sIgM+, kappa +, Ly-17(FcR)+, and Ly-5(B220)+. In addition, the J2C3 clones were ThB+ and contained subpopulations of Ia+ and Ly-1+ cells. Morphologically, these subclones resembled B lymphoblasts. None of the clones secreted Ig after stimulation with lipopolysaccharide and all were lysozyme and esterase negative.

The final category included subclones with characteristics of mature macrophages and monocytes. J2C3 clone 6 contained a subpopulation of cells that was Mac-1+, Mac-2+, Ly-17+, and Ia+, thus resembling mature, activated macrophages. Functional studies of this clone showed that the cells phagocytosed latex beads and produced alpha naphtyl esterase and lysozyme. J2C3 clone 3 did not express mature myeloid markers aside from Ly-17, but it was lysozyme positive and was therefore included in this category.

Taken together, these observations indicate that the sIg+, Mac-1+ phenotype obtained in the original parental J2C3 clone did not represent coexpression of cell surface markers, but rather the existance of discrete populations within a single cloned transformant.

Evaluations of Clonality.

The observations that the J-2 subclones included cells from different lineages raised the question of whether the initial lines may have been composed of a series of cells derived from independent transformation events. We therefore analyzed DNA prepared from the lines and subclones described above for the organization of Ig heavy and light chain genes and v-*myc* integrations (Fig. 2).

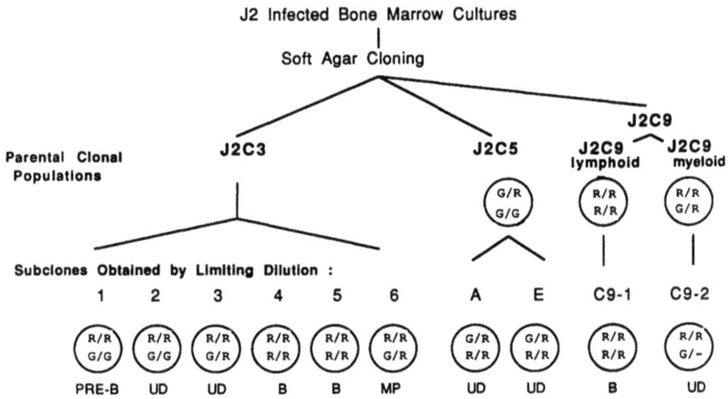

Cell Surface Antigens :

	1	2	3	4	5	6	A	E	C9-1	C9-2
Thy 1	-	-	-	-	-	-	-	-	-	-
IgM	-	-	-	+	+	-	-	-	+	-
K	-	-	-	+	+	-	-	-	+	-
Ia	-	-	-	-/+	-/+	-/+	-	-	-	-
ThB	-	-	-	+	+	-	-	-	-	-
B220	-/+	-	-	+	+	-	-	-	+	-
Lyb 2	-	-	-	-	-	-	-	-	-	-
Ly 17	+	-	-/+	+	+	-/+	-	-	+	-
6C3	-	-	-	-	-	-	-	-	-	-
PC 1	-	-	-	+	-	-	-	-	-	-
Ly 1	-	-	-	-/+	-/+	-/+	-	-	-	-
Mac 1	-	-	-	-	-	-/+	-	-	-	-
Mac 2	-	-	-	-	-	-/+	-	-	-	-
Ly 24	ND	+	+	ND	ND	ND	+	+	ND	+
Ly 5	ND	-	-	ND	ND	ND	-	-	ND	-

Fig. 1. FMF analyses of J2 subclones. Clones obtained by limiting dilution were analyzed by FMF with apanel of antibodies. The organization of Ig genes in each clone is indicated within the cells as germline (G), rearranged (R) or deleted (-) for alleles at the heavy chain (top line) or light chain loci (bottom line)

The organization of the Ig loci was determined by hybridization of EcoR1 digested genomic DNA to a JH specific probe. All of the J2C3 subclones contain identical JH rearrangements at both alleles as evidenced by the prescence of 9 kb and 1.7 kb fragments, and loss of the unrearranged germline 15 kb fragment found in NFS/N liver DNA. This observation would strongly suggest that all of the cells obtained, including pre-B, B, and myeloid lineages, were derived from a single transformed cell. A comparison of the J2C5 and J2C9 clones and their subclones also demonstrate identical rearrangements within each family. Thus, each clone and subclone presented identical rearrangements indicative of their clonal origin.

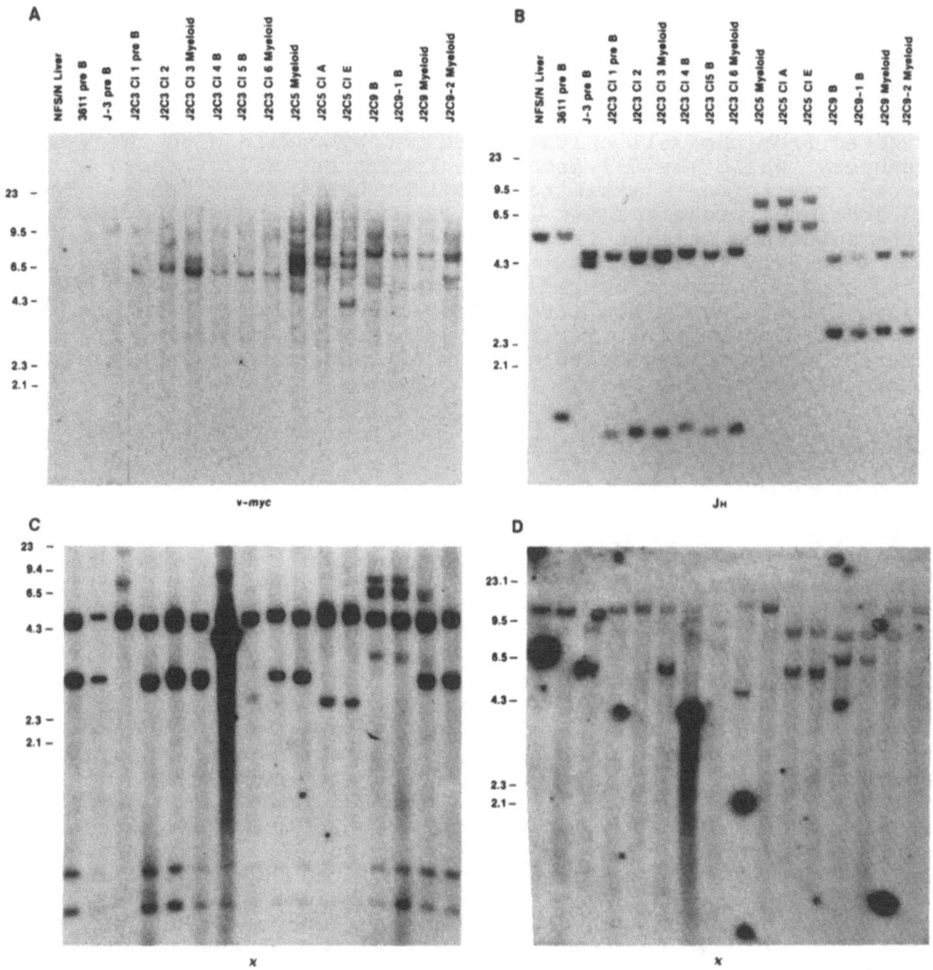

Fig. 2. Southern blot hybridization studies of v-myc integrations and Ig gene organization. 10 µg samples of high molecular weight DNA prepared from normal NFS/N liver or the indicated cell lines and clones was digested with appropriate enzymes, separated in 0.8% agarose gels, transferred to nitrocellulose and hybridized with nick-translated probes by established techniques. Panel A. BamH1 digests with a v-myc specific probe. Panel B. EcoR1 digested DNA hybridized with a JH specific probe (J11). Panel C. Hybridization pf HindIII digests with the pECk probe. Panel D. Bam H1 digests :Organization of kappa genes. Myeloid clones 3 and 6 had single rearrangements of 5.2 and 4.1 kb, repectively. The J2C5 parental population retained germline configuration while clones A and E had rearrangements of 7.5 and 5.5 kb. The J2C9 myeloid culture exhibited a 7.8 kb rearrangement. C9-2 retained one germline allele and appeared to have lost the second allele. In panels C and D, the intense hybridization seen with DNA from J2C3 clone 4 represents contamination with pBR DNA

In order to confirm the clonality of the transformants, DNA from the cultures was digested with BamH1 and hybridized to a v-myc probe which detects MC-29 v-myc sequences incorporated into the J-2 and J-3 retroviral constructs but does not cross hybridize with endogenous c-myc sequences. Thus, NFS/N liver DNA and DNA extracted from the 3611 culture does not hybridize with the v-myc sequences, while the J-3 induced cultures and all of the J-2 induced transformants exhibited distinct hybridizing bands. The J2C3 subclones presented two identical bands, suggesting that all of the subclones were in fact derived from a single transformed cell. The J2C5 family contained multiple bands, with a single shared band. A common hybridizing fragment is evident with the J2C9 family. Variations in the intensities of hybridizing bands in individual clones probably reflect reinfections occurring at different times during culture. Taken together, these data strongly support the clonality of the transformants.

Kappa Light Chain Gene Organization

The organization of kappa light chain genes was analyzed on Southern blots of DNAs digested with Hind III or Bam H1 and hybridized with the Jk-Ck probe, pECk (Fig. 2). The kappa genes of the 3611MSV pre-B cell line were in germline configuration, indicating that the cells were arrested at the large pre-B stage of differentiation. The finding that both alleles were rearranged in the J3 pre-B cell line permitted its designation as a small pre-B cell line.

The J2C3 mature B cell subclones 4 and 5 differed in having rearranged kappa genes, but unexpectedly the rearrangements were distinct for the two clones. Strikingly, analyses of the BamH1 digests revealed that the J2C3 myeloid clones 3 and 6 also had single kappa rearrangements that differed from each other and from those present in the B cell lines. The J2C5 family also exhibited some variability in kappa gene organization. The genes in the parental line were in germline configuration while both alleles were identically rearranged in the undifferentiated subclones. In the J2C9 family, the J2C9 lymphoid line and the derivative C9-1 subclone had two identical rearrangements, one of which was shared by the original myeloid line, but was deleted in the undifferentiated subclone C9-2.

Expression of Cellular and Viral Genes.

Our analyses of Ig gene organization demonstrated that identical and clearly functional rearrangements of heavy chain genes were shared by J2 subclones representative of each of the four general phenotypes, and that rearrangement at the kappa locus was an ongoing process in each cell type. Although FMF ad RIA studies had shown that expression of Ig proteins was restricted to the B and re-B cell lines and subclones, it was unclear if their absence in the myeloid and undifferentiated cell lines was determined at the level of transcription or translation. To answer this question, poly(A)+ RNA was prepared from cells and analyzed by Northern blot hybridizations for expression of mu heavy and kappa light chain genes (Fig. 3). The results of these studies showed that transcription of both loci was completely restricted to lines and subclones of the B cell lineage. The observation that the J-3 pre-B cell line had high levels of kappa transcripts although no protein was detected by RIA indicated that the transcripts were either sterile or functional but not translated.

The same RNA preparations were also analyzed for expression of c-myb, Ia, H-2K, and beta-2 microglobulin. c-myb transcripts were detected in the B lineage but not in the myeloid or undifferentiated lines or subclones. This was somewhat unexpected as many myeloid tumors and cells enriched for hematopoeitic precursors are characterized by high levels of myb expression (Duprey and Boettinger 1985). High levels of Ia mRNA were observed with the J-3 pre-B cell line and low levels with the J2C3 B cell clones 4 and 5.

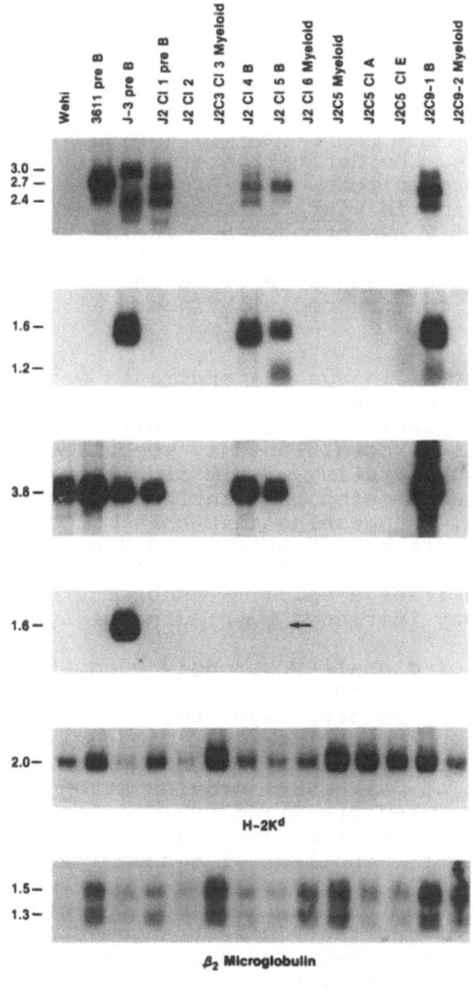

Fig.3 Northern blot analysis. Poly(A)+ mRNA was prepared from 3611MSV, J-3, and J-2 bone marrow transformants and NIH 3T3 fibroblasts infected with the same viruses and separated in 1% agarose/6 % formaldehyde gels by established techniques. Probes utilized for this analysis were purified inserts. The mu probe was a 1.96 kb BamH1/EcoR1 fragment, kappa light chain was detected using a 2.8 kb Hind III fragment of the pECk probe, a 0.5 kb EcoR1 fragment of murine c-myb, an 800 bp EcoR1 fragment of the IAd alpha probe, an 820 bp PvuII/PvuIII fragment of mouse H2-K cDNA, and a 500 bp Pst I fragment of human beta-2 microglobulin. Sizes of mRNAs are shown in kb and were determined from their migration relative to ^{32}P end labelled DNA fragment size standards and 28S and 18S ribosomal RNAs

Although there were variations in the levels of H-2K and B2M expression, no clear associations with the different phenotypes was observed. In further studies, it was found that none of the cells contained messages hybridizing with probes for IL-3, Il-4, IL-6, GM-CSF, or fms (data not shown). Finally, the mRNAs were examined for expression of v-myc, c-myc, and c-raf. All the tumors induced by J-2 or J-3 expressed high levels of genomic and lower levels of spliced messages hybridizing with the v-myc and c-raf probes. In keeping with earlier studies, high levels of v-myc expression were associated with complete suppression of c-myc transcription while c-raf transcripts were readily detected in cells expressing high levels of v-raf (data not shown).

DISCUSSION

We have demonstrated the derivation of clonally related undifferentiated progenitors, pre-B cells, B cells, and macrophages following in vitro infection of murine bone marrow cells with combined v-raf/v-myc oncogenes. The heterogeneous phenotypes obtained were not due to differences in target cell distribution based on LTR regulatory sequence differences, as the vectors in this analysis were constructed with identical LTR sequences (Rapp et al. 1985)

Phenotypically, the B lineage transformants were unusual in that they were negative for ThB and LyB-2, differing from established phenotypes characteristic of B cells. Although the significance of this observation is unknown, one may speculate that the combined oncogenes may be substituting within signal transduction pathways. Past reports have documented that antibodies to Lyb-2 stimulate B cell proliferation, suggesting that this molecule may be a receptor for a B cell growth factor (Subbarao and Mosier 1983). Since the pre-B cell lines induced by 3611-MSV or J-3 were LyB-2+, expression of both genes in B lineage cells may substitute for the signal transduction pathway initiated by binding of Lyb-2 to its ligand.

The demonstration of clonality strongly suggests that the target of v-raf/v-myc induced transformation was common to both B and myeloid lineages. Past analyses have suggested a close developmental relationship between these lineages. The derivation of myeloid cells from pre-B lymphoid HAFTL-3 and BAMC1 cell lines (Holmes et. al. 1986; Davidson et. al. 1988) provide strong support for this hypothesis. The related pre-B and myeloid lines obtained in this instance exhibited shared rearrangements of the immunoglobulin heavy chain locus. However, the lack of concommitant Ig expression by the pre-B lineage cells suggested that these rearrangements were aberrant. This implied that differentiation into the myeloid lineage was a consequence of aberrant Ig rearrangements and may have represented a salvage pathway for B lineage cells incapable of completing their normal differentiation scheme.

This was clearly not the case for the J2C3 B cell clones 4 and 5 and myeloid clone 6 as they exhibited identical JH rearrangents. The presence of different kappa rearrangements in the B cell and macrophage subclones suggests that the events determining final lineage commitment may have occurred only after these rearrangements were initiated.

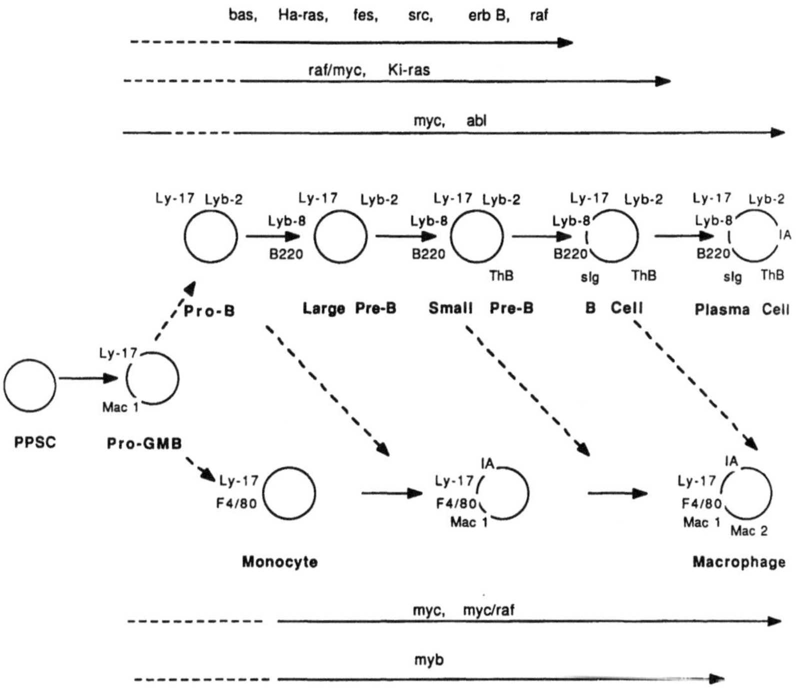

Fig. 4. Proposed pathway of B/myeloid differentiation. Established pathways of maturation are indicated by solid lines while pathways suggested by the present study are indicated by dashed lines. The ability of different oncogenes to transform cells arrested at various stages of differentiation is indicated by lines above or below the lineages, and is based on data presented here and elsewhere (Alt et al. 1981; Holmes et al., 1986; Lichtman et al., 1986; Pierce et al., 1984; Radke, et al., 1982; Rosenberg and Baltimore, 1976) The transition from the B cell to the monocyte/macrophage lineage may occur at any of several stages in the B cell pathway as indicated by cell lines transformed with raf/myc (this study), abl, bas, or Ha-ras (Holmes et al. 1986)

REFERENCES

Abramson S, Miller RG, Phillips RA (1977) The identification in adult bone marrow of pluripotent and restricted stem cells of the myeloid systems. J Exp Med 145: 1567-1579.

Alt F, Rosenberg N, et al. (1981) Organization and reorganization of immunoglobulin genes in A-MuLV-transformed cells: rearrangement of heavy but not light chain genes. Cell 27: 381-390.

Bauer SR, Holmes KL, Morse HC III, Potter M (1986) clonal relationship of the lymphoblastic cell line P388 to the macrophage cell line P388D1 as evidenced by immunoglobulin gene rearrangements and expression of cell surface antigens. J Immunol 136: 4695-4699.

Blasi E, Mathieson BJ, Varesio L, et al. (1985) Selective immortalization of murine macrophages from fresh bone marrow by a raf/myc recombinant murine retrovirus. Nature 318: 667-670.

Blasi, E, Radzioch D, Durum S, Varesio L (1987) A murine macrophage cell line, immortalized by v-raf and v-myc oncogenes, exhibits normal macrophage functions. Eur J Immunol 17: 1491-1498.

Boyd AW, Schrader JW (1982) Derivation of macrophage like lines from the pre-B lymphoma ABLS 8.1 using 5-azacytidine. Nature 297: 691-693.

Cleveland JL, Jansen HW, Bister K et al. (1986) Interaction between raf and myc oncogenes in transformation in vivo and in vitro. J Cell Biochem 30: 195-218.

Davidson WF, Fredrickson TN, Rudikoff EK et al. (1984) A unique series of lymphomas related to the Ly-1^{+} lineage of B lymphocyte differentiation. J Immunol 133: 744-753.

Davidson WF, Pierce JH, Rudikoff S, Morse HC III (1988) Relationships between B cell and myeloid differentiation: Studies with a B lymphocyte progenitor line, HAFTL-1. In Press.

Duprey SP, Boettinger D (1985) Developmental regulation of c-myb in normal myeloid progenitor cells. Proc Natl Acad Sci USA 82: 6937-6941.

Holmes KL, Morse HC III (1984) Coexpression of Lyb-2 and Mac-1 by murine lymphomas and subpopulations of normal spleen and bone marrow cells. Curr Top Microbiol Immunol 113: 224-230.

Holmes KL, Pierce JH, Davidson WF, Morse HC III (1986) Murine hematopoietic cells with pre-B or pre-B/myeloid characteristics are generated by in vitro transformation with retroviruses containing fes, ras, abl and src oncogenes. J Exp Med 164: 443-457.

Lichtman AH, Reynolds DS, et al. (1986) Mature murine B lymphocytes immortalized by Kirsten sarcoma virus. Nature 324: 489-491.

Morse HC III, et al. (1986) Recombinant murine retroviruses containing avian v-myc induce a wide spectrum of neoplasms in newborn mice. Proc Natl Acad Sci USA 83: 6868-6872.

Neame PB, et al. (1985) Simultaneous or sequential expression of lymphoid and myeloid phenotypes in acute leukemia. Blood 65: 142-148.

Pierce JH, Aaronson SA (1982) BALB and Harvey murine sarcoma virus transformation of a novel lymphoid progenitor cell. J Exp Med 156: 873-879.

Pierce JH, Aaronson SA, Anderson SM (1984) Hematopoietic cell transformation by a murine recombinant retrovirus containing the src gene of Rous sarcoma virus. Proc Natl Acad Sci USA 81: 2374-2376.

Radke K, Beug H, Kornfeld S, Graf T (1982) Transformation of both erythroid and myeloid cells by E26, an avian leukemia virus that contains the myb gene Cell 31: 643-653.

Rapp UR, Cleveland JL, Fredrickson TN, et al. (1985) Rapid induction of hematopoietic neoplasms in newborn mice by a raf(mil)/myc recombinant murine retrovirus. J Virol 55: 23-33.

Rosenberg N, Baltimore D (1976) A quantitative assay for transformation of bone marrow cells by Abelson murine leukemia virus. J Exp Med 143: 1453-1463.

Strong RC, Korsmeyer SJ, et al. (1985) Human acute leukemia cell line with the T(4;11) chromosomal rearrangement exhibits B lineage and monocytic characteristics. Blood 65: 21-31.

Subbarao B, Mosier DE (1983) Induction of B lymphocyte proliferation by monoclonal anti-Lyb2 antibody. J Immunol 130: 2033-2037.

Williams DA, Lemischka IR, Nathan DG, Mulligan RC (1984) Introduction of new genetic material into pluripotent hematopoietic stem cells of the mouse. Nature 310: 476-480.

Closely Related BCR/ABL Oncogenes Are Associated with the Distinctive Clinical Biologies of Philadelphia Chromosome Positive Chronic Myelogenous and Acute Lymphocytic Leukemia

O.N. WITTE

OVERVIEW

The study of oncogene activation in naturally occurring human leukemias provides a useful approach to define the range of growth regulatory mechanisms used by hematopoietic stem cells. Recent analysis of the structural changes created by the Philadelphia chromosome translocation [t(9;22)-(q34;q11)] shows that two alternative regions of the BCR gene on chromosome 22 can be used as the breakpoint site. This results in two alternative subsets of exons of the BCR gene remaining on the Philadelphia chromosome, which can be joined by RNA splicing to a common set of exons of the ABL oncogene derived from chromosome 9 sequences. The resulting chimeric mRNAs encode a 210,000 MW protein in CML (P210BCR/ABL) and a 185,000 MW protein in ALL (P185BCR/ABL). Both proteins retain an active tyrosine kinase activity derived from the ABL portion.

Although structurally related, these two alternative forms of the BCR/ABL oncogene are correlated with quite distinctive biologies. In CML most patients progress from a rather indolent or chronic stage of the disease towards a more aggressive blast crisis stage over a period of months to several years. During the chronic stage myeloid and erythroid elements capable of maturation but derived from the affected stem cell predominate in the peripheral blood. During the blast crisis phase a monomorphic outgrowth of an immature myeloid, erythroid, or lymphoid cell type without maturation dominates in the marrow and peripheral blood. In contrast, Ph1 positive ALL patients frequently present with very high peripheral white blood counts of leukemic lymphoblasts without maturation.

The major goals of this brief review are to describe the current status of our understanding of the structural rearrangements occurring in the Ph chromosome, the control of expression of the chimeric BCR/ABL oncogenes, the functional consequences of expression of BCR/ABL oncogenes in fibroblastic and hematopoietic cell types, and the connections between variation in clinical presentation and the structures of the oncogene products. The answers to many of these questions are incomplete at this time, but experimental strategies currently being used in attempts to clarify these issues will be described.

ALTERNATIVE BREAKPOINTS ON CHROMOSOME 22 RESULT IN DIFFERENT CONTRIBUTIONS OF BCR EXONS TO THE CHIMERIC ONCOGENES IN CML AND ALL

The important observation by Groffen and colleagues (Groffen et al 1984) that the breakpoint regions for the Ph1 chromosome in almost all

---

The Howard Hughes Medical Institute, Department of Microbiology and Molecular Biology Institute, UCLA, Los Angeles, CA, 90024, USA

Current Topics in Microbiology and Immunology, Vol. 141

CML patients could be localized to a small (5 Kb) segment (breakpoint cluster region) on chromosome 22, while the breakpoint on chromosome 9 upstream of the main body of ABL gene was less precisely localized can now be understood in terms of a splicing model to generate the chimeric BCR/ABL mRNA (see Figure 1). A subset of the exons of the gene surrounding the breakpoint cluster region on chromosome 22, now called BCR, can be spliced in-frame to the second coding exon of the ABL gene, resulting in a predicted large primary transcription product. By splicing out of the intronic regions this product results in an 8.7 Kb mRNA (Mes-Masson et al 1986; Shtivelman, Lifshitz, Gale & Canaani 1985) which encodes the P210BCR/ABL gene product (Daley, McLaughlin, Witte & Baltimore 1987; Konopka, Watanabe & Witte 1984). Depending on the precise location of the breakpoint on chromosome 22, the contribution of BCR exons at the BCR/ABL joint can vary subtly resulting in two alternative structures which are both in the same translational reading frame as the ABL segment (Shtivelman et al 1986). The resulting P210BCR/ABL protein is an active tyrosine kinase, with _in vitro_ enzymatic properties indistinguishable from the gag/abl form expressed by the Abelson murine leukemia virus (Davis, Konopka & Witte 1985). A cohesive and convincing body of evidence at the DNA, RNA and protein levels has correlated this set of molecular events with the clinical setting of CML and has proved useful in diagnosis, monitoring response to therapy and exploring the pathobiology of this disease (see Rosenberg & Witte 1988; Witte 1986 for recent reviews).

This condensation of cytogenetic and molecular data lead to a comparative study of leukemias related to CML either by a similar blood picture or a related cytogenetic event. One striking observation

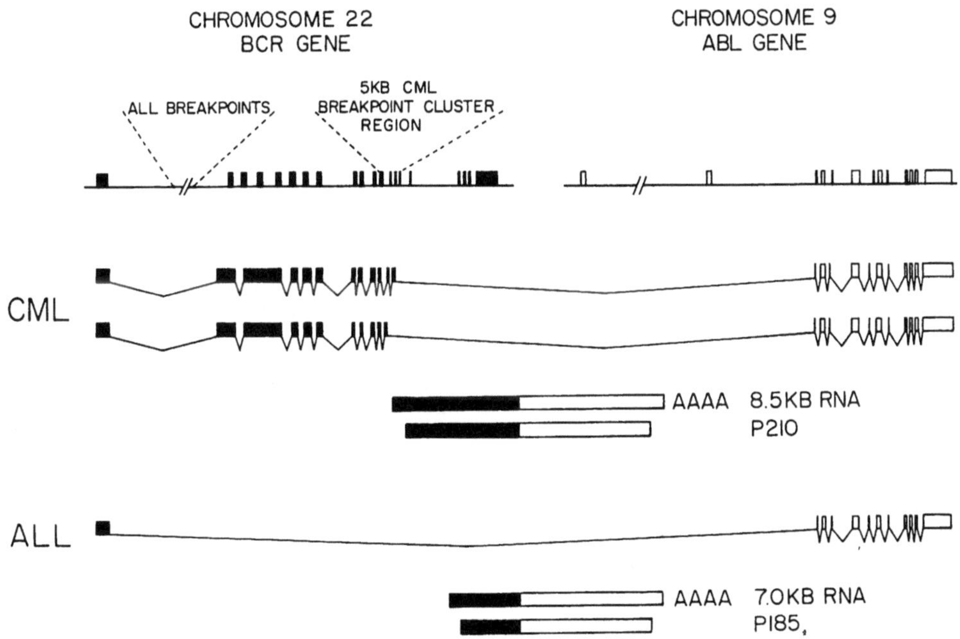

Fig. 1. Schematic representation of the molecular consequences of the Ph1 chromosome translation in CML and ALL. Distances are not to precise scale

was the discovery that an ABL related protein of 185,000 MW with active tyrosine kinase activity could be identified in acute lymphoblastic leukemia (Chan et al 1987; Clark et al 1987; Kurzrock et al, 1987). Initially it was not appreciated that the BCR gene was involved because the breakpoint cluster region seen in CML was not altered and fine structure cytogenetic mapping suggested that the site on chromosome 22 was more centromeric. However, recent data from several laboratories has clearly established that the P185 product is derived by a splicing mechanism that uses exons from both BCR and ABL (Clark et al 1988; Fainstein et al 1987; Hermans et al 1987). In this case the chromosome 22 breakpoints are found in the first intron of the BCR gene (Hermans et al 1987). To date, very few patients have been successfully mapped at the genomic level, but the complementary results from cDNA analysis and serological reactivities of the P185 protein unambiguously show that the amino terminal region of BCR is joined to precisely the same set of exons of the ABL gene as are used in the formation of P210. This structural information has been used to develop an ultrasensitive polymerase chain reaction technique starting with mRNA to detect leukemic cells in peripheral blood or bone marrow (Kawasaki et al 1988). Characterization of the <u>in vitro</u> enzymatic activity of the P185 product shows that it is indistinguishable from the viral abl forms or the P210 protein (A.M. Pendergast, S. Clark and O. Witte, unpublished observations).

BIOLOGICAL ACTIVITIES OF THE BCR/ABL GENE PRODUCTS

Although the correlation of the expression of P210 and P185 to CML and ALL provided a strong suggestion that they played a central role in the genesis of these leukemias, it was critical to test various <u>in vitro</u> tissue culture and <u>in vivo</u> animal models to assess their functional capabilities. Complete cDNA copies of the coding sequences for the P210 protein (Mes-Masson et al 1986) were constructed into retrovirus vectors and expressed in murine 3T3 type fibroblastic cell lines (Daley, McLaughlin, Witte & Baltimore 1987; McLaughlin, Chianese & Witte 1987) by transfection or infection. Surprisingly, unlike the Abelson murine leukemia virus gene product (P160v-abl) which is a potent transforming gene for 3T3 lines (Rosenberg & Witte 1988), the P210BCR/ABL expressing retrovirus did not induce morphological transformation even though high amounts of active tyrosine kinase were produced. Clearly the requirements for efficient fibroblast transformation were not met in the BCR/ABL chimeric protein.

During the course of these transfection experiments, recombinants formed between segments of the gag gene of the helper Moloney murine leukemia virus DNA included to provide viral packaging proteins and the BCR/ABL gene created a trimeric structure expressing a gag/BCR/ABL protein (named P220) with potent transforming activity for 3T3 lines (Daley, McLaughlin, Witte & Baltimore 1987). This strongly suggested that there was no defect in the tyrosine kinase activity of the BCR/ABL segment, but rather a cellular localization or perhaps protein association selectivity which the presence of an amino-terminal segment of the gag gene could overcome. It is likely that the amino-terminus of the gag gene directs the post-translational myristy- lation of the gag/BCR/ABL protein as it does for the native gag gene itself and the gag/abl fusion of the Abelson murine leukemia virus (Rosenberg & Witte 1988).

When the rescued stocks of P210BCR/ABL expressing retrovirus were tested in the standard Abelson murine leukemia virus (A-MuLV) bone marrow agar transformation assay, they showed no ability to induce the large transformed foci routinely seen with A-MuLV preparations. Occasionally, small foci (less than 500 cells) were detected with P210 expressing stocks, but they could not be explanted into continuous lines in vitro (unpublished observations). As an alternative approach, we exploited a long term bone marrow culture technique which can selectively enrich and replicate immature elements of the B cell lineage for many months (Whitlock & Witte 1982; Whitlock, Robertson & Witte 1984). When fresh bone marrow was infected with the P210 expressing retrovirus and plated under these selective conditions, we could demonstrate the outgrowth of clonal cell lines expressing P210 (McLaughlin, Chianese & Witte 1987). Such lines were clearly altered from the normal lymphocytes in the long term cultures in that they grew to higher densities, had a larger modal size, and had a lessened (but not zero) requirement for growth factor enriched media or stromal layers for continuous growth in vitro. Each of the clonal lines expressed high levels of the P210 tyrosine kinase, and showed one or a small number of retroviral vector integrations. Phenotypically, these clonal lines were in the early stages of the B cell lineage with immunoglobulin gene rearrangements of the DJ type, high levels of terminal deoxynucleotidyl transferase, and surface markers like B220 (MCLaughlin, Chianese & Witte 1987).

Most interestingly, although these clonal lines were very similar in differentiation phenotype to A-MuLV transformed cell lines, the P210 stimulated lines were in many cases non-tumorigenic in syngeneic animal challenges despite the presence of high levels of the P210 gene product. After continuous passage in culture for several months some of the low tumor potential lines could convert to a malignant phenotype, while others remained non-tumorigenic. Clearly, the expression of the P210 tyrosine kinase may be necessary, but it is in itself not sufficient for the full progression of the transformed phenotype in these cultured lymphocyte populations.

The dramatic difference between the clinical biology of Ph1 positive ALL and CML could possibly be due to a difference in efficiency of the growth stimulus of P185 compared to P210 in certain subpopulations of hematopoietic cells, such as immature lymphocyte precursors. Although this may be too simple, it is important to compare these gene products in the same quantitative transformation systems. We and others have recently described the molecular cloning of the unique BCR/ABL mRNA junction found in Ph1 positive ALL (Clark et al 1988; Fainstein et al 1987; Hermans et al 1987), and we have used this information to create a precise site directed deletion in our infectious P210 expressing clone which can now express the P185 protein (J. McLaughlin and O. Witte, manuscript in preparation). When this retrovirus is used in the same lymphoid transformation assay used for P210 we find that the P185 protein is an efficient growth stimulator for early B lineage cells. Clonal lines expressing P185 overgrow uninfected cells in the cultures. Our preliminary observations suggest P185 may be a stronger mitogenic signal than P210 for lymphoid cells but further work is needed to verify this point (J. McLaughlin, E. Chianese, and O. Witte, unpublished observations).

CURRENT STRATEGIES TO IDENTIFY GENES WHICH MAY COMPLEMENT THE FUNCTION OF P210 DURING PROGRESSION OF CML

The biological pattern of disease progression during CML is clear, but the molecular correlates that regulate these changes have not been identified. Two general mechanisms should be considered. The first would involve changes in the expression or regulation of P210BCR/ABL which would result in an altered level of the tyrosine kinase. This could putatively explain altered growth kinetics of the leukemic population, but it would be difficult to rationalize a selectivity for a specific cell type to grow out in blast crisis. The second general mechanism would require secondary genetic (mutations) or epigenetic changes (changes in gene expression which could cyclicly vary from an active to inactive state but do not involve changes in DNA sequence). Such secondary changes could be of two types: those which require the complementary expression of P210 for activity, and those which are transdominant to P210 expression and do not require continued expression of P210 or in fact the Philadelphia chromosome.

Although cell preparations from some patients in the blast crisis stage of CML show enhanced levels of the specific chimeric BCR/ABL mRNA, this is not a constant finding (Stam et al, 1985; Shtivelman et al 1987). Separate analyses of the levels of protein expression generally show a significant increase in the recovered amount of BCR/ABL kinase activity (Konopka et al 1985; Konopka et al 1986; Clark et al 1987). However, these are difficult studies to compare, since the cell population mixtures found in chronic phase vary between patients, and numerous other variables like treatment protocols could affect the relative messenger and protein levels. Little is currently known about the specific transcriptional regulation of the BCR gene. Genomic clones upstream of the first coding exon have been isolated but no functional demonstration of promotor or other regulatory elements has been worked out. Recently, work from this laboratory (A. Muller, J. McLaughlin, M. Timmons, and O. Witte, manuscript in preparation) has documented that the 5' untranslated region of the BCR gene contains sequences which are suppressive for translation. When synthetic mRNAs varying in the amount of 5' untranslated sequence are compared, only the species truncated to begin just above the AUG start site for P210 are strongly active in rabbit reticulocyte translation extracts. Even as few as 90 bases of 5' untranslated region will suppress the efficiency by 200 to 500 fold. The relationship of these <u>in vitro</u> observations to the clinical progression are difficult to define at this time, but it opens the area of translational regulation as a mechanism to vary the effective dosage of P210 in different cell types.

The possible activation of secondary oncogenes during blast crisis has been investigated in two alternative fashions. In the first, a specific search targeted at a known oncogene can be accomplished if there is a sensitive and practical screening method available. These criteria have been met for the ras oncogene family by searching for specific mutations in the cellular DNA with the use of the polymerase chain reaction technology and probing with oligonucleotides which vary at single base changes. Such an approach has revealed that many patients with acute myelogenous leukemia contain mutations in one of the cellular RAS alleles (Janssen et al 1987). A similar analysis of CML patients did not show an appreciable frequency of RAS mutations by this approach. An alternative approach is to search for activated oncogenes capable of inducing a transformed phenotype in rodent fibroblast cells following genomic DNA transfections. Mutated ras oncogenes would be expected to score well in such an assay due to their small size and potent effects on 3T3 type lines. In a recent paper (Liu, Hjelle & Bishop 1988) DNA from several independent blast crisis patients and one chronic phase patient could be shown to contain activated ras alleles of the N and H types, and in one case an unidentified fibroblast transforming gene. It will be interesting to see if these genes can work with P210 in hematopoietic transformation. It has already been shown that activated RAS oncogenes in retroviral form can abnormally growth stimulate multiple hematopoietic cell types.

Many oncogenes have been described which do not efficiently transform 3T3 lines, like some myc derived genes and P210 itself. Because of this we attempted to establish a system where a putative second oncogene of this type would co-operate with P210 to induce transformation of rodent fibroblast cell lines. Stocks of a P210 retrovirus rescued from a helper virus free packaging cell were used to create non-producer clones of NIH3T3 and Rat-1 fibroblast cell lines which stably produce high levels of the P210 protein (T. Gross-Lugo and O. Witte, unpublished observations). Interestingly, the Rat-1 derived line forms small colonies in agar under the growth stimulation of P210, while the NIH3T3 line did not form colonies. If the Rat-1/P210 line is superinfected with a murine retroviral construct expressing a form of the myc oncogene (Schwartz et al 1986) giant colonies are formed in agar. Nude mouse tumor challenge experiments show parallel changes. The Rat-1/P210/myc line causes rapidly growing tumors with a short latency period. We hope to use such systems to search for genes which by themselves cannot transform fibroblast cell lines, but in concert with P210 can induce growth changes.

PATHWAYS AND POSSIBLE MECHANISMS FOR THE ROLE OF P210 AND P185 IN HUMAN LEUKEMIAS

There is no clear evidence at this time for a single specific biochemical pathway on which members of the abl oncogene family function. However, the recent demonstration that A-MuLV can render myeloid and lymphoid cells growth factor independent suggests one possible mode of action (Cook et al 1987; Cook et al 1985; Pierce et al 1985). Similar results can be obtained with expression of P210 transferred by retroviral vector infection into myeloid cell lines dependent on either GM-CSF or IL-3 (G. Daley and D. Baltimore, personal communication; A. Muller, E. Chianese, and O. Witte, unpublished observations). Using a bcr/v-abl chimeric construct, S. Cory and colleagues (personal communication) have also shown IL-3 independence for a cultured cell line. Decreasing the requirement for growth factors naturally acting on progenitors of the myeloid and lymphoid pathways would certainly be a logical step in creating the cellular overgrowth common to both CML and ALL, but would not explain the variation in their biologies with regard to cell specificity unless P210 and P185 differ significantly in which growth factor pathways they can perturb.

Because such a variety of growth factor requirements can be reduced by the action of the abl oncogene, it is likely that its site of action may be further downstream, in a factor/receptor pathway common to a number of signaling systems. Searches for putative targets for the tyrosine kinase activity of abl oncogenes have turned up numerous candidates based on their containing detectable levels of phosphotyrosine (Bell et al 1987; Frackelton, Ross & Eisen 1983). However, at this time it is difficult to determine if these proteins form a link in a regulatory cascade or irrelevant phosphorylations of bystander proteins.

SUMMARY

Our understanding of the activation of the ABL oncogene in Ph1 positive human CML and ALL has dramatically increased in the last few years. The general mechanisms used to unite BCR and ABL segments has been deciphered, and critical biological, nucleic acid and serological reagents have now been prepared. Biological activities in an *in vitro* model system have been established for the P210 and P185 proteins. The next generation of experiments will likely focus on the difficult issues of relationship of gene expression to pathogenesis, and the intracellular pathways of action of these leukemia specific gene products.

ACKNOWLEDGEMENTS

Work in the authors' laboratory is supported by grants from the National Cancer Institute and the Howard Hughes Medical Institute.

REFERENCES

Bell JC, Mahadevan LC, Colledge WH, et al (1987) Abelson-transformed fibroblasts contain nuclear phosphotyrosyl-proteins which preferentially bind to murine DNA. Nature 325:552-554
Chan LC, Karhi KK, Rayter SI et al (1987) A novel *abl* protein expressed in Philadelphia chromosome positive acute lymphoblastic leukaemia. Nature 325:635-637
Clark SS, McLaughlin J, Crist WM, Champlin R & Witte ON (1987) Unique forms of the *abl* tyrosine kinase distinguish Ph^1-positive CML from Ph^1-positive ALL. Science 235:85-88
Clark SS, McLaughlin J, Timmons M et al (1988) Expression of a distinctive BCR-ABL oncogene in Ph^+ acute lymphocytic leukemia (ALL). Science 239:775-777
Cook WD, Fazekas de St. Groth B, Miller JFAP, MacDonald HR & Gabathuler R (1987) Abelson virus transformation of an interleukin 2-dependent antigen-specific T-cell line. Molec Cell Biol 7:2631-2635
Cook WD, Metcalf D, Nicola NA, Burgess AW & Walker F (1985) Malignant transformation of a growth factor-dependent myeloid cell line by Abelson virus without evidence of an autocrine mechanism. Cell 41:677-683
Daley GQ, McLaughlin J, Witte ON & Baltimore D (1987) The CML-specific P210 *bcr/abl* protein, unlike v-*abl*, does not transform NIH-3T3 fibroblasts. Science 237:532-535
Davis RL, Konopka JB & Witte ON (1985) Activation of the c-*abl* oncogene by viral transduction or chromosomal translocation generates altered c-*abl* proteins with similar *in vitro* kinase properties. Molec Cell Bio 5:204-213
Fainstein E, Marcelle C, Rosner A et al (1987) A new fused transcript in Philadelphia chromosome positive acute lymphocytic leukaemia. Nature 330:386-388
Frackelton Jr AR, Ross AH & Eisen H (1983) Characterization and use of monoclonal antibodies for isolation of phosphotyrosyl proteins from retrovirus-transformed cells and growth factor-stimulated cells. Molec Cell Biol 3:1343-1352
Groffen J, Stephenson JR, Heisterkamp N et al (1984) Philadelphia chromosomal breakpoints are clustered within a limited region, bcr, on chromosome 22. Cell 36:93-99
Hermans A, Heisterkamp N, von Lindern M et al (1987) Unique fusion of *bcr* and *c-abl* genes in Philadelphia chromosome positive acute lymphoblastic leukemia. Cell 51:33-40
Janssen JW, Steenvoorden ACM, Lyons J, Anger B, Bohlke JU, Bos JL, Seliger H, Bartram CR (1987) RAS gene mutations in acute and chronic myelocytic leukemias, chronic myeloproliferative disorders, and myelodysplastic syndromes. Proc Natl Acad Sci USA 84:9228-9232
Kawasaki ES, Clark SS, Coyne MY et al (1988) Diagnosis of chronic

myelogenous and acute lymphocytic leukemias by detection of leukemia-specific mRNA sequences amplified in vitro. Proc Natl Acad Sci in press

Konopka JB, Clark S, McLaughlin J, Nitta M, Kato Y, Strife A, Clarkson B & Witte ON (1986) Expression of the translocated c-abl gene in Ph1-positive B lymphoid cell lines from chronic myelogenous leukemia patients. Proc Natl Acad Sci 83:4049-4052

Konopka JB, Watanabe SM & Witte ON (1984) An alteration of the human c-abl protein in K562 leukemia cells unmasks associated tyrosine kinase activity. Cell 37:1035-1042

Konopka JB, Watanabe SM, Singer J, Collins S & Witte ON (1985) Cell lines and clinical isolates derived from Ph1-positive chronic myelogenous leukemia patients express c-abl proteins with a common structural alteration. Proc Natl Acad Sci 82:1810-1814

Kurzrock R, Shtalrid M, Romero P et al (1987) A novel c-abl protein product in Philadelphia-positive acute lymphoblastic leukaemia. Nature 325:631-635

Liu E, Hjelle B & Bishop JM (1988) Transforming genes in chronic myelogenous leukemia. Proc Natl Acad Sci USA 85:1952-1956

McLaughlin J, Chianese E & Witte ON (1987) In vitro transformation of immature hematopoietic cells by the P210 BCR/ABL oncogene product of the Philadelphia chromosome. Proc Natl Acad Sci USA 84:6558-6562

Mes-Masson A-M, McLaughlin J, Daley GQ, Paskind M & Witte ON (1986) Overlapping cDNA clones define the complete coding region for the $P210^{c-abl}$-gene product associated with chronic myelogenous leukemia cells containing the Philadelphia chromosome. Proc Natl Acad Sci USA 83:9768-9772

Pierce JH, Di Fiore PP, Aaronson, SA et al (1985) Neoplastic transformation of mast cells by Abelson-MuLV: abrogation of IL-3 dependence by a nonautocrine mechanism. Cell 41:685-693

Rosenberg N & Witte ON (1988) The viral and cellular forms of the Abelson (abl) oncogene. Adv Virus Res in press

Shtivelman E, Gale RP, Dreazen O, Berrebi A, Zaizov R, Kubonishi I, Miyoshi I & Canaani E (1987) bcr-abl RNA in patients with chronic myelogenous leukemia. Blood 69:971-973

Shtivelman E, Lifshitz B, Gale RP & Canaani E (1985) Fused transcript of abl and bcr genes in chronic myelogenous leukemia. Nature 315:550-554

Shtivelman E, Lifshitz B, Gale RP, Roe BA & Canaani E (1986) Alternative splicing of RNAs transcribed from the human abl gene and from the bcr-abl fused gene. Cell 47:277-284

Schwartz RC, Stanton LW, Riley SC, Marcu KB & Witte ON (1986) Synergism of v-myc and v-Ha-ras in the in vitro neoplastic progression of murine lymphoid cells. Mol Cell Biol 6:3221-3231

Stam K, Heisterkamp N, Grosveld G, de Klein A, Verma R, Coleman M, Dosik H & Groffen J (1985) Evidence of a new chimeric bcr/c-abl mRNA in patients with chronic myelocytic leukemia and the Philadelphia chromosome. N E J M 313:1429-1433

Whitlock CA, Robertson D & Witte ON (1984) Murine B cell lymphopoiesis in long term culture. J Immunol Meth 67:353-369

Whitlock CA & Witte ON (1982) Long Term culture of B lymphocytes and their precursors from murine bone marrow. Proc Natl Acad Sci USA 79:3608-3612

Witte ON (1986) Abl oncogene function. Cancer Surveys 5:183-198

The Pathogenesis of Tumors Induced by Helper Virus-Free Abelson Murine Leukemia Virus

P.L. GREEN, D. KAEHLER, and R. RISSER

INTRODUCTION

The Abelson murine leukemia virus (A-MuLV) rapidly induces pre-B cell lymphomas in susceptible mice (Risser et al., 1978) and transforms bone marrow lymphocytes *in vitro* (Rosenberg and Baltimore, 1976) when inoculated as a complex of virions containing A-MuLV genomes and Moloney MuLV (M-MuLV) genomes. Recently, we have analyzed the activity of helper virus-free A-MuLV pools harvested from supernatants of the psi2 retroviral packaging cell line which had been transfected with a cloned A-MuLV P160 provirus (Green et al., 1987a). The availability of helper virus-free A-MuLV has allowed us to follow the progression of transformed cells *in vivo* without the complications that result from virus replication (Green et al., 1987b; Whitlock et al., 1983). Individual transformed cells recovered from A-MuLV(psi2)-infected mice were analyzed molecularly and biologically to determine which properties correlate with the progression of the transformed cell to a fully malignant cell.

ISOLATION AND CLONALITY OF TRANSFORMED CELLS PRIOR TO TUMOR EMERGENCE

A-MuLV(psi2) efficiently induces monoclonal tumors in weanling BALB/cByJ mice (age 14-20 days). To follow the kinetics of the appearance of A-MuLV transformed cells *in vivo*, total bone marrow cells were removed from A-MuLV(psi2)-infected mice at various times after infection. These cells were plated in the *in vitro* agarose transformation assay and scored 10-14 days later as described (Green et al., 1987b). As early as three days after infection 0.2 transformed cells per 1×10^6 total bone marrow cells were present in infected mice. As expected, the number of transformed cells increased throughout the preleukemic phase of the disease and were at their maximum in tumor-bearing mice (Table 1).

Individual preleukemic transformed colonies (isolated from total bone marrow of A-MuLV(psi2)-infected mice 14 days after infection) were picked from agarose, and chromosomal DNA was isolated from each colony. Approximately 1 ug of DNA could be obtained from a single colony, and the proviral integration pattern of each colony was analyzed with an *abl*-specific probe by nucleic acid hybridization. Hybridization of colony DNAs digested with Hind III will detect an A-MuLV proviral/cellular DNA junction fragment for each independent proviral integration, in addition to the endogenous c-*abl* fragment. A total of seventy-five colonies from five infected mice were analyzed, and a representation of the hybridization pattern of eight colonies of one mouse is presented in Fig. 1. In addition to the c-*abl* endogenous band, a single A-MuLV provirus was detected in each of the 75 colony DNAs analyzed.

McArdle Laboratory for Cancer Research, University of Wisconsin, Madison, WI 53706

Current Topics in Microbiology and Immunology, Vol. 141

These data indicate that, within the limit of detection of this experiment, the population of preleukemic transformed cells present within an individual A-MuLV(psi2)-infected mouse is monoclonal by the midpoint of the tumor latent period.

Table 1. Transformed colonies from BALB/cByJ mice infected with helper virus-free A-MuLV

Source of Cells (Days post-infection)	Number of Colonies*	Range	Number of Mice
3	0.2	0-0.3	5
7	0.8	0-2.4	11
14	72	0-370	23
Leukemic BM (25-30 days)	270	$1\text{-}1\times10^3$	13
Leukemic Tissue	270	$1\text{-}1\times10^3$	15

Total bone marrow (BM) cells or a suspension of tumor cells were obtained from BALB/cByJ mice infected with A-MuLV(psi2). The cells were plated in the agarose bone marrow transformation assay, and colonies were scored 10-14 days after plating.

*The number of colonies are per 1×10^6 total bone marrow cells plated.

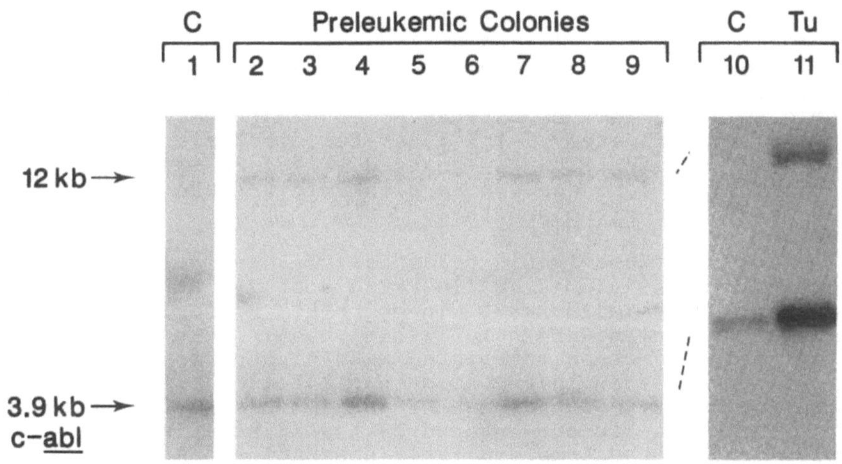

Fig. 1. Southern blot of DNAs from individual preleukemic colonies and a single tumor induced by transplantation of the same preleukemic colonies. Preleukemic colonies were isolated by the agarose transformation assay from total bone marrow of a 5 week old BALB/cByJ mouse infected 2 weeks earlier with 5×10^3 FFU of A-MuLV(psi2). Fourty individual colonies were picked from agarose. Chromosomal DNAs were prepared from 20 colonies (approximately 1 ug) and the other 20 colonies were transplanted into newborn BALB/cByJ mice. One tumor was induced by colony transplantation and chromosomal DNA was prepared. The DNAs were digested with Hind III, electrophoresed through a 0.7% agarose gel, and transferred to nitrocellulose. The

nitrocellulose filter was hybridized with 2 x 10^7 CPM of ^{32}P-labeled abl-specific probe pAB Bgl II and washed as described (Green et al., 1987a). The filters were exposed for 7 days with two intensifying screens at -70°C. Lanes: 1 (C), 1 ug BALB/cByJ liver DNA; 2-9, individual preleukemic colony DNAs; 10 (C), 5 ug BALB/cByJ liver DNA; 11 (Tu), 10 ug of DNA from the tumor induced by transplantation of a preleukemic colony. Molecular weights were determined by comparison with a Hind III digest of lambda DNA (23.1 kb, 9.42 kb, 6.68 kb, 4.36 kb, 2.32 kb, 2.03 kb, and 0.564 kb). The c-abl hybridizing fragment (3.9 kb) and the proviral junction fragment (12 kb) are indicated.

PROPERTIES OF INDIVIDUAL PRELEUKEMIC AND LEUKEMIC COLONIES

The results presented in the previous section indicate that A-MuLV-transformed cells present in the bone marrow of an individual preleukemic mouse are descendants of a single infected cell. We next determined whether the preleukemic colonies from a single infected mouse were qualitatively the same by studying their plating efficiencies *in vitro* and their tumorigenic potentials. Leukemic colonies were also analyzed in this manner as a basis for comparison.

Individual preleukemic and leukemic colonies were picked from agarose and were dispersed and counted. Each colony was divided into three equal parts (Fig. 2) and the cells were: a) replated in agarose with no feeder layer; b) replated in agarose in the presence of mouse bone marrow feeder layers; c) injected intraperitoneally into a newborn BALB/cByJ mouse. Replating efficiency was heterogeneous, but on the average leukemic colonies plated with approximately 15-fold higher efficiency than preleukemic colonies in the absence of feeder layers (Figs. 2C and 2A, respectively). The replating efficiency of preleukemic colonies in the presence of bone marrow feeder layers was increased approximately 8-fold; however, leukemic cell plating efficiency was not significantly changed (Figs. 2B and 2D, respectively) indicating a dependence of preleukemic transformed cells on feeder layers for *in vitro* growth.

The analysis of the tumorigenic potential of these colonies revealed that greater than 84% (16/19) of leukemic colonies were tumorigenic (Figs. 2C and 2D, shaded boxes). In contrast, approximately 3% (2/63) of preleukemic colonies were tumorigenic when injected into animals. This tumorigenic phenotype correlated with high plating efficiencies in the absence of feeder layers among preleukemic colonies (Figs. 2A and 2B, shaded boxes). Comparison of the A-MuLV proviral integration pattern of one of the tumors induced by transplantation of a preleukemic colony showed the same proviral integration pattern as the primary colony, thus indicating the tumor was an outgrowth of the inoculated preleukemic colony (Fig. 1, compare lanes 2-9 with lane 11). Transplantation of four individual leukemic colonies which arose from total bone marrow of the same tumor-bearing mouse resulted in the induction of four tumors. The A-MuLV proviral integration patterns of these tumors were identical to each other (data not shown). This result is consistent with the hypothesis that the primary leukemic colonies were monoclonal in origin and indicates that the initial tumor was a result of the expansion of these clones.

The data presented so far indicate that A-MuLV transformed cells present in an individual mouse during the preleukemic phase of the disease are monoclonal with respect to A-MuLV integration. However, analysis of their tumorigenic potential revealed qualitative differences between preleukemic colonies, suggesting that additional changes are necessary for a preleukemic transformed cell to progress to a malignant cell.

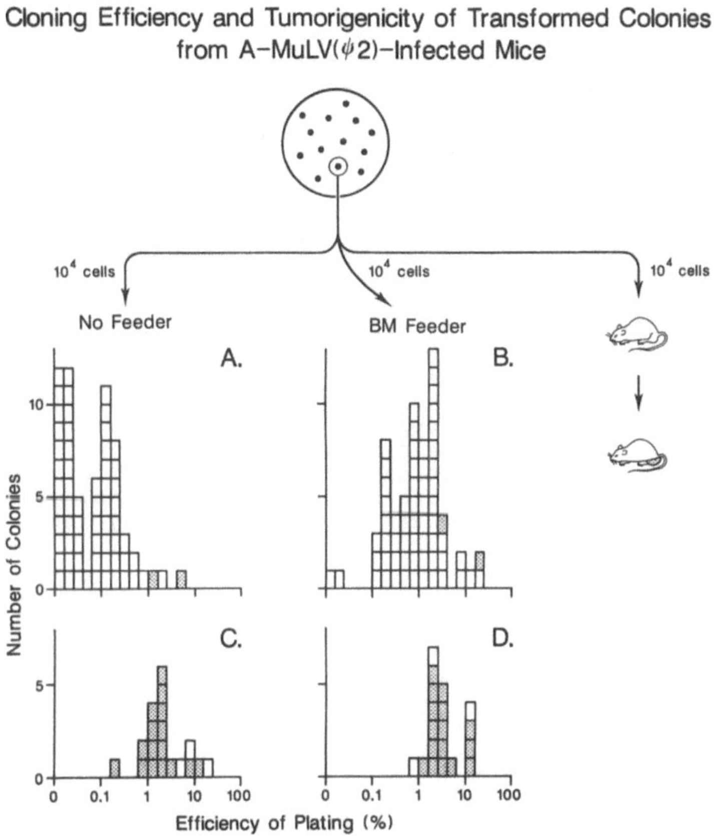

Fig. 2. Individual preleukemic colonies were isolated from total bone marrow of preleukemic mice (14 days post-infection) and individual leukemic colonies were isolated from total bone marrow of tumor-bearing mice (25-30 days post-infection). Each colony was divided into three equal parts (approximately 1 x 10^4 cells/part). Part 1 was replated in agarose over no feeder layer, part 2 was replated in agarose over bone marrow (BM) feeder layers, and part 3 was transplanted into newborn BALB/cByJ mice. Macroscopic colonies were scored 10-14 days post-plating, and mice were observed for three months after inoculation. Each box represents a single colony tested, and the shaded boxes represent colonies which were tumorigenic. A and B are plating efficiencies of preleukemic colonies in the absence and presence of feeder layers, respectively. C and D are plating efficiencies of leukemic colonies in the absence and presence of feeder layers, respectively. Tumorigenic preleukemic colonies/total preleukemic colonies tested (2/63). Tumorigenic leukemic colonies/total leukemic colonies tested (16/19).

THE EXPRESSION OF V-ABL IN PRELEUKEMIC AND LEUKEMIC COLONIES

The v-*abl* P160 protein is a member of the tyrosine kinase family of oncogenes, and this kinase activity is required for its transforming function. The v-*abl* P160 protein phosphorylates itself as well as other cellular proteins. To determine if different levels of the v-*abl* oncogene product correlated with the qualitative differences between preleukemic and leukemic colonies, the level of v-*abl* protein product expressed in individual preleukemic and leukemic colonies was measured by *in vitro* kinase assays.

Preleukemic and leukemic colonies were picked from agarose, and cell lysates were made from 5 x 10^4 cells of each colony. These cell lysates were immunoprecipitated and the immune complexes subjected to an *in vitro* protein kinase assay. Immediately following the *in vitro* kinase assay, the phosphorylated proteins were identified by SDS-PAGE and autoradiography. The autophosphorylation activity of the P160 v-*abl* transforming protein was determined in the individual colonies by aligning the exposed film with the dried gel, excising the P160 band, and determining the amount of ^{32}P present by liquid scintillation. The data represented in Fig. 3 indicate that the autophosphorylation activity of v-*abl* (^{32}P CPM present in the P160 v-*abl* protein) in preleukemic colonies does not differ significantly from that in leukemic colonies. These results indicate that the changes that occurred within a clonal population of preleukemic colonies to make them leukemic cannot be attributed to selection for a specific level of v-*abl* kinase activity. Therefore, progression of a preleukemic cell to a leukemic cell most likely involves an additional event(s).

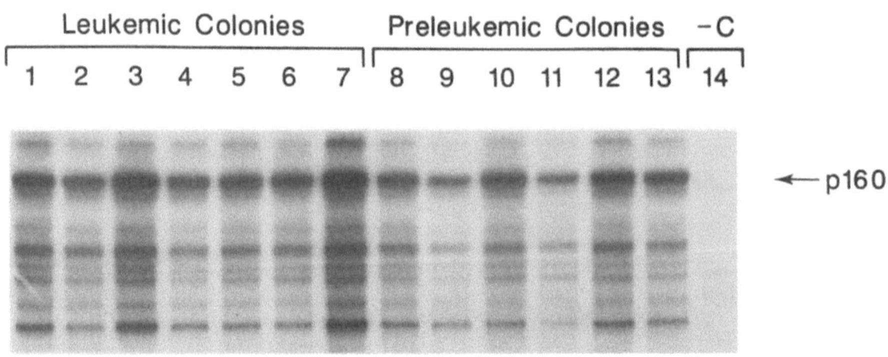

Fig. 3. Comparison of the autokinase activity of the p160 v-*abl* protein in preleukemic and leukemic colonies. Individual colonies were isolated from A-MuLV(psi2)-infected mice by an agarose bone marrow transformation assay. Individual preleukemic colonies were isolated from total bone marrow of preleukemic mice (14 days post-infection) and individual leukemic colonies were isolated from total bone marrow of tumor-bearing mice (25-30 days post-infection). Individual colonies were picked from agarose, and the viability (85%) and cell number (5 x 10^4-1 x 10^5) was determined. Lysates were made

from 5 x 10^4 viable cells and immunoprecipitated with p15^{gag} antisera. The *in vitro* protein kinase assay was performed on the Staph A-immune complexes, and the eluted proteins were immediately subjected to SDS-PAGE. All lanes contain 5 x 10^4 cell equivalents of lysate except lanes: 3 (8.8 x 10^4); 7 (7 x 10^4); 11 (2.6 x 10^4), Lanes: 1-7, individual leukemic colonies; 8-13, individual preleukemic colonies; 14 (-C), normal bone marrow. The exposed film was aligned with the dried gel, and the region containing the p160 protein was excised, and the amount of ^{32}P was determined by liquid scintillation counting. The raw counts in CPM (parentheses) normalized to 5 x 10^4 cell equivalents are: Lanes: 1 (159); 2 (158); 3 (141); 4 (117); 5 (125); 6 (129); 7 (205); 8 (117); 9 (102); 10 (130); 11 (188); 12 (154); 13 (119); 14 (20).

RESCUE OF TUMORIGENIC POTENTIAL OF A-MuLV(psi2) FOLLOWING SUPERINFECTION OF RESISTANT MICE WITH A-MuLV(psi2)

Previous data indicated that adult BALB/c mice and weanling C57BL/6 (B6) mice were 30- or 100-fold less sensitive to tumor induction by A-MuLV(psi2) than weanling BALB/c mice (Green et al., 1987a). To determine if this quantitative resistance could be overcome by superinfection of the mice with helper virus, we infected adult BALB/c mice with a subtumorigenic dose of A-MuLV(psi2), and then infected some of the mice with M-MuLV helper virus. The results of that experiment, summarized in Table 2, indicate that superinfection of A-MuLV-(psi2)-infected mice with helper virus can rescue the tumorigenic potential of A-MuLV(psi2), even when the helper virus is inoculated as late as 11 days after infection with A-MuLV(psi2). The latent period, but not the final incidence, is affected by the dose of helper virus used to rescue tumor formation. Less extensive data with B6 mice indicate that tumor formation can be induced by superinfection of A-MuLV(psi2)-infected mice with M-MuLV helper.

Table 2. Rescue of tumorigenic potential of A-MuLV(psi2) by superinfection of adult BALB/c mice with helper virus

A-MuLV(psi2) Dose (FFU)	M-MuLV Dose (PFU)	Day of M-MuLV Infection	Tumors/Total	Latent Period
10^4	0	-	0/30	90
10^4	3 x 10^5	0	7/15	36
10^4	2 x 10^6	0	7/9	34
10^4	2 x 10^5	8-11	8/17	47
10^4	6 x 10^5	8-11	7/12	42
10^4	2 x 10^6	8-11	6/17	34

These results suggest that A-MuLV-infected cells persist in resistant mice for at least 11 days after infection with A-MuLV(psi2), and the cells (or virus within them) are activated to induce tumors following infection with helper virus.

To determine if A-MuLV(psi2)-infected resistant mice contained transformed cells, we infected the mice with subtumorigenic doses of A-MuLV(psi2), removed the bone marrow 14 days later, and quantified the number of transformed cells present in the bone marrow. The results of that experiment, presented in Table 3, indicated that resistant mice contain significantly fewer colony-forming cells than susceptible mice. The morphology of the cells is more myeloid, and individual colonies are smaller (50-200 cells) than those found in preleukemic mice.

Table 3. Transformed colonies in bone marrow of A-MuLV(psi2)-infected mice

Mouse	Age (d)	Virus Dose FFU	Colonies/10^6 BM cells	Tumors
BALB/c	14	5×10^3	23	25/27
BALB/c	60	5×10^3	2	0/30
BALB/c	14	2×10^5	70	17/17
C57BL/6	14	2×10^5	3	2/20
BALB/c	14-60	(2×10^6 M-MuLV)	0.05	ND

DISCUSSION

We followed the progression of Abelson disease in vivo by comparing growth and tumorigenic properties of preleukemic and leukemic cell clones isolated from the bone marrow of mice infected with helper virus-free A-MuLV. Our results indicate that only 3% of preleukemic colonies formed tumors on transplantation in vivo, and these cells also had high replating efficiencies in the absence of feeder layers. Greater than 84% of leukemic colonies formed tumors on transplantation in vivo, and the replating efficiencies of these cells were not dependent on feeder layers. Preleukemic and leukemic colonies were monoclonal populations of cells identified by single proviral integration events, and they showed similar levels of the v-abl protein as determined by protein kinase assays. Therefore, it is reasonable to infer that preleukemic colonies give rise to leukemic colonies. From the results presented, we conclude that an event(s) unrelated to the expression of A-MuLV occurs within a transformed cell clone and provides that cell with a selective growth advantage characteristic of a malignant cell.

These results are consistent with the theory that a cellular event(s) in addition to the expression of v-abl is required for a cell to be tumorigenic. At least two hypotheses can be considered for the nature of this event. One hypothesis is that the target cell for tumorigenesis generates progeny cells with differing self-renewing capabilities, but no stable genetic differences distinguish these progeny. Thus, the process of clonal selection results in the enrichment of cells with greater self renewing capabilities. The second hypothesis is that leukemic cells differ from preleukemic cells as a consequence of a genetic mutation which is selected for during tumorigenesis. This mutation could involve the activation of a cellular proto-oncogene. If a second genetic event is required, experiments could be designed to try to identify that event.

The role of the helper virus in resistant mice may be to increase the dose of A-MuLV *in vivo* and thereby effect transformation of additional or more sensitive target cells, or the helper virus may mediate an essential step in tumor induction, such as the activation of a proto-oncogene. Experiments in progress with self-inactivating A-MuLV vectors may help to answer these questions.

REFERENCES

Green PL, Kaehler DA, Risser R (1987a) Cell transformation and tumor induction by Abelson murine leukemia virus in the absence of helper virus. Proc Natl Acad Sci USA 84:5932-5936

Green PL, Kaehler DA, Risser R (1987b) Clonal dominance and progression in Abelson murine leukemia virus lymphomagenesis. J Virol 61:2192-2197

Risser R, Potter M, Rowe WP (1978) Abelson virus induced lymphomagenesis in mice. J Exp Med 148:714-726

Rosenberg N, Baltimore D (1976) A quantitative assay for transformation of bone marrow cells by Abelson murine leukemia virus. J Exp Med 143:1453-1463

Whitlock CA, Ziegler Z, Witte ON (1983) Progression of the transformed phenotype in clonal lines of Abelson virus-infected lymphocytes. Mol Cell Biol 3:596-604

The Development of Three Distinct Avian B Cell Lymphomas

E.H. HUMPHRIES, C. FRISCH BARTH, and E. PIZER

INTRODUCTION

The infection of 1 day old SC Hyline chicks with avian leukosis virus (ALV) induces a bursal-dependent B cell lymphoma in approximately 50% of the population (Baba and Humphries 1984). This disease has provided a useful model for the examination of multistep neoplastic development (Purchase and Burmester 1978). Progression to terminal metastatic disease occurs only after 4 to 6 months of incubation and follows the appearance of distinct preneoplastic and neoplastic lesions (Cooper et al. 1968; Neiman et al. 1980). Both of these earlier stages are dependent upon the bursal microenvironment for their development. Progression from preneoplasia to neoplasia appears to be stochastic in nature (Baba and Humphries 1985). Examination of both the tumor and cell lines derived from the tumor has demonstrated that they synthesize and may secrete IgM (Cooper et al. 1974; Chen et al. 1983; Baba et al. 1985). Extensive analysis has demonstrated that ALV integration and formation of the provirus disrupts one allele of the c-*myc* locus (Hayward et al. 1981; Fung et al. 1981; Payne et al. 1982). This insertional mutagenesis results in elevated transcription of c-*myc* mRNA and appears to be responsible for the development of the preneoplastic lesion (Neiman et al. 1985; Thompson et al. 1987). The nature of the subsequent genetic alterations required for progression from preneoplasia to neoplasia and later to metastatic disease have not been identified.

Recent data from studies both in our laboratory and elsewhere have demonstrated that at least two other genetic alterations can lead to the development of an avian B cell tumor. The initial oncogenic events associated with the development of these tumors result in the expression of c-*myb* and v-*rel*. Characterization of these newly described lymphomas indicates that, while the tumor cells are phenotypically similar to those of the ALV-induced bursal-dependent B-cell lymphoma, their biological development is distinct.

THE EXPRESSION OF V-*REL* INDUCES A B CELL LYMPHOMA

Reticuloendotheliosis virus [REV-T(REV-A)] induces a poorly defined tumor originally described as reticuloendothelial (Theilen et al. 1966; Olson 1967). It has been suggested that cell lines developed following *in vitro* infection of spleen cells are early lymphoid in nature (Beug et al. 1981; Lewis et al. 1981). In contrast to these observations, we have recently shown that infection of 1 day old chicks with REV-T and the non-immunosuppressive helper virus, chicken syncytial virus [REV-T(CSV)], leads to acute development of B cell lymphomas (Barth and Humphries 1988). The development of these tumors is extremely rapid, killing the chick within 7 to 14 days. Tumor development appears to be independent of the bursal

Department of Microbiology, Southwestern Medical Center
University of Texas, 5323 Harry Hines Blvd.
Dallas, Texas 75235

Current Topics in Microbiology and Immunology, Vol. 141

environment and occurs in multiple tissues. Further, the establishment of cell lines from tumor tissue does not require extended *in vivo* transfer or prolonged *in vitro* culture indicating that additional genetic changes are not required for *in vitro* proliferation. In contrast to the isolation of cell lines derived from the ALV-induced lymphoma, isolation of cell lines expressing v-*rel* is a highly efficient process.

We have continued our analysis of the REV-T(CSV)-derived cell lines in order to further define their B cell character. To assay for the secretion of IgM, the lines were labeled with ^{35}S-methionine and the culture media collected and used for immunoprecipitation with antisera directed against chicken Ig. All 10 cell lines examined thus far have released IgM. The analysis of several cell lines is shown in Fig. 1. The variation in the amount of IgM secreted was 50 to 100-fold.

Cell lines were also examined for the continued expression of the oncogene v-*rel* using a Northern analysis of total cellular RNA (Fig. 2). Both the genomic and the spliced subgenomic mRNAs are expressed abundantly. While some variation was noted, the level of expression was consistent with that expected from a viral LTR. The analysis also indicates that c-*rel* mRNA is expressed in these lines. Further analysis has thus far failed to detect elevated expression of any additional oncogenes.

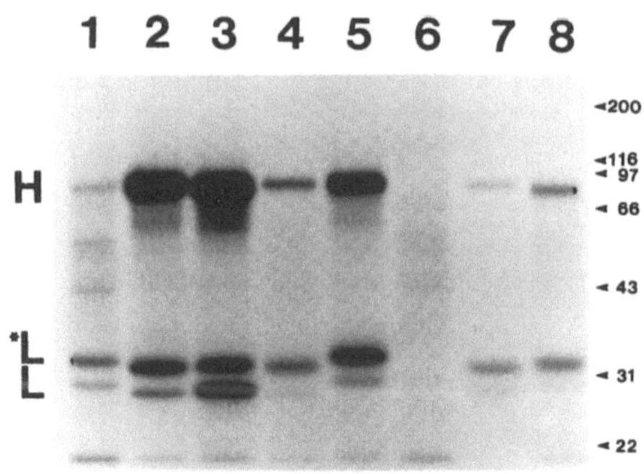

Fig. 1. Secretion of IgM by REV-T(CSV) cell lines. Cell lines were incubated with ^{35}S-methionine and the culture media immunoprecipitated with a rabbit polyclonal antisera against chicken Ig. Immunoprecipitates were collected using protein A-Sepharose and subjected to polyacrylamide gel electrophoresis. Lanes 1,3,4,5,7 and 8 are independently isolated REV-T(CSV) cell lines. Lane 2 is an IgM secreting bursal lymphoma cell line (Baba et al. 1985); Lane 6 is an REV-T(CSV) derived cell line that does not express IgM (Barth and Humphries 1988).
H and L indicate the position of heavy and light chain, respectively; *L indicates the position of glycosylated light chain. Molecular mass markers are given in kilodaltons

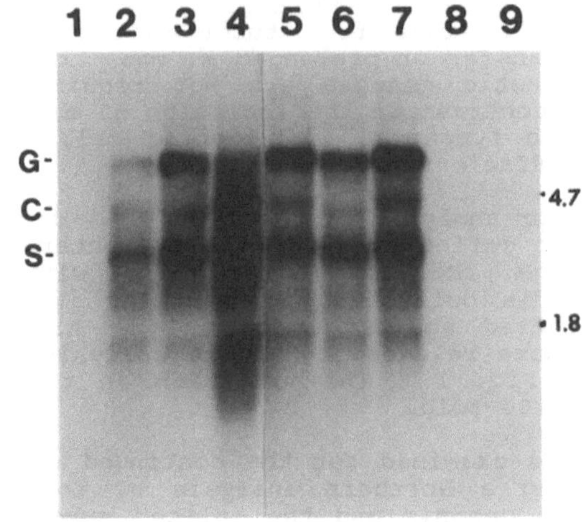

Fig. 2. Expression of v-*rel* in REV-T(CSV) cell lines. Total cellular RNA was prepared from REV-T(CSV) cell lines using guanidine isothiocyanate and examined by Northern analysis. The probe in the hybridization was a 967bp *Eco*RI fragment from v-*rel* (Wilhelmsen and Temin 1984). Lanes 2,3,4,5,6 and 7 are RNAs from REV-T(CSV) cell lines. Lanes 1, 8 and 9 are RNAs from normal fibroblasts, a cell line derived from an ALV-induced bursal lymphoma and an AMV-derived myeloid cell line respectively. G:genomic(5.9kb) mRNA; C:c-*rel*(4.0kb) mRNA; S:spliced(2.9kb) mRNA. Molecular mass markers are given in kilobase pairs(kb)

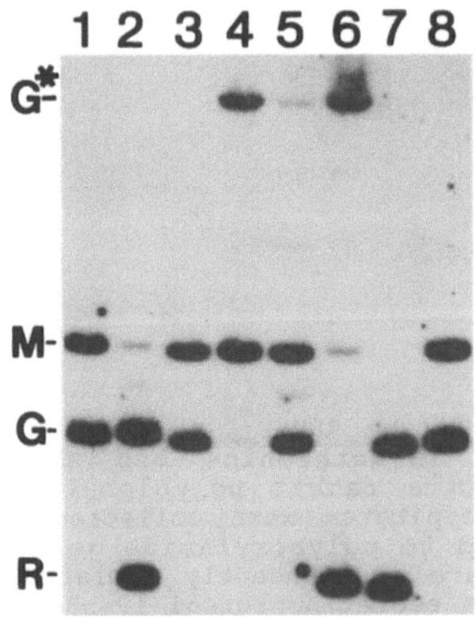

Fig. 3. Southern analysis of DNA from REV-T(CSV) lines. DNAs were digested with *Sal*I and *Sca*I and analyzed by electrophoresis, capillary transfer and hybridization to a 1.2Kb *Eco*RI-*Sal*I DNA fragment from the avian light chain constant region (Thompson and Neiman 1987). Lanes 1 through 8 contain DNA from independently derived cell lines. M, G and R designate the migration positions of the diversified, germ line and rearranged (non-diversified) alleles of the light chain locus. The large fragment [G*] in lanes 4 and 6 is the methylated form of the germ line allele

REV-T(CSV) cell line DNA was doubly digested with *Sca*I/*Sal*I and examined by Southern analysis to establish the configuration of the two alleles of the light chain Ig locus (Fig. 3). This analysis distinguishes between germ line, rearranged and diversified alleles (Reynaud et al. 1985; Thompson et al. 1987). The analysis of 26 IgM positive cell lines revealed that 24 of the lines had one allele in germ line configuration. Of the 24 active alleles, 11 were rearranged and diversified while 13 were rearranged without diversification. In no instance was the rearranged allele partially diversified. The presence of active alleles that are exclusively rearranged or rearranged and diversified suggests that the expression of v-*rel* inhibits the process of diversification of the active allele. As further evidence in support of this hypothesis, it was noted that the ratio of rearranged alleles to diversified alleles in the B cell population of a 1 day old bursa was the same as the ratio in the cell lines derived from tumors of chicks infected at day 1 of age. (Thompson et al. 1987; and data not shown).

The data presented above further substantiate that the expression of v-*rel* induces an IgM positive B cell lymphoma. These lymphomas secrete IgM and contain rearranged light chain genes. While the active light chain allele may have undergone diversification, further diversification appears to be inhibited by v-*rel*. In addition, we have examined the lymphomas for the expression of a variety of surface markers that are present on B cells. Our results indicate that, while the development of this B cell tumor is very different from that of the c-*myc*-derived lymphoma described above, we can not distinguish the two tumors at the cellular level by expression of surface antigens.

INTEGRATION OF ALV WITHIN THE C-*MYB* LOCUS RESULTS IN THE DEVELOPMENT OF A THIRD B CELL LYMPHOMA.

In studies that were designed to increase the frequency of preneoplasia in the SC Hyline chick, we had infected 12 day old embryonated eggs with the RAV-1 strain of ALV. To our surprise, 13 of 34 chicks developed disseminated lymphoma 4 to 6 weeks after hatching. The metastatic tumor was characterized by *in situ* immunoperoxidase analysis using a monoclonal that detects avian IgM. The results demonstrated that all 13 chicks had developed a rapid metastatic B cell lymphoma. In our previous studies of the bursal-dependent B cell lymphoma, in which all the chicks were infected as 1 day old hatchlings, the earliest metastatic disease we observed was detected at 15 weeks of age. Further, while histological analysis revealed aberrant expression of IgM, which may mark a population of malignant cells, there was no obvious primary tumor in the bursa. These observations suggested that this tumor was a B cell lymphoma distinct from both the c-*myc* and the v-*rel*-derived lymphomas described above.

Using Southern analysis, we examined the c-*myc* locus in 9 of these apparently novel B cell lymphomas. In one tumor there was a fragment diagnostic of an altered c-*myc* locus. The yield of this fragment, however, was well below the molar yield expected from a clonal tumor. The remainder of the tumors contained apparently normal c-*myc* loci. In contrast, Southern analysis of the c-*myb* loci revealed that this locus had been altered in all 9 tumors. We have analyzed these tumors further and have identified integrated ALV proviral sequences located at different sites within the c-*myb* locus (Fig. 4).

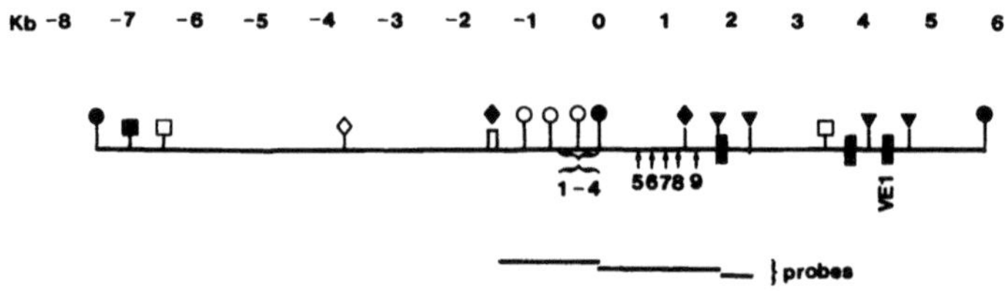

Fig. 4. The sites of 9 RAV-1 insertions are located (↑) within the c-*myb* locus. [-█-] c-*myb* exons; VE1 indicates the first exon transduced into v-*myb*. Restriction endonuclease sites: *Eco* RI (●); *Bam*HI (■); *Not*I (○); *Hind*III (□); *Sma*I (◆); *Pst*I (▼); *Bgl*II (◇); Nucleic acid fragments used as probes in mapping these insertions are as indicated and are derived from a molecular clone of c-*myb* (Perbal et al. 1983)

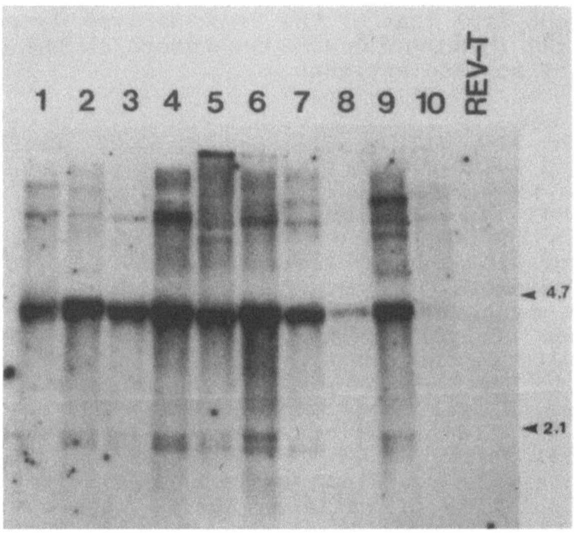

Fig. 5. Northern analysis of RNA isolated from B-cell lymphomas that contain altered c-*myb* loci. Total RNA was isolated from tumor tissue and subjected to agarose gel electrophoresis, capillary transfer and hybridization to a DNA probe containing v-*myb* sequences. Lanes 1 to 10 contain RNA from 10 different tumors. The lane marked REV-T contains RNA from a cell line derived from one of the REV-T(CSV)-induced B cell lymphomas described above

Northern analysis of RNA isolated from 8 of these tumors demonstrated that c-*myb* mRNA was 50 to 100-fold more abundant than in normal B cells (Fig. 5). Further analysis has shown that an LTR probe hybridizes with RNA at the same location on the gel suggesting that both c-*myb* and viral RNA sequences are part of a single mRNA species. It would appear, therefore, that the transcription from the c-*myb* locus may be influenced by integrated LTR sequences using mechanisms similar to those described previously in ALV-induced deregulation of the c-*myc* locus.

The data presented in Figs. 4 and 5 indicate that elevated expression of c-*myb* can induce a biologically distinct B cell lymphoma. We do not yet know whether the protein product of c-*myb* is structurally altered in these tumors. Again, using monoclonal antibodies that detect surface proteins, we have been unable to distinguish this B cell tumor from the two B cell tumors described above (data not shown). Similar findings have recently been reported by Kanter et. al. (1988).

THE THREE B CELL LYMPHOMAS ARE BIOLOGICALLY DISTINCT

Table 1 summarizes the biological properties of the three tumors described in this work. These observations indicate that these tumors, while all B cell tumors, develop along distinct biological pathways. The most striking difference is the multistep nature of the c-*myc*-derived tumor. While there is clear evidence for progression in the development of this tumor, the rapid dissemination of metastatic disease in both the v-*rel* and c-*myb*-derived tumors suggest that the initial oncogenic event in these tumors is sufficient for the appearance of malignant disease. This property also suggests that microenvironmental factors play little if any role in the establishment of specific stages required for tumor development. In contrast, the critical role that the bursal environment plays in the development of the c-*myc*-derived tumor has been well documented.

Perhaps the most interesting difference among the tumors is the marked *in vitro* proliferative capacity of the v-*rel*-induced tumors. While only preliminary data are available at present, it appears as though the c-*myb* tumors resemble the c-*myc* tumors in that both are difficult to adapt to culture as are normal avian B cells. These three tumors provide an opportunity to compare significant differences in B cell proliferative capability and experimental comparison should yield important information regarding B cell growth requirements.

Table 1. A Comparison of the Development of Three Avian B cell Lymphomas[a]

	Initial Oncogenic Event		
Property	ALV\c-*myc*	ALV\c-*myb*	v-*rel* Expression
specific target cells[b]	yes	yes	no
a stochastic multistage process	yes	probably not	no
a specific micro-environment is critical for tumor development	yes	probably not	no
in vitro proliferation of the tumor	no	probably not	yes
1° tumor	localized and easily defined	diffuse/not identifiable	diffuse/not identifiable
metastatic tumor	normal but not always present	yes	yes
somatic diversification of the Vλ locus	delayed	apparently normal	blocked by tumor development
average time until death	4-6 months	6-8 weeks	7-14 days

[a]The initial oncogenic event has been designated by ALV\c-*myc* (ALV insertion within the c-*myc* locus): ALV\c-*myb* (ALV insertion within the c-*myb* locus): v-*rel* expression (REV-T(CSV) infection)

[b]Discrete cell populations at different stages of avian development that are susceptible to tumor formation

SUMMARY

We have described the development of three distinct avian B cell lymphomas. Characterization of the tumor cells has shown them to have all the well recognized markers of this lineage. At present, we are unable to distinguish them by surface antigen profile. In contrast, their development as tumors can be uniquely identified at the biological level. These data lead us to propose that the genetic pathways by which these tumors develop are also unique. The evidence in support of this hypothesis at present is that the three genetic alterations believed to be responsible for initiation of the neoplastic process are different.

ACKNOWLEDGMENTS

We thank Ed Filardo and Gary Huffnagle for stimulating discussions, Rae Allen for excellent technical assistance, and Brenda West for superb typing. The work in our laboratory was supported by Public Service awards CA32295 and CA41450 from the National Institute of Health.

REFERENCES

Baba TW, Humphries EH (1984) Differential response to avian leukosis virus infection exhibited by two chicken lines. Virol. 135:181-188

Baba TW, Humphries EH (1985) Formation of a transformed follicle is necessary but not sufficient for development of an avian leukosis virus-induced lymphoma. Proc. Natl. Acad. Sci. USA 82:213-216

Baba TW, Giroir BP, Humphries EH (1985) Cell lines derived from avian lymphomas exhibit two distinct phenotypes. Virology 144:139-151

Barth CF, Humphries EH (1988) A nonimmunosuppressive helper virus allows high efficiency induction of B-cell lymphomas by reticuloendotheliosis virus strain T. J. Exp. Med. 167:89-108

Beug H, Muller H, Grieser S, Doederlein G, Graf T (1981) Hematopoietic cells transformed *in vitro* by REV-T avian reticuloendotheliosis virus express characteristics of very immature lymphoid cells. Virology 115:295

Chen LC, Courtneidge SA, Bishop JM (1983) Immunological phenotype of lymphomas induced by avian leukosis viruses. Mol. Cell. Biol. 3:1077-1085

Cooper MD, Payne LN, Dent PB, Burmester BR, Good RA (1968) Pathogenesis of avian lymphoid leukosis. I. Histogenesis. J. Natl. Cancer Inst. 41:373-389

Cooper MD, Purchase HG, Bochman DE, Gathings WE (1974) Studies on the nature of the abnormality of B-cell differentiation in avian lymphoid leukosis: Production of heterogeneous IgM by tumor cells. J. Immunol. 113:1210-1222

Fung YK, Fadly AM, Crittenden LB, Kung HJ (1981) On the mechanisms of retrovirus-induced avian lymphoid leukosis:

Deletion and integration of the proviruses. Proc. Natl. Acad. Sci. USA 78:3418-3422
Hayward WS, Neel BG, Astrin SM (1981) Activation of a cellular *onc* gene by promoter insertion in ALV-induced lymphoid leukosis. Nature (London) 290:475-480
Kanter MR, Smith RE, Hayward WS (1988) Rapid induction of B-cell lymphomas: Insertional activation of c-*myb* by avian leukosis virus. J. Virol. 62:1423-1432
Lewis RB, McClure J, Rup B, Niesel DW, Garry RF, Hoelzer JD, Nazerian K, Bose HR (1981) Avian reticuloendotheliosis virus: Identification of the hematopoietic target cell for transformation. Cell 25:421
Neiman PE, Jordan L, Weiss RA, Payne LN (1980) Malignant lymphoma of the Bursa of Fabricius: Analysis of early transformation. In: Essex M, Todaro G, zur Hausen H (eds) Viruses in Naturally Occurring Cancers, Cold Spring Harbor Laboratory, New York, pp 519-528
Neiman P, Wolf C, Enrietto PJ, Cooper GM (1985) A retroviral *myc* gene induces preneoplastic transformation of lymphocytes in a bursal transplantation assay. Proc. Natl. Acad. Sci. USA 82:222-226
Olson LD (1967) Histopathologic and hematologic changes in moribund stages of chicks infected with T-virus. Am. J. Vet. Res. 28:1501
Payne GS, Bishop JM, Varmus HE (1982) Multiple arrangements of viral DNA and an activated host oncogene in bursal lymphomas. Nature (London) 295:209-214
Perbal B, Cline JM, Hillyard RL, Baluda MA (1983) Organization of chicken DNA sequences homologous to the transforming gene of avian myeloblastosis virus. II. Isolation and Characterization of λ *proto-amv* DNA recombinant clones from a library of leukemic chicken DNA. J. Virol. 45:925-940
Purchase HG, Burmester BR (1978) In: Hofstad MS, Calnek BW, Helmboldt DF, Reid WM, Yoder HM (eds) Diseases of Poultry, Iowa State Univ. Press, Ames, pp 418-468
Reynaud CA, Anquez V, Dahan A, Weill JC (1985) A single rearrangement event generates most of the chicken immunoglobulin light chain diversity. Cell 40:283-291
Theilen GH, Zeigel RF, Twiehaus MJ (1966) Biological studies with RE virus (strain T) that induces reticuloendotheliosis in turkeys, chickens, and Japanese quail. J. Natl. Cancer Inst. 37:731
Thompson CB, Humphries EH, Carlson LM, Chen CLH, Neiman PE (1987) The effect of alterations in *myc* gene expression on B-cell development in the Bursa of Fabricius. Cell 51:371-381
Thompson CB, Neiman PE (1987) Somatic diversification of the chicken immunoglobulin light chain gene is limited to the rearranged variable gene segment. Cell 48:369-378
Wilhelmsen KC, Temin HM (1984) Structure and dimorphism of c-*rel* (turkey), the cellular homolog to the oncogene of reticuloendotheliosis virus strain T. J. Virol. 49:521-529

Bursal Stem Cells as Targets for *myc*-Induced Preneoplastic Proliferation and Maturation Arrest

P. E. NEIMAN[1], E. B. GEHLY[1], L. M. CARLSON[2], R. C. COTTER[1], and C. B. THOMPSON[2]

INTRODUCTION

Induction of lymphomas in the bursa of Fabricius of chickens by replication competent, non-transforming avian retroviruses (Avian Leukosis Virus, ALV) is both an economically significant, naturally occurring disease and an experimental model system which has provided seminal insights into the role of the myc oncogene in B-cell neoplasia. Infection of the bursa of susceptible birds results in the development of lymphomas in a series of specific stages beginning with a preneoplastic proliferation of pyroninophilic B-lymphoblasts within individual bursal follicles, a lesion termed a transformed follicle (Cooper et al. 1968, Neiman et al. 1980a). Transformed follicles appear to be precursors of discrete bursal nodules and, later, metastasizing lymphomas (Neiman et al. 1980a, 1980b, Baba and Humphries 1985). These advanced lesions are clonal neoplasms (Neiman et al. 1980b) in which the expression of the c-myc proto-oncogene has been deregulated as a result of the nearby integration of an ALV provirus (Hayward et al. 1981).

We recently explored further the role of myc in this process by exploiting a bursal transplantation technique (Neiman et al. 1985). Embryonic bursal lymphocytes were infected exvivo with a replication defective transforming virus, HB1, which carries a v-myc oncogene (Enrietto et al. 1983) and infused into recipient birds whose bursal lymphocytes had been ablated by treatment with a cytotoxic drug, cyclophosphamide. Bursal follicles reconstituted by HB1-infected cells, and expressing v-myc, were indistinquishable from the transformed follicles observed in the early stages of ALV induced lymphomagenesis. Based on these observations we postulated in these proceedings (Thompson and Neiman 1986) that the effect of myc deregulation was to fix bursal lymphocytes in an activated cycling state represented by the cell population of the transformed follicle. The inability to turn off myc expression in these cells, we supposed, prevented the normal accumulation of small resting lymphoctes which make up the majority of the post hatching bursal lymphocyte population. Apparently maintenance of cell proliferation by v-myc leads to a high probability of further changes characteristic of lymphoid malignancy. In this paper we modify and extend this model to take into account the evidence that myc-mediated bursal lymphomagenesis is a developmentally regulated process.

DEVELOPMENTAL ASPECTS OF BURSAL LYMPHOMAGENESIS

A distinctive feature of ALV induced lymphomagenesis is the fact that

[1]Division of Basic Sciences, Fred Hutchinson Cancer Research Center and the Departments of Medicine, Pathology and Laboratory Medicine, University of Washington, 1124 Columbia St., Seattle, Washington 98104.
[2]Howard Hughes Medical Inst., University of Michigan, Ann Arbor, MI

Current Topics in Microbiology and Immunology, Vol. 141

the virus induces this disease in the bursa only if infection takes place during bursal organogenesis in the embryo or a few days after hatching (Burmester et al. 1960, Maas et al. 1982). This limited window of sensitivity exists despite the fact that resting and cycling bursal lymphocytes are present for several months after hatching and might be expected to serve as targets for tumor induction. It was intriguing to us that this window in development corresponded to the period when bursal stem cells have been detected in the bursa (Toivanen and Toivanen 1973, Pink et al. 1985, Ratcliffe 1985, Weill et al. 1986). This stem cell has been characterized by these investigators as a bursal derived sIg+ cell with the specific ability to repopulate cyclophosphamide depleted bursal follicles. As pointed out previously in these proceedings (Thompson et al. 1986, Weill et al. 1987) our initial results transplanting HB1 infected embryonic bursal lymphocytes (Neiman et al. 1985) were consistent with the notion that such bursal stem cells might be selective targets for the action of myc oncogenes.

The cells of HB1 v-myc induced transformed follicles have the cardinal properties of bursal stem cells.

We tested HB1 induced transformed follicle cells for the principal phenotypic properties of bursal stem cells, namely their ability to repopulate ablated bursal follicles and to diversify their immunoglobulin genes (Weill et al. 1986, Reynaud et al. 1985, 1987, Thompson and Neiman et al. 1987). The main experimental strategy was the use of serial bursal transplants as outlined in Figure 1.

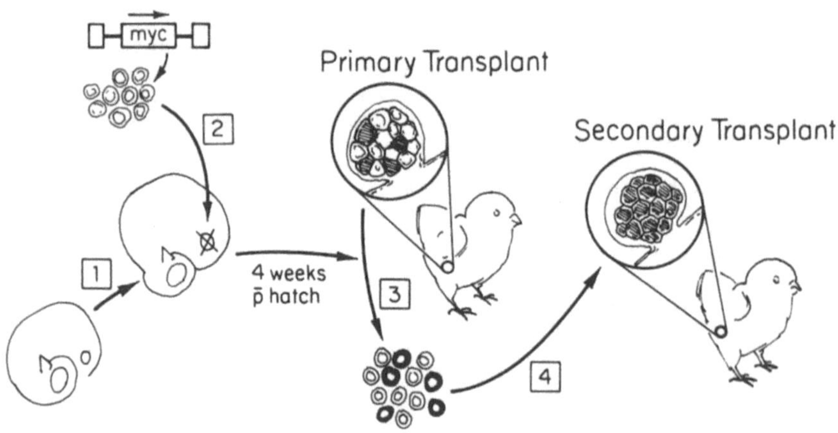

Figure 1. Schematic representation of the serial bursal transplant system (Thompson et al. 1987). The steps are: 1) Recipient embryos (days 15, 16, 17) are treated with cyclophosphamide to ablate lymphocytes in bursal follicles. 2) Bursal donor lymphocytes from 18 day embryos are infected with HB1 virus bearing a v-myc gene and transplanted into primary recipients. 3) Four weeks after hatching . reconstituted follicles are mixtures of normal and transformed (darkly shaded) follicles. Cell suspensions from these bursas are transplanted into secondary cyclophosphamide treated recipients. 4) Four weeks after hatching these secondary transplants contain only transformed reconstituted follicles. Cell suspensions from these secondary transplants can be transferred to a third recipient (see Table 1)

As displayed in the figure, bursas reconstituted with HB1 infected embryonic lymphocytes contain a mixture of transformed and normal follicles by four weeks after hatching (Neiman et al. 1985). The difference between transformed and normal follicle morphology was determined by the presence or absence respectively of the HB1 v-myc gene expressed at high levels in the follicle lymphoid cell population (Neiman et al. 1985, Thompson et al. 1987). Normal follicle cell populations lack bursal stem cells detectable in the transplantation assay by four weeks after hatching, even when the proliferative cell compartment is specifically isolated and tested (Thompson et al. 1987). We found, however, that four week bursal follicle populations from primary recipients of HB1 infected transplants could repopulate ablated follicles in secondary recipients when transformed follicles were present in the primary transplant and that all follicles in secondary recipients had transformed morphology (Fig. 1, Thompson et al. 1987). Figure 2 shows myc RNA levels by Northern blotting analysis in small resting and large cycling lymphocyte populations from normal 4 week bursas compared with myc RNA levels in nearly pure populations of transformed follicle cells from secondary transplants (Fig. 1)

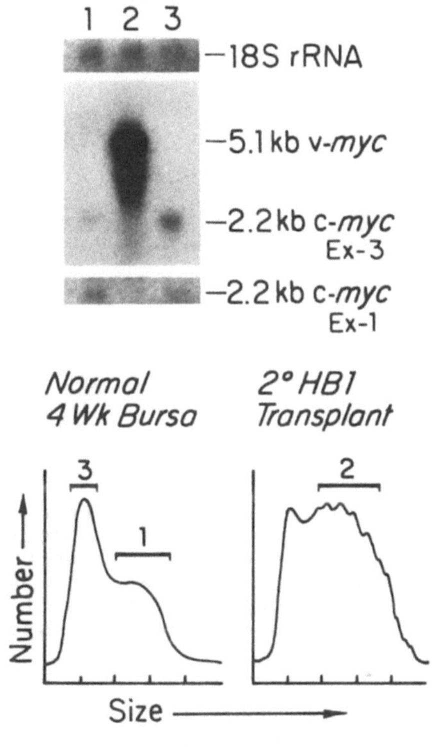

Figure 2. Myc RNA levels were assayed by Northern blotting in bursal lymphocyte populations fractionated by size by counter flow centrifugation as shown in the lower panels. Fractions 1 and 3 are respectively the large cycling and small resting populations of normal 4 week bursas. Fraction 2 is the large transformed follicle cell population of a secondary bursal transplant (see Fig. 1). RNAs detected include 18S ribosomal RNA as a loading control, 5.1Kb v-myc and 2.2Kb c-myc in RNAs detected with a c-myc exon 3 probe, and the same 2.2 Kb c-myc RNA detected with c-myc exon 1 probe (not present in HB1 v-myc RNA)

The data in Figure 2 show that the 5.1 Kb HB1 v-myc mRNA is present at levels at least 20 fold higher than those of either normal bursa lymphocyte population [or of normal embryonic bursal lymphocytes (not shown)]. In addition 2.2 Kb c-myc mRNA levels were reduced below detectable levels in these transformed follicle cells. Furthermore, transformed follicle cells from primary recipients (i.e. four weeks after introduction of the HB1 v-myc gene) were at least 100 fold more efficient at producing detectable repopulation than were normal bursal cell populations from 18 day embryos (Table 1, Thompson et al. 1987). Finally, we found by measuring the disappearance of germ line restriction endonuclease sites from the rearranged light chain variable region gene (Thompson and Neiman 1987) that HB1-induced transformed follicle cells were as efficient as the progeny of normal bursal stem cells in diversification of immunoglobulin genes (Thompson et al. 1987).

Interpretation: HB1 v-myc permits proliferation of and induces a selective maturation block in bursal stem cells.

A reasonable interpretation of these observations is outlined in Figure 3.

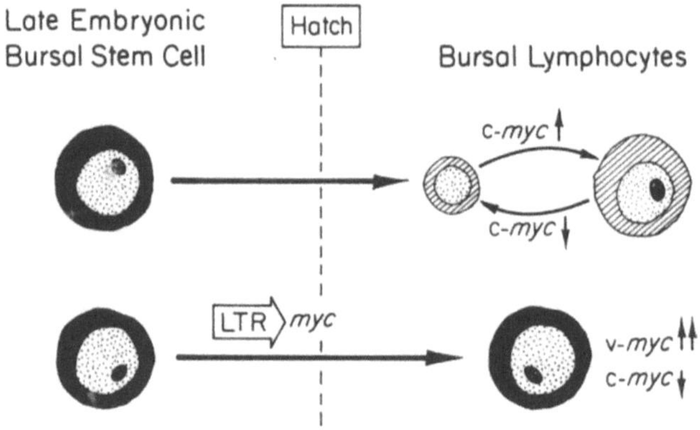

Figure 3. Schematic interpretation of the effect of unregulated HB1 v-myc expression on bursal stem cell differentiation. The left side depicts a pyroninophilic embryonic stem cell which differentiates into resting and activated population after hatching, shown on the right-side. The lower segment depicts effects of arrested maturation in the presence of a deregulated myc gene expressed from a viral promoter-enhancer system (LTR)

Normal bursal stem cells present among the proliferating cells of embryonic bursal follicles give rise to the population of small resting lymphocytes which dominate the post-hatching bursal follicles. These cells in turn are activated, at least transiently, to produce the proliferative large cell compartment of the bursal cell population. None of these post hatching populations retain stem cell activity detectable in transplantation assays. Although the purpose of this transition is at this point unclear, we postulate that a period of quiescence is required for some aspect of bursal lymphocyte maturation. If a deregulated myc gene is introduced into bursal stem cells they fail to differentiate into the small resting and activated large cell populations of normal post hatching follicles but, instead, remain arrested in the stem cell stage and form the pyroninophilic blast cell population of the transformed follicle. For convenience Figure 3 depicts the normal stem cell target of the embryonic bursa as having the same morphology as the transformed follicle cell, but, in fact, the morphology of such cells has not yet been determined. The availability of nearly pure populations of transformed follicle cells in this experimental system may permit the generation of monoclonal reagents which will recognize normal stem cell specific antigens and permit the identification of normal bursal stem cells. In any case these myc induced transformed follicle cells retain the capacity to reconstitute bursal follicles and appear to continue to require association with the bursal epithelium for proliferation. Furthermore, they continue to diversify the germline sequence of rearranged immunoglobulin light chain variable region genes at a rate indistinquishable from that of normal bursal cells (Thompson et al. 1987). Therefore their maturation arrest is selective in that it involves the persistence of stem cell function, but does not block diversification of Ig light chain genes.

FUNCTIONAL INSTABILITY AND NEOPLASTIC PROGRESSION IN V-MYC INDUCED TRANSFORMED FOLLICLE CELLS

The foregoing discussion emphasizes the normal stem cell-like properties of myc oncogene induced transformed follicle cells, but these cells are clearly phenotypically unstable and subject to functional change in two important respects. First, their efficiency in repopulating ablated bursal follicles, as measured by the minimal number of transformed follicle cells required for histologically detectable repopulation, deteriorates over time without obvious morphologic change in the cell population. This conclusion is supported by the data in table 1.

As shown in the table the minimal reconstituting dose of transformed follicles cells in a secondary transplant (see Fig. 1) is about 10^4. In a third round of transplantation, which is about 8 weeks after the original infection with HB1, transformed follicle cells are about 100 fold less efficient (minimal reconstituting dose around 10^6 transformed follicle cells) in reconstituting cyclophosphamide ablated bursal follicles than similar donor cells in the secondary transplants. The nature of the changes which underlie this deterioration of function are not presently known, but could be related to neoplastic progression since more advanced bursal neoplasms also are inefficient in homing to the bursa (Neiman, unpublished data).

The second change in transformed follicle cell populations with time is gross neoplastic progression. About 5% of primary bursal transplants with HB1 infected donor stem cells contain obvious bursal lymphomas (spreading neoplasms not confined to the follicular architecture). As shown in Table 1, in secondary transplants this number has increased to about 20%, a proportion which persists in tertiary transplants. These data are consistent with the interpretation that v-myc transformed stem cells gradually both lose the ability to act as stem cells and undergo full neoplastic conversion at a fairly high rate. The tumors that we detected after each transfer probably represent the progeny of residual functional reconstituting stem cells in the *myc* transformed population which underwent further genetic change mediating full neoplastic conversion during the four week period between transplantation and histologic analysis of reconstituted bursas.

TABLE 1. Efficiency of normal and v-*myc* transformed bursal stem cells in reconstituting cyclophosphamide ablated bursal follicles.

Source of bursal stem cells	Numbers of recipients analysed	Minimum reconstituting dose [a]	Percent of recipients with bursal tumors
Normal 18 day embryos	> 100	$0.5 - 1 \times 10^6$	--
Total normal 4 week bursa	37	No reconstitution	--
Proliferating fraction 4 week bursa (normal)	5	No reconstitution	--
T.F cells from primary transplants [b]	40	$1 - 3 \times 10^4$	21%
T.F. cells from secondary transplants [c]	11	2×10^6	19%

a. Minimal reconstituting dose is the number of donor cells required to repopulate at least 10% of ablated bursal follicles in at least 1/3 of recipient birds.

b. T.F. cells are transformed follicle cells from primary recipients of HB1 infected embryonic donor bursal lymphocyte analysed four weeks after hatching as depicted in Figure 1.

c. These are pure transformed follicle cell preparations from secondary transplants (described in Fig. 1) transplanted to tertiary cyclophosamide treated recipients as analysed four weeks after hatching.

CONCLUSION

The principal conclusion from these studies is that deregulation of *myc* expression in the chicken B-cell lineage is tumorigenic principally when it occurs in a particular window of development represented by the bursal stem cell. The phenotype of such *myc* transformed stem cells involves a partial maturation arrest which doesn't seem to affect adversely the processing of immunoglobulin genes. Such transformants are, however, phenotypically unstable and both lose stem cell function and undergo neoplastic conversion at a high rate. Thus, while *myc* oncogenes do not seem sufficient for full neoplastic change in this system, their expression in bursal stem cells appears to place these target cells at high risk for further genetic change, the precise nature of which remains to be determined. The availability of reasonable numbers of myc transformed stem cells in this experimental system may provide an opportunity for further insights into oncogenesis in the B-cell lineage.

ACKNOWLEDGEMENTS

We thank our colleagues at the Hutchinson Center Mark Groudine and Bob Eisenman for thoughtful discussions and Jan Smith for assistance in preparation of the manuscript. This work was supported by grants CA 20068 and CA 28151 from the National Cancer Institute.

REFERENCES

Baba TW, Humphries EH (1985) Formation of a transformed follicle is necessary but not sufficient for development of an avian leukosis virus-induced lymphoma. Proc Natl Acad Sci USA 82:213-216

Burmester BR, Fontes AK, Walter WG (1960) Pathogenicity of a viral strain (RPL 12) causing avian visceral lymphomatosis and related neoplasms. III. Influence of host age and route of inoculation. J Natl Cancer Inst 24:1423-1442.

Cooper MD, Payne LN, Dent PB, Burmester BR, Good RA (1968) Pathogenesis of avian leukosis I. Histogenesis. J Natl Cancer Inst 41:373-389.

Enrietto PJ, Payne LN, Hayman MJ (1983) A recovered avian myelocytomatosis virus that induces lymphomas in chickens: Pathogenic properties and their molecular basis. Cell 36:369-379.

Hayward WS, Neel B, Astrin SM (1981) Activation of a cellular onc gene by promoter insertion in ALV-induced lymphoid leukosis. Nature 290:475-480.

Maas HJL, Debber GF, Groenend JE (1982) Age-related resistance to avian-leukosis virus. 3. Infectious virus neutralizing antibody and tumors in chickens innoculated at various ages. Avian Path 11:309-327.

Neiman PE, Jordan L, Weiss RA, Payne LN (1980a) Malignant lymphoma of the bursa of Fabricius: Analysis of early transformation. In: Viruses in Naturally Occuring Cancers, Cold Spring Harbor Laboratory, Cold Spring Harbor, New York, p 519-527.

Neiman P, Payne LN, Weiss RA (1980b) Viral DNA in bursal lymphomas induced by avian leukosis viruses. J Virol 34:178-186.

Neiman P, Wolf C, Enrietto PJ, Cooper GM (1985) A retroviral myc gene induces preneoplastic transformation of lymphocytes in a bursal transplantation assay. Proc Natl Acad Sci USA 82:222-226

Pink JRL, Ratcliffe MJH, Vainio O (1985) Immunoglobulin-bearing stem cells for clones of B (bursa-derived) lymphocytes. Eur J Immunol 15:617-620.

Ratcliffe MJF (1985) The ontogeny and cloning of B-cells in the bursa of Fabricus. Immunol Today 6:223-227.

Reynaud CA, Anquez V, Dahan A, Weill J-C (1985) A single rearrangement event generates most of the chicken immunoglobulin light chain diversity. Cell 40:283-291.

Reynaud CA, Anquez V, Grimal H, Weill J-C (1987) A hyperconversion mechanism generates the chicken light chain preimmune repertoire. Cell 48:379-388.

Thompson CB, Challoner PB, Neiman PE (1986) Normal and neoplastic B-cell development in the bursa of Fabricius. In: Current Topics in Microbiology and Immunology, vol 132, Springer-Verlag Berlin-Heidlberg, pp 209-214.

Thompson CB, Neiman PE (1987) Somatic diversification of the chicken immunoglobulin light chain gene is limited to the rearranged variable gene segment. Cell 48:369-378.
Thompson CB, Humphries EH, Carlson LM, Chen C-LH, Neiman PE (1987) The effect of alterations in myc gene expression on B-cell development in the bursa of Fabricius. Cell 51:371-381.
Toivanen P, Toivanen A (1973) Bursal and post-bursal stem cells in chickens. Funtional characteristics. Eur J Immunol 3:585-595.
Weill J-C, Reynaud CA, Lassila O, Pink JRL (1986) Rearrangement of chicken immunoglobulin genes is not an ongoing process in the embryonic bursa of Fabricius. Proc Natl Acad Sci USA 83:3336-3340.
Weill J-C, Leibowitch M, Reynaud C-A (1987) Questioning the role of the embryonic bursa in molecular differentiation of B lymphocytes. In: Current Topics in Microbiology and Immunology, vol. 135, Springer-Verlag Berlin-Heidlberg, pp 111-124.

Molecular Analysis of an AIDS-Associated Burkitt's Lymphoma: Near-Identity with Endemic Cases

F. G. HALUSKA[1], G. RUSSO[1], M. ANDREEFF[2], and C. M. CROCE[1]

INTRODUCTION

The most completely studied B-cell malignancy is Burkitt's lymphoma, an understanding of which has been facilitated by a cytogenetic and molecular analyses of the chromosome translocations which are consistently exhibited by these tumors (reviewed in Haluska et al., 1987a). The translocations seen in Burkitt's lymphomas abnormally juxtapose one of the immunoglobulin loci with the c-myc proto-oncogene. A small proportion of these chromosome translocations involve the light chain loci. Thus, the t(2;8)(p11;q24) translocation approximates the κ chain genes on chromosome 2 with the c-myc gene on chromosome 8; similarly, the t(8;22)(q24;q11) translocation juxtaposes the c-myc gene and the λ chain genes. But in over 80% of cases, the relevant translocation is the t(8;14)(q24;q32). In this instance, the c-myc gene is translocated into the immunoglobulin heavy chain region. Consequently, normal regulation of c-myc expression is abrogated, and malignancy ensues.

Extensive study of the molecular structures of translocated t(8;14) chromosomes has revealed that the chromosome breakpoints are dispersed throughout both the IgH and c-myc regions. Despite this apparent heterogeneity, the distribution of breakpoints is not random. The locations of translocation breakpoints have been shown to correlate with the origin of the tumor from which they derive. Burkitt's lymphomas of endemic origin generally carry 14q+ chromosomes which join regions far 5' of c-myc (Haluska et al., 1986; Pelicci et al., 1986) to Ig J_H (Haluska et al., 1986) or D_H (Haluska et al., 1987b) regions. This suggests that these translocations arise early in B-cell ontogeny through mistakes in the function of the immunoglobulin recombinase system (reviewed in Haluska et al., 1987c). In contrast, tumors which arise sporadically most commonly exhibit junction of regions immediately 5' of c-myc, or within the first exon or intron of the gene, to Ig μ, γ, or α switch sequences. These tumors apparently arise later in B-cell development due to translocations catalyzed by Ig isotype switching enzymes. These molecular differences between endemic and sporadic tumor-specific translocations therefore suggest an explanation for well-described clinical, epidemiologic, and phenotypic differences between these subtypes of Burkitt's lymphoma (Klein and Klein, 1985).

Burkitt's lymphomas also occur in patients suffering from the acquired immune deficiency syndrome (AIDS) (Ziegler et al., 1982). These tumors share several features with endemic Burkitt's lymphomas. Many of them occur in conjunction with infection by the Epstein-Barr virus (Ziegler et al., 1982), while EBV is found in over 95% of endemic Burkitt's lymphoma cases (Klein and Klein, 1985); they arise in the clinical setting of severe systemic immunosuppression, thought

[1]The Wistar Institute, 3601 Spruce Street, Philadelphia, PA 19104 USA
[2]Memorial Sloan-Kettering Cancer Center, 1275 York Avenue, New York, NY 10021

Current Topics in Microbiology and Immunology, Vol. 141

to be an important co-factor for the development of endemic Burkitt's lymphoma (Klein and Klein, 1985); and they are often preceded by expansion of clonal B-cell populations (Pelicci et al., 1986), also regarded as a step in the pathogenesis of endemic Burkitt's tumors (Klein and Klein, 1985). These clinical parallels between AIDS-associated and endemic Burkitt's lymphomas suggested to us that molecular similarities might exist as well. To this end we analyzed a case of Burkitt's lymphoma diagnosed in a patient infected with HIV.

RESULTS

Clinical Presentation

The patient, a 26 year old male homosexual, first came to medical attention six years prior to being seen by us when he experienced generalized lymphadenopathy and was told he had contracted mononucleosis. Four years ago he suffered an enlarged right axillary lymph node and night sweats; biopsy of the node demonstrated reactive hyperplasia, and the patient was informed that he had "chronic mononucleosis". Four weeks prior to his presentation he noticed a painful mass in his right axilla, which subsequently enlarged rapidly. Biopsy of the mass disclosed a Burkitt's lymphoma. Cytogenetic analysis revealed a t(8;14) translocation in the tumor. The patient also was shown to harbor the HIV.

Molecular Analysis

Genomic DNA was extracted from the biopsy tissue and digested with various restriction enzymes as previously described (Haluska et al., 1986). Southern blot experiments (not shown) revealed rearrangement of both Ig J_H alleles, germline Ig κ and λ loci, and germline c-myc and bcl-2 configurations, utilizing previously described probes for these genes (Haluska et al., 1986; Tsujimoto et al., 1985; Erikson et al., 1983; Croce et al., 1983). By analogy with the observations we have made in endemic Burkitt's lymphomas, we hypothesized that the t(8;14) translocation breakpoint would lie far 5' of c-myc and within the J_H region of the IgH locus. We thus undertook to isolate clones containing the Ig J_H region. We constructed a genomic library using biopsy DNA and screened approximately 800,000 recombinant clones with a DNA probe for the J_H region (as per Haluska et al., 1986). Clones were obtained which, when physically mapped, corresponded to both rearranged J_H alleles. Some of the clones are illustrated in Fig. 1.

A unique 1.9kb Hind III fragment was purified and subcloned from the region 5' of the region of homology to the J_H probes (see Fig. 1). This probe, when hybridized to Southern-blotted DNAs from a panel of somatic cell hybrids containing segregated human chromosomes, hybridized only to hybrids containing human chromosome 8 (not shown). As Fig. 2 illustrates, the genomic DNA including this probe is amplified in the HL 60 and COLO 320 cell lines, both of which carry amplification units which encompass regions surrounding c-myc. These lines of evidence demonstrate that the clones illustrated in Fig. 1 indeed include the t(8;14) junction, and that the Hind III 1.9kb fragment derives from chromosome 8.

We obtained the nucleotide sequence of the translocation breakpoint (see Fig. 3). Using the dideoxynucleotide chain termination method on M13 clones, we found that the breakpoint on chromosome 14 is upstream of the J5 segment.

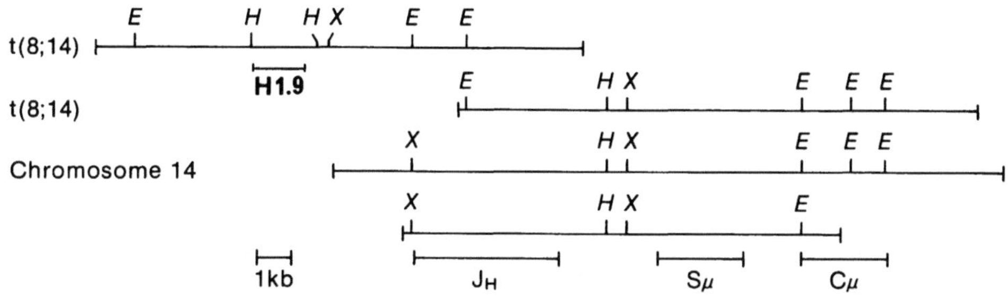

Fig. 1. Restriction map of FID recombinant clones. Two rearranged clones encompassing the t(8;14) chromosomal translocation are shown on the top. On the bottom, two germ line clones encompassing Cμ, Sμ and J_H on chromosome 14 are represented. The probe H1.9 used for mapping is also indicated

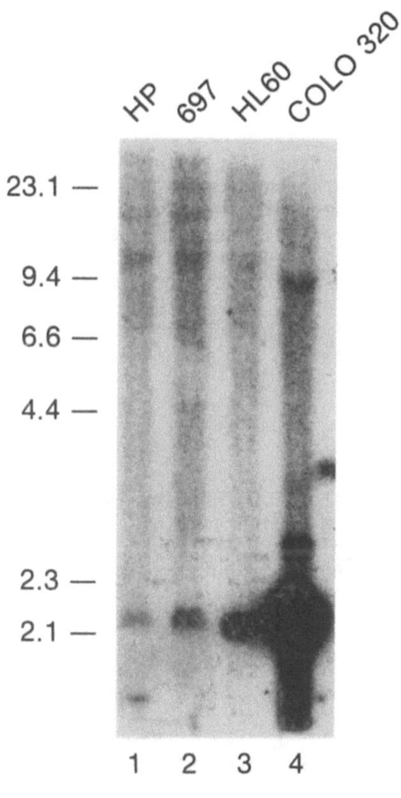

Fig. 2. Southern blot analysis of Hind III-digested DNAs with the probe H1.9. HP is human placental DNA, 697 is a pre-B leukemia cell line, HL60 is a promyelocytic cell line, and COLO 320 is a moderately undifferentiated adenocarcinoma of the sigmoid colon. HL60 and COLO 320 carry an amplification unit which encompasses regions surrounding c-myc (see text for explanation)

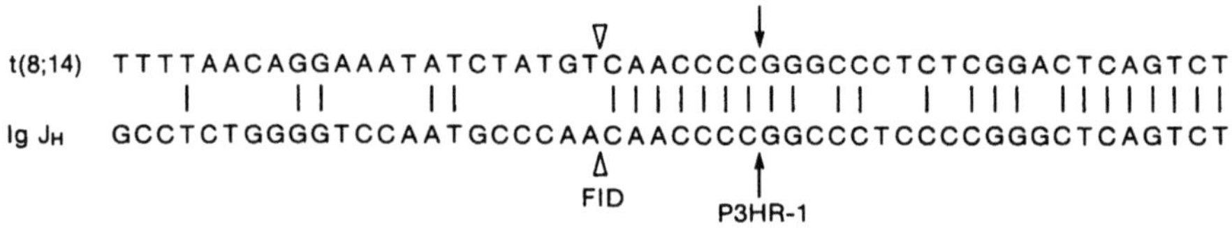

Fig. 3. Nucleotide sequences of the FID t(8;14) breakpoint and the normal 14 J_H region. The breakpoint of FID is indicated by △. The P3HR1 breakpoint is indicated by ↑

Interestingly, the breakpoint is seven nucleotides from the previously described P3HR-1 endemic Burkitt's lymphoma breakpoint (Haluska et al., 1986). Upstream of the breakpoint on chromosome 14 there is a heptamer which matches at 5 out of 7 positions the heptamer recombination signal which lies apposed to the J5 segment: CAATGCC is found near the breakpoint, whereas CAATGTG is observed at J5. Located further upstream, 23 base pairs from the heptamer, is a nonamer also previously described for the P3HR-1 breakpoint. In short, the breakpoint localization on chromosome 14 is nearly identical to that of the endemic Burkitt's lymphoma P3HR-1.

Where does the breakpoint reside on chromosome 8? The Hind III 1.9kb fragment hybridizes to somatic cell hybrid DNA whose only human chromosome is the 14q+ These data also suggest that the common features of these otherwise very different diseases are central to their common molecular pathogenesis. EBV infection, clonal B-cell expansion, and immunosuppression appear to increase the likelihood of the occurrence of a pathogenic recombinase error. In this patient, as well as in endemic Burkitt's lymphoma patients, all of these factors are operative. As a result, the recombinase mistakenly joins regions near c-myc to the J_H (or D_H) region. Consequently, c-myc deregulation and tumorigenesis follow.

ACKNOWLEDGEMENTS

We thank Ms. C. Long for preparation of this manuscript. F.G. Haluska is a trainee of the Medical Scientist Training Program at the University of Pennsylvania School of Medicine, NIH T32 GM 07170. This work was supported in part by a grant from the National Cancer Institute, CA 39860, Outstanding Investigator Award (C.M. Croce).

REFERENCES

Croce CM, Thierfelder W, Erikson J, Nishikura K, Finan J, Lenoir G, Nowell PC (1983) Transcriptional activation of an unrearranged and untranslocated c-myc oncogene by translocation of a C λ locus in Burkitt lymphoma cells. Proc Natl Acad Sci USA 80: 6922-6926

Erikson J, Nishikura K, ar-Rushdi A, Finan J, Emanuel B, Lenoir G, Nowell PC, Croce CM (1983) Translocation of an immunoglobulin κ locus to a region 3' of an unrearranged c-myc oncogene enhances c-myc transcription. Proc Natl Acad Sci USA 80: 7581-7585

Haluska FG, Finver SN, Tsujimoto Y, Croce CM (1986) The t(8;14) chromosomal translocation occurring in B-cell malignancies results from mistakes in V-D-J joining. Nature 324: 158-161

Haluska FG, Tsujimoto Y, Croce CM (1987a) Oncogene activation by chromosome translocation in human malignancy. Ann Rev Genet 21: 321-345

Haluska FG, Tsujimoto Y, Croce CM (1987b) The t(8;14) translocation of the Daudi endemic Burkitt's lymphoma occurred during immunoglobulin gene rearrangement and involved the D_H region. Proc Natl Acad Sci USA 84: 6835-6839

Haluska FG, Tsujimoto Y, Croce CM (1987c) Mechanisms of chromosome translocation in B- and T-cell neoplasia. Trends Genet 3: 11-15

Klein G, Klein E (1985) Evolution of tumors and the impact of molecular oncology. Nature 315: 190-195

Pelicci P-G, Knowles DM II, Arlin ZA, Wieczorek R, Luciw P, Dina D, Basilico C, Dalla-Favera R (1986) Multiple monoclonal B cell expansions and c-myc oncogene rearrangements in acquired immune deficiency syndrome-related lymphoproliferative disorders: implications for lymphomagenesis. J Exp Med 164: 2049-2076

Pelicci P-G, Knowles DM II, Magrath I, Dalla-Favera R (1986) Chromosomal breakpoints and structural alterations of the c-myc locus differ in endemic and sporadic forms of Burkitt lymphoma. Proc Natl Acad Sci USA 83: 2984-2988

Tsujimoto Y, Cossman J, Jaffe E, Croce CM (1985) Involvement of the bcl-2 gene in human follicular lymphoma. Science 228: 1440-1443

Ziegler JL, Miner RC, Rosenbaum E, Lennette ET, Shillitoe E, Casavant C, Drew WL, Mintz L, Gershow J, Greenspan J, Beckstead J, Yamamoto K (1982) Outbreak of Burkitt's-like lymphoma in homosexual men. Lancet ii: 631-633

Transgenic Mouse Models for Hematopoietic Tumorigenesis

A. W. Harris, W. Y. Langdon, W. S. Alexander, I. K. Hariharan, H. Rosenbaum, D. Vaux, E. Webb, O. Bernard, M. Crawford, H. Abud, J. M. Adams, and S. Cory

The development of transgenic mice carrying specific cancer-promoting genes represents a major advance in molecular oncology (for reviews, see Cory and Adams, 1988; Hanahan, 1987). In principle, the effect of a particular oncogene can be assessed for every cell type within the animal by this approach. The transgene can either be linked to a catholic regulatory sequence and expressed throughout all tissues, or its expression can be confined to a particular cell lineage by fusing it to a tissue-specific promoter/enhancer control element. Our own interests center on hematopoietic neoplasia, so we have produced transgenic mice harboring oncogenes targeted for constitutive expression in hematopoietic cells. The principal regulatory sequence we have used is the immunoglobulin (Ig) heavy chain enhancer (Eμ). While the function of this sequence is to direct expression of Ig heavy chain (*IgH*) genes in B lymphoid cells, it may also be active in at least some T lymphoid and even myeloid cells since the *IgH* locus is transcriptionally active in certain T and myeloid cell lines (Kemp et al, 1980a) and thymocytes (Kemp et al, 1980b) and an immunoglobulin μ transgene can be expressed in T as well as B cells (Grosschedl et al, 1984). By linking an oncogene to Eμ, we therefore expected to be able to assess its transforming potential for B lymphoid cells and perhaps also other hematopoietic cells. The oncogenes introduced into mice were selected principally on the basis of known associations with hematologic malignancies and include the *myc*, N-*myc*, *abl* and N-*ras* genes. We here discuss our most recent results with the well-established Eμ-*myc* line and our preliminary findings with new lines.

Eμ-*myc* MICE

Deregulated expression of the c-*myc* gene has been implicated in several types of neoplasm. Most human Burkitt lymphomas, murine plasmacytomas and rat immunocytomas carry a chromosome translocation which has activated c-*myc* by linking it to the *IgH* locus (Cory, 1986). The plasmacytoma ABPC17 proved to have undergone a more complex recombination which resulted in removal of a 2.3 kb segment of the enhancer region from one *IgH* allele and its insertion at a site 361 bp 5' to the c-*myc* gene (Corcoran et al, 1985). Injection of the Eμ-*myc* gene cloned from this tumor into fertilized oocytes produced a unique strain of mice predestined to develop B lymphoid tumors (Adams et al, 1985).

The Walter and Eliza Hall Institute of Medical Research, Royal Melbourne Hospital, Melbourne, 3050, Australia.

Current Topics in Microbiology and Immunology, Vol. 141

B Cell and Plasma Cell Tumors

Pathology of Eμ-*myc* Disease

Autopsies on animals with advanced disease, combined with histologic analysis of affected tissues and phenotypic characterization of the tumors have established a detailed picture of the pathology of Eμ-*myc* disease (Harris et al, 1988). Most mice develop a massive enlargement of all lymph nodes and/or the thymus. The tumors are highly malignant by transplantation assay and are categorized morphologically as lymphoblastic lymphomas, usually accompanied by lymphoid leukemia and granulocytosis. They thus clearly differ from the reticulum cell type B tumors that arise much later in some non-transgenic mice of the same (C57BL/6 x SJL)F2 genetic background. The Eμ-*myc* tumors appear to arise at random from a population of pre-B cells expanded by constitutive expression of the *myc* transgene (see below). Some tumors arise as early as 3 weeks, others as late as 9 months. The rate at which cycling benign pre-B cells spontaneously convert to malignancy is estimated to be $\sim 10^{-10}$ per cell per generation.

All of the more than 50 tumors analyzed to date have proved to be of B lymphoid origin, even those predominantly involving the thymus (Harris et al, 1988; Cory, Harris, Langdon and Adams, in preparation). They exhibit B lineage surface markers and Ig gene rearrangements, none carry Thy 1, and only one or two have undergone rearrangement of the γ or β T cell receptor loci. A minority (19%) of the tumors express surface Ig and are thus B cell lymphomas, another 29% contain varying mixtures of pre-B and B cells, while the major class can be assigned to various pre-B stages exhibiting either DJ or VDJ heavy chain rearrangement. The apparent restriction of expression of the transgene to B lymphoid cells (Alexander et al, 1987) would account for the absence of T or myeloid tumors in Eμ-*myc* mice.

Most of the primary tumors can readily be established as permanent lines in culture. Significantly, many of the pre-B tumors yield cultured lines of sIg^+ B cells. The Ig gene rearrangement patterns and the occurrence of mixed pre-B/B tumors suggest that maturation of some malignant pre-B cells can proceed *in vivo*. Thus *myc*-induced tumorigenesis need not prevent some subsequent differentiation.

Influence of Genetic Background

One factor contributing to the variation in the rate of onset of tumors in Eμ-*myc* mice is their variable genetic background. Because transgenic mice are produced most efficiently from hybrid eggs, the founder mouse was an F2 derived from the C57BL/6 (B6) and SJL/J strains and the colony has been maintained as an F2 pool by successive matings of male transgenic mice with multiple normal (B6 x SJL)F1 hybrid females. At the same time, we have begun to directly test the influence of genetic background by serial backcrossing with three inbred strains, SJL, B6 and BALB/cAn (Table 1). The SJL line was lost after 3 backcrosses because it developed lymphomas too rapidly. The B6 line, by contrast, develops tumors relatively slowly and has now reached generation 11. Transgenic females of this line transmit a higher predisposition to lymphomagenesis in their progeny than do males - an example of gender-specific imprinting of transgene activity. The tumor onset kinetics are different again in the BALB/c line. While the first lymphomas appear very early in life, the subsequent onset rate is fairly low. How the genetic background influences the rate of tumorigenesis is unclear, but it is interesting to note that the longest-surviving B6SJL Eμ-*myc* mice had a lower increase in the number of pre-B cells (Langdon et al, 1986).

Table 1. Lymphoma onset in sublines of Eμ-*myc* 292-1 mice

Eμ-*myc* subline[a]	Tumor onset		
	lag period[b] (wk of age)	linear rate (% per wk)	median age (wk)
B6SJL	5	17 and 6	9
SJL N3	6	36	7
B6 N5	5	6	16
BALB/c N6	0	6	11

[a]Groups of mice were progeny of appropriate backcross transgenic males mated to normal females of indicated strain
[b]Obtained by extrapolating log-linear decline in % healthy mice back to 100%

Plasmacytoma Induction

To date, no plasmacytomas have arisen spontaneously in Eμ-*myc* mice, even though most conventional plasmacytomas carry a *myc* gene deregulated by recombination with the *IgH* locus (reviewed by Cory, 1986). This paradox might have reflected the genetic background of the mice, since plasmacytoma induction is readily accomplished only in BALB/c or NZB mice. However, the BALB/c Eμ-*myc* line also produces lymphomas rather than plasmacytomas. The limiting factor may be the presence of an adequate milieu for the progressive growth of plasma cells. When Eμ-*myc* mice from the B6SJL and the B6 lines are injected with pristane, which provokes the formation of a granuloma, those which do not succumb to B lymphoid tumors of the "spontaneous" type by about 7 months of age develop plasmacytomas in association with the granuloma tissue in the peritoneal cavity. The role of the granuloma may be to provide a sustained local source of IL-6, a factor that stimulates plasma cell growth (Nordan and Potter, 1986; O'Garra et al, 1988). Since the transgene allows plasmacytomas to develop in strains that are normally resistant to pristane induction, plasmacytoma susceptibility in conventional mice may be determined primarily by genes that affect the propensity to undergo a *myc*-Ig locus translocation.

Tumor Onset Requires Additional Genetic Change

Several observations argue that lymphomagenesis in Eμ-*myc* mice requires more than expression of the transgene. (a) Lymphoid tissues in young Eμ-*myc* mice lacking any overt sign of disease express the transgene at the same level as do Eμ-*myc* tumor cells (Alexander et al, 1987), but do not provoke malignancy on transplantation (Langdon et al, 1986). (b) The time of onset of tumors in individual mice varies widely (Harris et al, 1988). (c)

The tumors are clonal (Adams et al, 1985), even though probably all B lineage cells in the mice express the transgene.

To test directly the hypothesis that tumor onset requires the activation of additional oncogene(s), we are screening Eμ-*myc* tumor DNA for the presence of genes capable of transforming NIH-3T3 fibroblasts to grow as fibrosarcomas in nude mice. DNA from one Eμ-*myc* tumor provoked particularly rapid fibrosarcomas which were found to harbor an amplified N-*ras* gene. Since integration of transfected DNA frequently involves amplification and/or rearrangement, we inferred that the amplified N-*ras* gene derived from the Eμ-*myc* tumor. An N-*ras* cDNA has now been cloned from the Eμ-*myc* tumor and shown to carry a mutation at amino acid 61, one of the two residues most frequently implicated in *ras* activation. The mutated N-*ras* gene has recently been cloned into a retroviral vector and shown to transform non-tumorigenic pre-B cells from Eμ-*myc* bone marrow. These results clearly establish that collaboration of at least two oncogenes (*myc* and N-*ras*) was responsible for development of one B lymphoid tumor (Alexander, Bernard and Cory, in preparation).

The Pre-Neoplastic State in Eμ-*myc* mice

To investigate how constitutive *myc* expression establishes a high predisposition to malignancy, we have analyzed in some detail the pattern of B lymphoid differentiation in young pre-tumorous animals (Langdon et al, 1986). A remarkable polyclonal expansion of pre-B cells becomes evident as early as 18 days of gestation. Young adult Eμ-myc mice contain some 5 times as many pre-B cells as their normal littermates, and the pre-B cells are found in the spleen as well as the bone marrow. Significantly, the thymus and lymph nodes are essentially free of pre-B cells, even though these are prominent sites of tumor growth in older animals.

The number of mature (sIg^+) B cells in the animals is somewhat depressed, and fewer splenic B cells express surface IgD. Nevertheless, the mice remain healthy and their immunologic function is not grossly impaired (Vaux et al, 1987). Immunization provoked specific antibody responses, although some were delayed. Moreover, isotype switching was not grossly impaired because both IgM and IgG antibody-forming cells were generated in the spleen. Stimulation *in vitro* with either mitogens or T-independent antigens enabled some of the Eμ-*myc* B cells to proliferate and differentiate into antibody-secreting cells, albeit at lower frequency than in normal controls. The level of Eμ-*myc* expression in the LPS-stimulated B cells was comparable to that in the bone marrow pre-B cells. In sum, deregulated *myc* expression within the B lymphoid lineage appears to favor self-renewal over maturation, resulting in an increased proportion of pre-B cells. B cell differentiation, while retarded, is not prevented.

Both pre-B and B cells in the Eμ-*myc* mice are notably larger than most equivalent cells in normal mice. Indeed, the typical resting small pre-B and B cell populations are undetectable. At least 2.5 times as many of the pre-B and B cells are in the G_2+S phases of the cell cycle and the large size of the nuclei of cells in the G_0/G_1 fraction suggests that they are in G_1 rather than G_0 (Langdon, Harris and Cory, submitted). Thus constitutive c-*myc* expression may prevent B lymphoid cells from entering the G_0 state.

Eμ-*myc* Pre-B Cells Progress Towards Autonomy upon Prolonged Culture

To analyze further the effect of constitutive *myc* expression on cells in the B lineage we have characterized the growth of Eμ-*myc* bone marrow cells in culture (Langdon, Harris and Cory, submitted). The Eμ-*myc* B lineage cells are not autonomous, since they die rapidly when cultured without feeder cells. Long-term cultures can be readily established on a feeder layer of bone marrow stromal cells. The proliferating Eμ-*myc* cells are markedly larger than normal pre-B cells grown under these conditions and show much more cell cycle activity. Since the Eμ-*myc* cultures initially grow to only slightly higher densities than normal cultures, the rapidly cycling Eμ-*myc* cells must die more rapidly than normal cells, perhaps due to an inability to regulate their *myc* expression in an environment in which growth is limited by the availability of some growth factor.

A striking feature of the Eμ-*myc* cultures was their dramatic change to much higher density growth after about 15 weeks. Despite this change, the cells still died rapidly when removed from their feeder layers, and they were not tumorigenic when inoculated into histocompatible recipients. The Eμ-*myc* population that survived 23 weeks of culture underwent a further change and became stromal cell independent and tumorigenic. This approach has thus allowed us to monitor stepwise growth changes that occur during the progression to a fully malignant cell.

The Eμ-*myc* Mouse as a Test-Bed for Cooperating Oncogenes

The pre-neoplastic B lineage cells in young Eμ-*myc* mice provide an ideal system in which to investigate the ability of other oncogenes to collaborate with deregulated *myc* expression to transform B lineage cells. We have used retroviruses to deliver the second oncogene, both *in vivo* and *in vitro*, and have compared the growth characteristics of cells expressing two oncogenes with those expressing a single one.

Transgenic animals infected with helper virus-free stocks of viruses carrying the v-H-*ras* or v-*raf* oncogenes were found to develop advanced lymphomas by 3 weeks of age, while infected non-transgenic and non-infected transgenic animals did not. Thus, both of these oncogenes can individually cooperate with the activated *myc* gene to induce lymphoid malignancy. Abelson virus did not enhance the onset of pre-B lymphomas, raising the possibility that v-*abl* cannot collaborate with *myc* in the transformation of pre-B cells or that the co-expression of these genes is usually lethal for pre-B cells. An alternative hypothesis, which we currently favor, is that the primary target for transformation by Abelson virus is a primitive cell in which the heavy chain enhancer is not yet active (Langdon, Harris and Cory, unpublished).

In vitro infection of bone marrow cells from Eμ-*myc* and normal mice with the same viruses has confirmed and extended the *in vivo* work (Dyall-Smith and Cory 1987; Alexander and Cory, in preparation). Uninfected Eμ-*myc* B lineage cells did not proliferate in soft agar or liquid culture and were not tumorigenic. Infection of Eμ-*myc* bone marrow with either the Harvey or 3611 murine sarcoma virus produced 50-100 times more colonies in the agar assay than similarly infected normal marrow cells, a striking illustration of the synergy between *myc* and either v-H-*ras* or v-*raf*. Furthermore, in liquid culture, B lymphoid lines emerged from v-H-*ras* or v-*raf* virus-infected Eμ-*myc* marrow cultures at higher frequency and with shorter latency than from infected normal marrow. The B lymphoid cells expressing v-*raf* alone were clearly only partially transformed; they required feeder cells for growth *in vitro*, did not form colonies in soft agar and were non-tumorigenic. While tumorigenic lines did grow out from the H-*ras* virus-infected normal marrow, the early cultures were only oligoclonal and most of the cells were unable to grow in soft agar. The low frequency of transformed cells implies that these clones harbored other oncogenic changes and that v-H-*ras* expression is insufficient for complete transformation of B lineage cells.

With the Abelson virus, Eμ-*myc* bone marrow cultures produced somewhat fewer colonies than normal cultures, even though the target for Abelson virus is thought to be the pre-B cell, and pre-B cells are 4-fold more frequent in Eμ-*myc* than in normal bone marrow. The newly derived Abelson/Eμ-*myc* lines, like normal Abelson lines, were not all tumorigenic in nude mice. However, when tumorigenic lines were compared, the Eμ-*myc*/v-*abl* lines plated more efficiently in soft agar than the Abelson lines derived from normal bone marrow. Taken together, these results suggest that co-expression of v-*abl* and a deregulated *myc* gene offers certain growth advantages but is insufficient for complete transformation of early B lineage cells.

PRODUCTION OF NEW TRANSGENIC LINES

To expedite the introduction of other oncogenes into mice, we have developed two cassettes containing the heavy chain enhancer (Eμ) and the RNA splice signals and polyadenylation sequence from the SV40 T antigen gene. One vector (EμSV) contains the SV40 early region promoter. The other (EμV_H) contains the immunoglobulin variable region promoter. A polylinker inserted after the promoter allows a choice of restriction sites for cloning the gene of interest. The assembled cassette can be removed from plasmid sequences by digestion with the restriction endonuclease Not I, which cuts extremely rarely within eukaryotic sequences. Hence the gene can readily be delivered to the mouse eggs in a form free of bacterial sequences, which sometimes prevent proper tissue-specific expression of a transgene.

Eμ-N-*myc* Mice

The vertebrate genome contains several *myc*-related genes, one of which, N-*myc*, is homologous to c-*myc* both in organization and sequence. The amplification and elevated expression of N-*myc* in many human neuroblastomas, retinoblastomas and small cell lung carcinomas (reviewed in Schwab, 1986) suggested that abnormal expression of the gene was instrumental in the progression of these tumors. Its validation as an oncogene was obtained by *in vitro* experiments showing that, like c-*myc*, N-*myc* can transform established lines of rat fibroblasts (Small et al, 1987) and can collaborate with an activated *ras* gene to transform embryonic fibroblasts (Schwab et al, 1985; Yancopoulos et al, 1985). Low level N-*myc* expression has been noted in pre-B cell (but not B cell or plasma cell) lines (Zimmerman et al, 1986), but the oncogenic potential of N-*myc* for hematopoietic cells was unknown.

To directly assess *in vivo* the consequences of constitutive N-*myc* expression for B lymphoid development, we have produced two lines of transgenic mice carrying N-*myc* in the EμSV cassette. Strikingly, both lines show a high predisposition to lymphoma development. The disease profile in the first line is very similar to that in Eμ-*myc* mice; the animals develop clonal disseminated pre-B (and B) lymphomas involving lymph nodes and thymus, mostly at 8 to 18 weeks of age (Fig. 1). Tumor development within the second line is considerably slower, the mice being free of tumors until 17 weeks or more after birth.

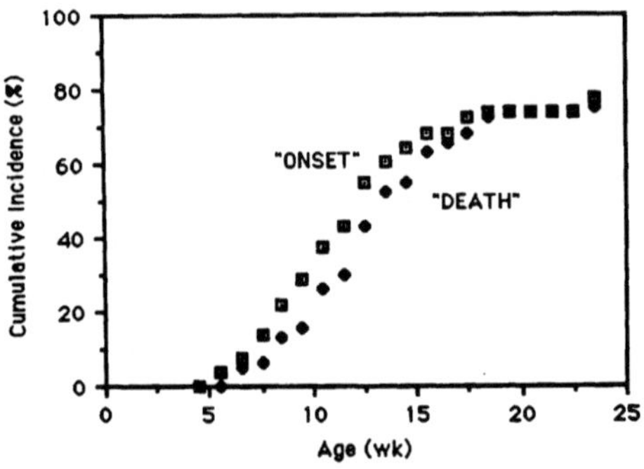

Fig. 1. Incidence of lymphoma in a group of 76 EμSV-N-myc mice. Onset times were assessed by weekly examination for enlarged lymph nodes. Mice were killed when they showed general signs of ill health

Eμ-N-*ras* Mice

Although the three closely related *ras* genes - H-*ras*, N-*ras* and K-*ras* - are activated by similar mutations, there appears to be a bias in certain cancers towards mutation of a particular member of the *ras* family. More specifically, N-*ras* mutations predominate in human hematologic malignancies, particularly acute myeloid leukemia (Bos et al, 1985; Farr et al, 1988). In the mouse, treatment with N-methylnitrosourea induces thymic lymphomas which frequently harbor a mutated N-*ras* gene (Guerrero et al, 1984).

To produce N-*ras* transgenic mice, a mutated human N-*ras* cDNA was inserted into the EμSV vector. The mutation was effected *in vitro* and the cDNA encodes an aspartic acid residue at amino acid 12 instead of glycine. The two female primary transgenic Eμ-N-*ras* mice developed T lymphomas, one at 10 weeks and the other at 17, and a line was established from the second. To our surprise, no B lymphoid tumors have developed within this line. Instead, the mice succumb to either T lymphomas or tumors that appear to be of myeloid derivation, the latter being classified morphologically as type A reticulum cell sarcomas (Table 2). Strikingly, when an Eμ-SV-N-*ras* mouse was bred to Eμ-*myc* mice, offspring carrying both transgenes rapidly developed very early B lymphoid tumors. The cells in these tumors express the B lineage cell surface marker Ly5(B220), but most have not yet undergone *IgH* recombination. The profile of tumor development within the Eμ-N-*ras* line suggests that the EμSV vector can promote expression within the B, T and myeloid lineages. It is not clear why the tissue-specificity of expression is broader than for Eμ-*myc* mice, but it presumably involves the apposition of the Eμ and SV40 sequences. At face value, the data argue that a mutated N-*ras* gene strongly predisposes to development of T lymphoid and myeloid rather than B lymphoid neoplasia and that co-expression with a deregulated *myc* gene renders it a potent transforming agent for early B lymphoid cells.

Table 2. Tumors of EμSV-N-*ras* mice in the first 6 months of life

Tumor type[a]	No. of animals with tumor[b] females	males
Thymic T lymphoma	9	3
Reticulum cell sarcoma type A	10	17
TTL and RCS-A together	10	3

[a]The first 15 of each type were diagnosed by cell surface markers and/or histology; the remainder by distinctive gross pathology
[b]From a monitored group of 85 transgenic B6SJL animals

Eμ-v-*abl* Mice

The *abl* oncogene was first identified as the transforming gene of the acutely transforming Abelson murine leukemia virus. When injected intravenously or intraperitoneally, this virus transforms pre-B cells (Rosenberg, 1982), but intrathymic injection provokes T lymphomas (Cook, 1982). Moreover, BALB/c mice treated with pristane develop plasmacytomas at a faster rate after Abelson virus injection (Potter et al, 1983).

We have produced three lines of mice transmitting a v-*abl* gene linked to Eμ and the SV40 promoter. Surprisingly, none of the transgenic offspring have developed pre-B lymphomas. Perhaps, as suggested above, the primary target for Abelson pre-B disease is a very early cell in which the Eμ enhancer is not yet active. To date, two of the Eμ-v-*abl* lines have produced plasmacytomas with quite long latent periods, while the third has produced T lymphomas (Fig. 2). Preliminary results from a cross between one of the former and the Eμ-*myc* strain indicate that double-transgenic Eμ-*myc*/Eμ-v-*abl* mice can develop plasmacytomas very early in life without a requirement for pristane treatment.

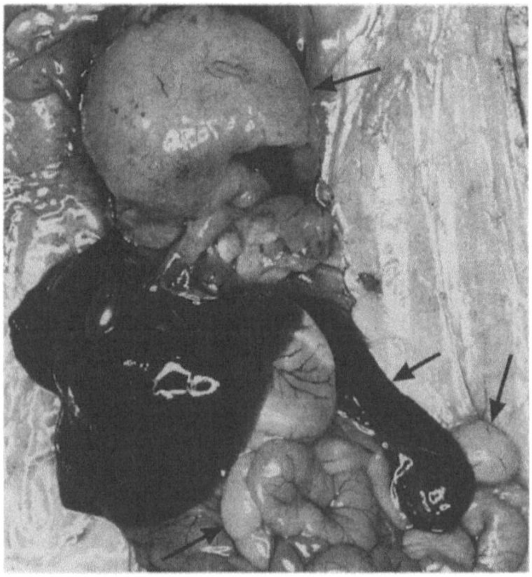

Fig.2. T lymphoma in 13-wk mouse of the Eμ-SV-v-*abl* 1 line. Autopsy shows a very large thymoma, an enlarged spleen and expanded inguinal and mesenteric lymph nodes

Eμ-*bcr*-v-*abl* Mice

The hallmark of chronic myeloid leukemia is the small 'Philadelphia' chromosome produced by a 9;22 chromosome translocation which fuses most of the c-*abl* gene to the 5' portion of a gene termed *bcr* (Shtivelman et al, 1985; Grosveld et al, 1986). The hybrid gene encodes a chimeric *bcr-abl* protein with increased tyrosine kinase activity (Konopka and Witte 1985). We have constructed a facsimile of the *bcr*-c-*abl* gene by fusing the relevant *bcr* sequences from a *bcr* cDNA clone (Hariharan and Adams, 1987) to v-*abl* sequences. Although v-*abl* bears a few amino acid substitutions and a deletion with respect to c-*abl*, these alterations are unlikely to play a major role in activating c-*abl* as a transforming gene (Ben-Neriah and Baltimore, 1986).

The *bcr*-v-*abl* gene was inserted into the EμVH vector and several primary transgenic mice have been born carrying multiple copies of the hybrid gene. One mouse died of a pre-B lymphoma only 17 days after birth, while two others developed T lymphomas after 7 and 11 weeks. A line was successfully established from one of the T lymphoma-bearing mice and from another Eμ-*bcr*-v-*abl* mouse which is still healthy after 1 yr. Tumor incidence in both the lines is very low but three T lymphomas have been characterized. The predominant occurrence of T lymphomas was most unexpected, since CML usually develops into an acute myeloid leukemia or B cell ALL, not a T lymphoid neoplasm.

FUTURE DIRECTIONS

We are currently analyzing the pre-tumor phase in each of the new transgenic strains, hoping to gain insight into the effect of the different oncogenes on proliferation and differentiation of cells within the various hematopoietic lineages. The timing, distribution and level of expression of the transgenes may provide important clues to the cellular origins of the disease profiles in the different strains. The mice provide a unique resource for dissecting the role of individual oncogenes,and their potential for cooperativity, in the generation of hematologic malignancies.

ACKNOWLEDGEMENTS

We thank Drs F. Alt, D. Baltimore and A. Hall for clones used to construct transgenes, Craig Hyland, Tracey Watson, Linda Byrt, Maureen Stanley, Gina Bari and Steven Mihajlovic for technical assistance, and Iras Collins for preparing the manuscript. This work was supported by the National Health and Medical Research Council (Canberra), the Drakensberg Trust and the US National Cancer Institute (CA 12421 and CA 43540).

REFERENCES

Adams JM, Harris AW, Pinkert CA, Corcoran LM, Alexander WS, Cory S, Palmiter RD and Brinster RL (1985) The c-*myc* gene driven by immunoglobulin enhancers induces lymphoid malignancy in transgenic mice. Nature 318: 533-538

Alexander W, Schrader JW and Adams JM (1987) Expression of the c-*myc* oncogene under control of an immunoglobulin enhancer in Eμ-*myc* transgenic mice. Mol Cell Biol 7: 1436-1444

Ben-Neriah Y and Baltimore D (1986) Structural modification of c-*abl* in lymphoma and leukemia. Curr Topics Microbiol Immunol 132: 81-89

Bos JL, Toksoz D, Marshall CJ, Verlaan-de Vries M, Veeneman GH, van der Eb AJ, van Boom JH, Janssen JWG and Steenvorden ACM (1985) Amino-acid substitutions at codon 13 of the N-*ras* oncogene in human acute myeloid leukaemia. Nature 315: 726-730

Cook WD (1982) Rapid thymomas induced by Abelson murine leukemia virus. Proc Natl Acad Sci USA 79: 2917-2921

Corcoran LM, Cory S and Adams JM (1985) Transposition of the immunoglobulin heavy chain enhancer to the *myc* oncogene in a murine plasmacytoma. Cell 40: 71-79

Cory S (1986) Activation of cellular oncogenes in hemopoietic cells by chromosome translocations. Adv Cancer Res 47: 189-234

Cory S and Adams JM (1988) Transgenic mice and oncogenesis. Ann Rev Immunol in press

Dyall-Smith D and Cory S (1987). Transformation of bone marrow cells from Eμ-*myc* transgenic mice by Abelson murine leukemia virus and Harvey murine sarcoma virus. Oncogene Res in press

Farr CJ, Saiki RK, Erlich HA, McCormick F and Marshall CJ (1988) Analysis of *RAS* gene mutations in acute myeloid leukemia by polymerase chain reaction and oligonucleotide probes. Proc Natl Acad Sci USA 85: 1629-1633

Grosschedl R, Weaver D, Baltimore D and Costantini F (1984) Introduction of a μ immunoglobulin gene into the mouse germ line: Specific expression in lymphoid cells and synthesis of functional antibody. Cell 38: 647-658

Guerrero I, Villasante A, D'Eustachio P and Pellicer A (1984) Isolation, preliminary characterisation, and chromosomal localization of the activated mouse N-*ras* gene from a carcinogen-induced thymic lymphoma. Science 225: 1041-1043

Hanahan D (1986) Oncogenesis in transgenic mice. In: Kahn P and Graf T (eds) Oncogenes and Growth Control. Springer, Berlin, pp 349-363

Hariharan IK and Adams JM (1987) cDNA sequence for human *bcr*, the gene that translocates to the *abl* oncogene in chronic myeloid leukaemia. EMBO J 6: 115-119

Harris AW, Pinkert CA, Crawford MC, Langdon WY, Brinster RL and Adams JM (1988) The Eμ-*myc* transgenic mouse. A model for high incidence spontaneous lymphoma and leukemia of early B cells. J Exp Med in press

Kemp D, Harris AW, Cory S and Adams JM (1980a) Expression of the immunoglobulin Cμ gene in mouse T and B lymphoid and myeloid cell lines. Proc Natl Acad Sci USA 77: 2876-2880

Kemp DJ, Wilson A, Harris AW and Shortman K (1980b) The immunoglobulin μ constant region gene is expressed in mouse thymocytes. Nature 286: 168-170

Konopka JB, Watanabe SM and Witte ON (1984) An alteration of the human c-*abl* protein in K562 leukemia cells unmasks associated tyrosine kinase activity. Cell 37: 1035-1042

Langdon WY, Harris AW, Cory S and Adams JM (1986) The c-*myc* oncogene perturbs B lymphocyte development in Eμ-*myc* transgenic mice. Cell 47: 11-18.

Nordan RP and Potter M (1986) A macrophage-derived factor required by plasmacytomas for survival and proliferation *in vitro*. Science 233: 566-569

O'Garra A, Umland S, DeFrance T and Christiansen J (1988) 'B-cell factors' are pleiotropic. Immunol Today 9: 45-54

Potter M, Sklar MD and Rowe WP (1973) Rapid viral induction of plasmacytomas in pristane-primed BALB/c mice. Science 182: 592-594

Rosenberg N (1982) Abelson leukemia virus. Curr Topics Microbiol Immunol 101: 95-126

Schwab M (1986) Amplification of proto-oncogenes and tumor progression. In: Kahn P and Graf T (eds) Oncogenes and Growth Control. Springer, Berlin, pp332-339

Schwab M, Varmus HE and Bishop JM (1985) Human N-*myc* gene contributes to neoplastic transformation of mammalian cells in culture. Nature 316: 160-162

Shtivelman E, Lifshitz B, Gale RP and Canaani E (1985) Fused transcript of *abl* and *bcr* genes in chronic myelogenous leukaemia. Nature 315: 550-554

Small MB, Hay N, Schwab M and Bishop JM (1987) Neoplastic transformation by the human gene N-*myc*. Mol Cell Biol 7: 1638-1645

Vaux DL, Adams JM, Alexander WS and Pike BL (1987) Immunologic competence of B cells subjected to constitutive c-*myc* oncogene expression in immunoglobulin heavy chain enhancer *myc* transgenic mice. J Immunol 139: 3854-3860

Yancopoulos GD, Nisen PD, Tesfaye A, Kohl NE, Goldfarb MP and Alt FW (1985) N-*myc* can cooperate with *ras* to transform normal cells in culture. Proc Nat Acad Sci USA 82: 5455-5459

Zimmerman KA, Yancopoulos GD, Collum RG, Smith RK, Kohl NE, Davis KA, Nau MM, Witte ON, Toran-Allerand D, Gee CE, Minna JD and Alt FW (1986) Differential expression of *myc* family genes during murine development. Nature 319: 780-783

Genetic Studies on Eμ-*myc* Transgenic Mice

C.L. SIDMAN[1], J.D. MARSHALL[1], and A.W. HARRIS[2]

INTRODUCTION

Transgenic mice carrying the *c-myc* oncogene under the control of the Ig heavy chain enhancer ("Eμ-*myc*" mice)(Cory and Adams, 1988; Harris et al., 1988a) develop B lineage tumors within the first year of life, and offer a reproducible and attractive system with which to dissect the genetic and cellular processes involved in tumorigenesis. Because the founder mouse of this strain was a (B6xSJL)F2 animal, the primary transgenic mouse stock has been maintained by continual crossing of transgenic animals to (B6xSJL)F1 hybrids. The line so maintained (genetically equivalent to an F2, and thus termed FE2) continously segregates the B6 and SJL genomes. Although over time the amount of each component averages 50%, any locus in any individual mouse could have 2 B6 alleles, 2 SJL alleles, or one of each. Thus, it is not surprising that considerable heterogeneity has been observed in the rate with which individual (B6xSJL)FE2-Eμ-*myc* transgenic mice develop visible tumors and die (Harris et al 1988b). In fact, three phases of tumor onset and death have been described in this stock; a lag phase, a rapid onset phase for some animals, and a slower onset phase for the remaining mice (Harris et al, 1988b). With the aim of reducing this genotypic and phenotypic variability, Eμ-*myc* mice have been repeatedly backcrossed to the parental B6 and SJL backgrounds. After only a few generations, the SJL-backcrossed stock developed tumors with essentially only the rapid onset pattern (Harris, 1987; Harris et al., 1988a), and could no longer be propagated. In contrast, backcrossing to B6 yielded a stable congenic Eμ-*myc* transgenic line that developed tumors predominantly with the slower onset pattern. This stock has now reached the tenth generation (N10) of backcrossing, and forms the basis for the genetic and functional studies reported here.

BACKGROUND GENE INFLUENCES

To emphasize the influence of background genes on the phenotypic expression of the Eμ-*myc* transgene, Table 1 shows the mean lifespan and average percent SJL genome in several Eμ-*myc* transgenic stocks. At the least severe extreme, B6(N10)-Eμ-*myc* mice, with only a nominal 1% of SJL genome remaining, had an average lifespan of 134±11 days. At the other extreme, the SJL(NE3)-Eμ-*myc* stock, with an average of 87.5% SJL genome, only lived 82±5 days. Genetically intermediate (average 50% SJL genome) (B6xSJL)FE2-Eμ-*myc* mice were also phenotypically intermediate, with a lifespan of 103±15 days.

[1]The Jackson Laboratory, 600 Main Street, Bar Harbor, ME 04609, USA, and [2]The Walter and Eliza Hall Institute of Medical Research, Royal Melbourne Hospital, Melbourne, 3050, Australia.

Current Topics in Microbiology and Immunology, Vol. 141

Table 1. Genetic background and Eμ-*myc* lifespan

Background	% SJL	Lifespan (days)
(B6xSJL)FE2	50	103±15
SJL(NE3)	87.5	82±5
B6(N10)	1	134±11
(B6xBALB)F1	.5	99±9

Since the genetic context in which the Eμ-*myc* construct is placed exerts such a profound effect on the rate of tumor development, we have begun to backcross the slowly affected B6-Eμ-*myc* transgenic strain to various other backgrounds, to identify and locate genes that interact with the Eμ-*myc* transgene. The BALB/cByJ strain was included in the panel of genetic backgrounds thus tested because of the well-known plasmacytoma susceptibility of closely related BALB/cAn mice. Table 1 also shows that the single (haploid) copy of BALB/c genome in (B6xBALB)F1-Eμ-*myc* mice made these mice as severely affected as the original (B6xSJL)FE2-Eμ-*myc* animals. Thus, at least one gene (or combination of genes) in the BALB/c background can act in a dominant fashion to exacerbate the tumorigenesis arising from the Eμ-*myc* transgene. Further F2, backcross, and recombinant inbred (RI) analyses are underway in order to determine the number and location of such tumorigenesis modifying genes in the BALB/cBy genome.

Transgenic Eμ-*myc* mice can be distinguished at weaning from their non-transgenic littermates by the size of their peripheral blood cell nuclei (Harris, 1987; Harris et al., 1988b). We have adapted this technique to the fluorescence activated cell sorter (FACS) in order to be able to easily type mice at a young age, and then follow them individually thoughout life. The FACS technique also allows us to determine the cellular nature of the lymphocyte populations present during this process. Figure 1 shows as examples of such analysis one each transgenic and non-transgenic littermate from the B6(N10)-Eμ-*myc* stock at one week of age. On the left, the scatter plots have been gated to show only lymphoid cells, and the percentages of those that fall into the large lymphocyte category (the enclosed areas) are indicated. Mouse #2 was clearly a transgenic animal, since over twice as many of its lymphocytes (46%) fell into the large lymphocyte category as did those from mouse #1 (with 18% large lymphocytes). On the right in Fig. 1, fluorescence plots show the FITC-anti-Ig (horizontal) and PE-anti-Thy-1 (vertical) staining of the large lymphocytes identified in the left-hand scattergrams. Non-transgenic mouse #1 showed 11%, 4%, and 85% of its large lymphocytes as T cells (Thy-1^+,Ig^-), B cells (Thy-1^-,Ig^+), and null or pre-B cells (Thy-1^-, Ig^-), respectively. In contrast, transgenic animal #2 had an increased percentage (94%) of null, presumably pre-B, cells. Either these null (or pre-B) cells were present in greatly increased numbers and so diluted out the T cells, or the Eμ-*myc* transgene directly down-regulated the development of T cells in this mouse.

Applying this early FACS typing technique to the same genetic stocks as were studied for lifespan in Table 1 gave rise to the percentages of large lymphocytes shown in Table 2. At three weeks of age, the

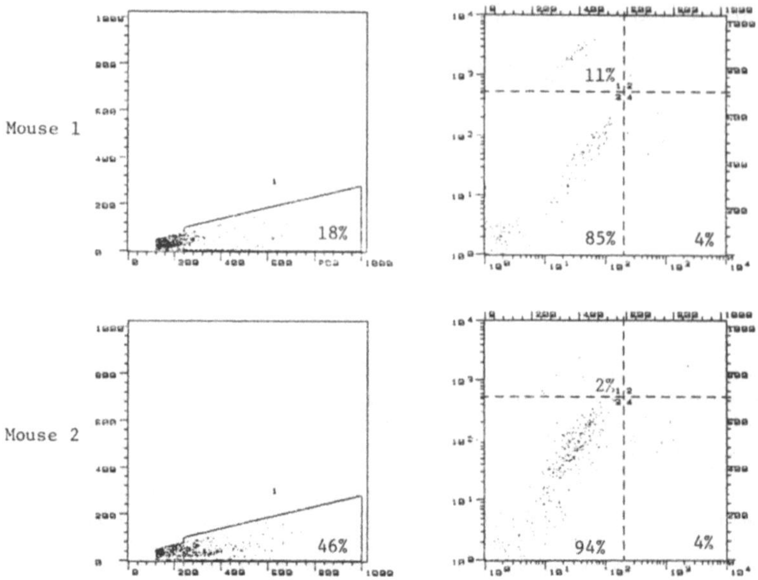

Fig. 1. FACS analysis of peripheral blood lymphocytes from one week old B6(N10) mice. On the left, scatter plots of gated lymphocytes, with forward scatter (horizontal) and right-angle scatter (vertical). The boxed area indicates large lymphocytes, with the number showing the percentage of total lymphocytes included as large lymphocytes. On the right, fluorescence plots of the large lymphocytes, with FITC-anti-Ig (horizontal) and PE-anti-Thy-1 (vertical). The numbers indicate the percentage of large lymphocytes included within the quadrants shown

Table 2. Genetic background and Eμ-*myc* percentage of large lymphocytes

Background	- Eμ-*myc*	+ Eμ-*myc*
(B6xSJL)FE2	9±1	39±3 (22,23,24,26,30,31,32,34,38 39,43,44,49,51,52,55,57,59)
SJL(NE3)	13±1	41±3 (33,39,41,41,51)
B6(N10)	12±1	25±3 (16,17,20,20,21,27,33,33,39)
(B6xBALB)F1	20±1	49±3 (41,52,54)

shorter-lived transgenic backgrounds (B6xSJL)FE2, SJL(NE3), and (B6xBALB)F1 showed much higher percentages of large lymphocytes than the longer-lived B6(N10) stock. Values from individual animals are also shown in Table 2 for Eμ-*myc* transgenic mice. While far from conclusive, 4/18 or 22% of the (B6xSJL)FE2-Eμ-*myc* transgenics had values within the statistical range (less than 28%) of the B6(N10)-Eμ-*myc* stock, suggesting the possibility of a single gene whose B6 allele is recessive to the SJL form.

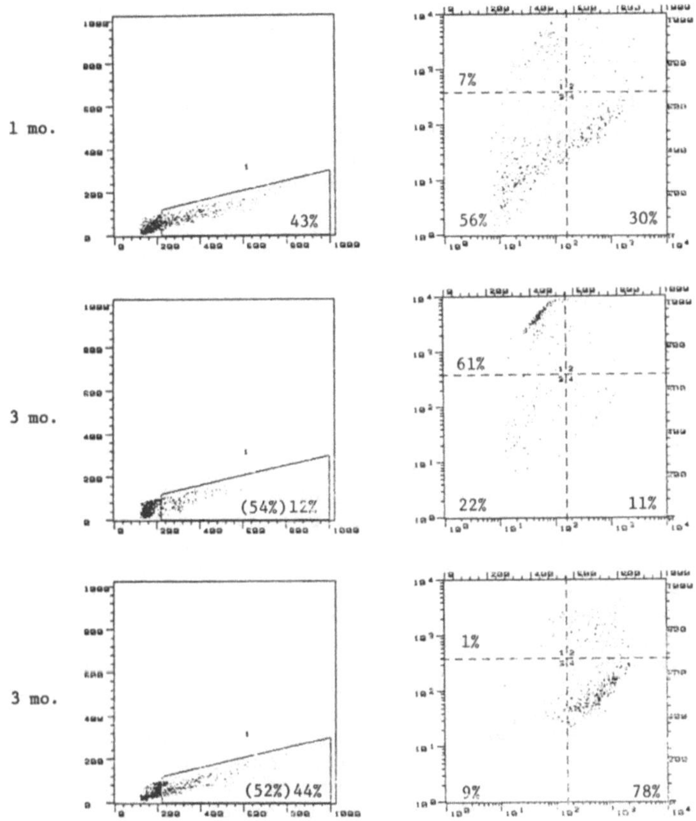

Fig. 2. FACS analysis of (B6xBALB)F1-E*μ*-*myc* mice. Figure organized like Fig. 1

TRANSIENT STATE OF BLOOD LYMPHOCYTE ENLARGEMENT

Having developed the assay to initially type transgenic animals, we began to serially follow individual mice throughout their lifespan. Figure 2 illustrates some of the patterns observed, using three animals from the (B6xBALB)F1-E*μ*-*myc* stock as examples. In the top panels of Fig. 2, typical patterns for a one month old transgenic animal are seen. Forty-three percent of its lymphocytes were large, and of these, the greatest percentages were null (or pre-B) (56%) and B (30%) cells. In the middle of Fig. 2, a transgenic animal is shown which, when analyzed at one month of age, had shown 54% large lymphocytes. At three months of age, however, large cells accounted for only 12% of its blood lymphocytes. Of these, the percentages of null/pre-B and B cells had dropped to more normal levels (22% and 11%, respectively), while the T cell compartment had increased to 61%. The repeated finding of increased numbers of large T cells in mice whose percentages of large lymphocytes had regressed suggests the possibility that T cell activity may play a role in such regression. Furthermore, the percentage of animals experiencing this regression, and its timing, appear to vary among the different backgrounds of transgenic mice. While all B6(N10)-E*μ*-*myc* mice have been observed to undergo regression in their percentages of large peripheral blood lymphocytes, some (B6xSJL)FE2- and (B6xBALB)F1-E*μ*-*myc* animals have not been seen to do so. Either these latter mice have experienced only brief and undetected remissions, or else they have gone directly from early leukemia into the terminal leukemic plus lymphomatous phase. In any case, all Eμ-*myc* mice (of all backgrounds) in our experience so far have shown easily detectable excesses of large peripheral blood lymphocytes immediately before death. At the bottom of Fig. 2 is an example of a transgenic mouse (a littermate of that shown in the middle of Fig. 2) which never appeared to lose its large peripheral blood lymphocyte population (52% at one month, and 44% at three months). This mouse was so ill that it had to be sacrificed one day after the analysis shown, but as frequently seen in terminally ill Eμ-*myc* transgenics, its large peripheral blood lymphocytes were predominantly of the B cell type.

CONCLUSIONS

Although the transgenic Eμ-*myc* oncogene guarantees that mice carrying it will develop B lineage tumors, for some months the animals appear healthy and are immunocompetent (Vaux et al. 1987). The activated Eμ-*myc* transgene by itself thus does not preclude any aspect of B lymphocyte development, differentiation, or function. Secondary events are likely to be required to complete the tumorigenic process that the Eμ-*myc* transgene sets in motion. The ability to serially follow individual animals throughout their development, plus readily manipulated genetic differences that affect the process, offer powerful tools with which to dissect the multiple steps involved in tumorigenesis. Preliminary data presented here indicate that a limited number of dominant-acting genes may exacerbate the development of this model of B cell tumorigenesis, and suggest that they may do so by affecting the animal's attempts to (immunologically?) control the developing tumor population. Exciting information should emerge during the coming year from the further genetic experiments currently underway to define the number and nature of the background genes affecting this tumorigenesis model, and to probe the role of other immune cells and reactivities in this process.

ACKNOWLEDGEMENTS

We thank B. Gott, M. Crawford and T. Watson for excellent technical assistance in these studies, and Drs. P. Hoppe and R. Evans for critical reading and comments on the manuscript. This work was supported by grants CA-35845, CA-43540, AI-20232 and AI-25765 from the National Institutes of Health. The contents of this publication are solely the responsibility of the authors and do not necessarily represent the official views of the NIH.

REFERENCES

Cory S, Adams JM (1988) Transgenic Mice and Oncogenesis. Ann Rev Immunol 6:25-48

Harris AW (1987) In: Melchers F, Potter M (eds) Mechanisms of B Cell Neoplasia. Editiones Roche, Basel, pp 344-353

Harris AW, Langdon WY, Alexander WS, Hariharan IK, Rosenbaum H, Vaux D, Webb E, Bernard O, Crawford M, Abud H, Adams JM, Cory S (1988a) Transgenic Mouse Models for Hematopoietic Tumorigenesis. Curr Topics Microbiol Immunol in press (this volume)

Harris AW, Pinkert CA, Crawford M, Langdon WY, Brinster RL, Adams JM (1988b) The E*μ-myc* Transgenic Mouse: A Model for High-Incidence Spontaneous Lymphoma and Leukemia of Early B Cells. J Exp Med 167:353-371

Vaux DL, Adams JM, Alexander WS, Pike BL (1987) Immunologic Competence of B Cells Subjected to Constitutive c-myc Oncogene Expression in Immunoglobulin Heavy Chain Enhancer myc-Transgenic Mice. J Immunol 139:3854-3860

Differential Expression of *myc*-family Genes During Development: Normal and Deregulated N-*myc* Expression in Transgenic Mice

R. DILDROP, K. ZIMMERMAN, R. A. DEPINHO, G. D. YANCOPOULOS, A. TESFAYE, and F. W. ALT

INTRODUCTION

The myc-family of cellular oncogenes is a dispersed multi-gene family that includes the c-, N- and L-myc genes (For review see Alt et al., 1986). The human and murine c-, N, and L-myc genes encode related but distinct nuclear proteins and have a similar overall organization (Kohl et al., 1986; DePinho et al., 1986, Stanton et al., 1986; DePinho et al., 1987; Legouy et al., 1987; Kay et al., 1988); all have three exons with the first encoding a potentially untranslated leader sequence. All three genes also cooperate similarly with an activated Ha-Ras oncogene to transform primary rat embryo fibroblasts (REFs) (Yancopoulos et al., 1985; DePinho et al., 1987). Despite striking similarities of regions of the myc proteins and the in vitro transforming activities, the three genes are conserved as distinct sequences throughout vertebrate species suggesting unique functional roles. This possibility is supported by findings that the genes are differentially expressed in a stage- and tissue-specific manner during human and murine differentiation (Zimmerman et al., 1986). In addition, deregulated c-myc expression has been causally implicated in the genesis of a wide variety of different tumor types and occurs by a variety of different mechanisms; deregulation of the N- and L-myc genes has been clearly implicated only in a few naturally occurring tumors (eg. human neuroblastomas and small cell lung carcinomas) and only by the mechanism of gene amplification (reviewed by Alt et al., 1986). Together, these findings suggest that the myc genes play related but distinct roles in growth and/or differentiation of mammalian cells. This report summarizes our recent efforts to develop model systems to help define differential functions of these genes and to elucidate molecular mechanisms involved in the differential control of their expression.

TISSUE AND STAGE SPECIFIC EXPRESSION OF MYC-FAMILY GENES

High level expression of the N- and L-myc genes is restricted with respect to tissue and stage in developing mice and humans, while that of c-myc is more generalized (Zimmerman et al., 1986; DePinho et al., in prep.). In general, all three myc-family genes show stage-specific expression in most tissues or cell lineages in which they are expressed; the highest expression levels of N- and L-myc always seems to occur at earlier stages. Previous studies implied that c-myc expression has a role in cell growth or differentiation (reviewed by Marcu, 1987). In situ hybridization analyses of N-myc expression in certain cell lineages within differentiating mouse brain and kidney suggest that N-myc expression is a feature of early differentiation stages, regardless of cell proliferative state; for example, post-mitotic but not yet differentiated neuroblasts in the brain express high N-myc RNA levels (Maugrauer, Alt, and Ekblom, J. Cell. Biol., in press).

The Howard Hughes Medical Institute and Departments of Biochemistry and Microbiology, College of Physicians and Surgeons of Columbia University, 630 W. 168th St. New York, New York, 10032

Myc-gene Expression During Lymphocyte Differentiation

Analyses of B lymphocyte differentiation demonstrated that myc-family genes are differentially expressed during the progression of cells through this pathway (Fig. 1; Zimmerman et al., 1986). Significant levels of L-myc expression have not been observed in any known stage of B lymphocyte differentiation, although some preliminary results suggest the possibility of expression in tumors representing very early precursors (DePinho et al., unpublished observations). Expression of c-myc and N-myc occurs in the earliest defined B cell precursors and continues in pre-B cells--which express heavy (H) but not light (L) immunoglobulin (Ig) chains. However, N-myc expression ceases but c-myc expression continues when pre-B cells progress to the B-cell stage-which involves expression of complete (H plus L chains) Ig surface molecules (Figure 1). Recently, we have isolated a novel sequence (PB.10) that is potentially myc-related (Oltz, Kaplan, and Yancopoulos, unpublished); expression of this sequence appears to decline at the B cell activation stage of the pathway while that of c-myc continues (Figure 1). Thus, differential and/or combinatorial expression of myc-family genes or potentially-related genes correlates with the three distinct stages of this B cell pathway.

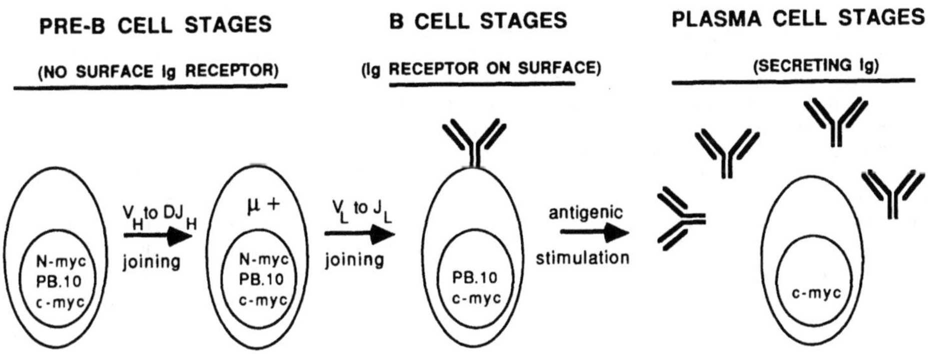

Figure 1. Myc-Gene Expression During B-cell Differentiation. Details are in the text.

Analysis of T cell tumor analogs or purified normal T cells suggest that L-, N-, and c-myc are differentially expressed in that pathway (DePinho et al., in preparation). T cells can be purified into surface receptor-positive and negative subsets based on expression of the receptor-associated molecule T3; receptor-negative cell populations express both N- and c-myc, whereas receptor-positive populations express equally high c-myc levels but greatly reduced N-myc levels. Thus, N-myc expression declines at a similar stage in B and T cell differentiation, at the point where precursor lymphocytes acquire their surface receptor and become functional cells poised to react with cognate antigen.

In general the results with the differentiating lymphocyte populations, coupled with results from differentiating organ systems (Zimmerman et al., 1986; Maugrauer, Alt, and Ekblom, In Press) support the notion that differential or perhaps combinatorial expression of these genes may be important in normal development (Alt et al., 1986).

REGULATION OF N-MYC GENE EXPRESSION

Transfection of N-myc Constructs into Cultured Cells

To identify molecular elements that mediate tissue and developmental stage specific expression of myc genes, we introduced a mouse N-myc genomic clone containing the promoter and the three exons into various cell lines (Legouy et al., 1987b). These experiments demonstrated that transiently and stably introduced human and murine N-myc genes are expressed both in human and murine cell lines that express the endogenous N-myc gene, as well as in cell lines which do not express the endogenous gene. As an alternative approach to define N-myc regulatory elements, we fused portions of the N-myc promoter region to reporter genes; constructs with as little as 600bp or as much as 3kb 5' of the human N-myc exon 1 sequence gave reporter gene expression in cells that did or did not express the endogenous N-myc gene (Legouy et al., 1987b). Several factors could lead to differential expression of introduced versus endogenous N-myc genes in the cell lines (see below).

Expression of the Human N-myc Gene in Transgenic Mice

As an alternative approach to elucidate mechanisms that regulate N-myc expression, we introduced the human N-myc clone employed in the transfection assays into the germline of transgenic mice (Figure 2). To date, three lines have been produced that carry from 1 to 5 copies of the human transgene. To monitor expression of endogenous murine N-myc and the human N-myc transgene, total RNA from various newborn and adult tissues was assayed for ability to differentially protect murine- or human-specific N-myc probes from digestion with S1 nuclease (Figure 2). As described (Zimmerman et al., 1986), endogenous N-myc was expressed differentially in tissues with highest levels in brain and kidney; expression in these tissues decreased in older mice. The human N-myc transgene had an expression pattern essentially identical to that of its murine counterpart--showing both the tissue- (high levels in brain and kidney) and developmental-stage specificity.

These findings clearly demonstrate that regulatory elements necessary for both tissue- and stage-specific expression of the human N-myc gene in mice are contained within the limited sequences surrounding the gene in the clone employed (3kb of 5' sequence and 2kb of 3'sequence). Thus, the lack of appropriately regulated expression in the transfection experiments apparently is not due to the lack of proper target sequences. Although other explanations are possible, these results suggest an interesting possibility regarding regulation of genes, such as N-myc, that are expressed early in multiple lineages and subsequently turned-off: factors that establish functional competence of negative regulatory sequences associated with these genes may only be present during a particular window of differentiation; therefore, the gene may need to go through normal development to be correctly regulated.

DEREGULATED N-MYC EXPRESSION IN TRANSGENIC B-LYMPHOID CELLS

The N-myc gene is expressed during the early stages of B cell differentiation and therefore may play some role in this pathway; in addition, despite significant N-myc expression in pre-B cells (as high as neuroblastomas with a single N-myc copy), deregulated expression of this gene has never been implicated in pre-B neoplasias. In contrast, translocation of the c-myc gene into an Ig locus is a common feature of human B cell lymphomas and murine plamacytomas (reviewed by Klein and Klein, 1985). Likewise, introduction of a c-myc gene that is deregulated by the presence of an Ig enhancer element into the germline of transgenic mice leads to the generation of reproducibly inherited pre-B and B-lymphoid malignancies (Cory et al., 1985). To study the role of N-myc in normal B cell differentiation, as well as to elucidate potential roles in generation of B-lineage neoplasias, we constructed N-myc transgenic mice carrying various N-myc constructs designed to deregulate expression in this lineage. All were based on incorporation of the Ig H chain enhancer element (Eu) into the construct; and were thus similar, in design, to the deregulated c-myc constructs previously employed by others (Cory et al, 1985). The Eu element is active throughout B cell differentiation and therefore its incorporation into N-myc constructs was used to force N-myc expression beyond the pre-B/B cell transition (Figure 1). Three such constructs were employed (Figure 3). One (EN) contained a complete copy of the N-myc gene including the promoter linked to an upstream copy of Eu. A second (ESN) contained the second and third exons of the N-myc gene (with the complete exon 2/3 open reading frame) downstream of Eu and an SV40 promoter. The third (EVN) contained the second and third N-myc exons downstream from Eu and an Ig V_H promoter.

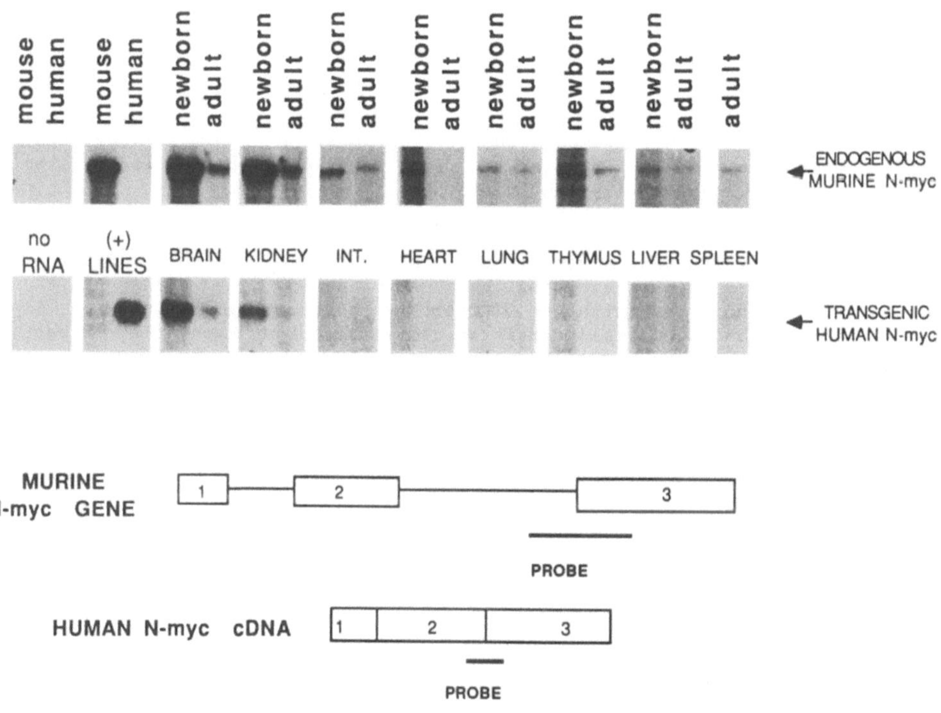

Figure 2. Expression of the human N-myc Gene in Transgenic Mice. Details are in the text.

Lymphoid Neoplasias in E_uN-myc Transgenic Mice

Five independent lines of EN-myc, 4 independent lines of ESN-myc, and 6 independent lines of EVN-myc transgenic mice were produced. Copy numbers of

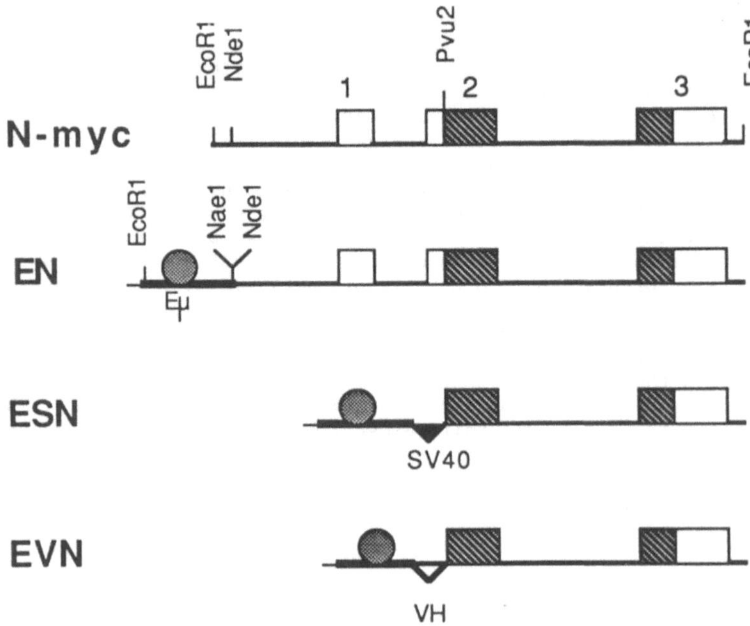

Figure 3. Deregulated E_uN-myc Constructs. The location of N-myc exons, the various promoters, the IgH enhancer, and some restriction endonuclease sites are indicated. Other details are in the text.

the transgenes ranged from a few to much greater than 10 depending on the line. Control transgenics carrying the human or murine N-myc transgenes that were expressed under the influence of their own controlling elements (eg. Fig. 2), did not exhibit an increased rate of neoplasia of any type. In contrast, at least 8 of 15 primary mouse line carrying the E_u-deregulated N-myc constructs reproducibly developed B lymphoid malignancies; with respect to onset and type of B cell malignancy, no clear differences were observed among the three constructs. Several lines that did not appear to express the deregulated construct did not develop tumors; although analyses are still preliminary, lines that develop B-lymphoid tumors often have high copy numbers and express the construct at high levels apparently in a tissue-specific fashion (see below).

Two EN-myc lines and four EVN-myc lines were chosen for more detailed analyses (both subsequently referred to as E_uN-myc because there were no definitive differences in their manifestations). Offspring from these lines reproducibly develop B-lymphoid neoplasias, usually during the first 10-20 weeks of life--although, as with E_uc-myc mice (Cory and Adams, 1988), time of onset appears random with respect to given individuals. The stage of tumors and their manifestations are also somewhat variable--although all of approximately 40 tumors analyzed (with one possible exception; see below), appear to be of B-lymphoid origin. Often, these tumors manifest as enlarged lymph nodes with involvement of the spleen and thymus. Another frequent manifestation is gross skull-deformation due to the accumulation of tumor cells without obvious invasion of the brain. Generally, the tumor cells can be characterized as plasmacytoid in gross morphology, but more detailed analyses demonstrate that they probably represent all major stages of the B cell pathway (see below). In general, as with tumor cells from E_uc-myc mice (Cory and Adams, 1988), they are readily transferable by injection into ascites or intraperitoneally into syngeneic recipients and are usually readily adaptable to growth in culture.

Staging of E_uN-myc Tumors

Initial analyses of Ig expression patterns in primary E_uN-myc tumors or cell lines indicated that individual tumors represent either the pre-B (u-only), B cell (H plus L chain), or in several cases plasma cell stage (evidenced by presence of serum paraproteins). With respect to cell lines, analyses of Ig H chain and kappa L chain rearrangement patterns suggested that 3 were at the pre-B stage (no L chain gene rearrangement) while 9 represented either the B cell or later stages of this pathway (rearrangement of both H and L chain genes). Detailed analyses of primary tumors and cell lines also suggested that some pre-B lines progressed to the B cell stage (rearrangment of L chain genes) during growth in vivo or in vitro and were thus similar to lines generated from E_uc-myc mice (Cory and Adams, 1988). Analyses of rearrangement patterns suggest that most tumors are monoclonal as previously noted for B cell neoplasias in E_uc-myc mice (Cory et al., 1985).

To date, most characteristics of the E_uN-myc disease appear quite similar to those of the E_uc-myc disease (the latter summarized by Cory and Adams, 1988). However, there were certain potential differences. Thus far, tissue-specificity of E_uc-myc activity was absolute; only B-lineage tumors were observed. This is also true for the E_uN-myc mice with the possible exception of one tumor that infiltrated both kidneys; the cell line established from this tumor had no Ig gene rearrangments and passaged tumor cells appear to home to the kidney. More analyses will be necessary to determine the nature of this unusual tumor; it may be non-lymphoid or may represent a very early B cell precursor that has not yet rearranged Ig genes. The latter type of B lymphoid tumor was not observed in E_uc-myc mice. In this context, it is also notable that although tumors representing the B cell stage (H plus L chain expression) are a minority in the E_uc-myc mice, so far they are a majority in E_uN-myc mice. Finally, no tumors representing the plasma cell stage were observed in the E_uc-myc mice; at least several of E_uN-myc appear to represent this stage. However, as noted previously (Cory and Adams, 1988), the stage of tumors generated could simply reflect strain specific-differences; in fact this may account for any of the potential differences we have observed since the E_uN-myc mice were of different genetic background than used for generation of E_uc-myc mice. The latter possibility is particularly notable since E_uN-myc mice produced in the same strain as the E_uc-myc mice appear to develop a disease that in most respects outlined above appears indistinguishable from the E_uc-myc disease (Suzanne Cory, personal communication). However, one striking difference between tumors that arise in E_uc-myc as opposed to E_uN-myc mice that seems quite significant is that all EuN-myc tumors expressed exceptionally high levels of the N-myc gene compared to expression levels in normal or tumor counterparts derived from normal mice (see below); E_u-c-myc expression levels in tumors were found to be comparable to c-myc expression levels in normal cells.

B-cell Differentiation in E_u-N-myc Transgenic Mice

A major goal of the generation of E_uN-myc transgenics is to determine how disruption of normal N-myc expression affects the B cell pathway. The E_uc-myc transgenics have already been quite instructive in this context. Mice carrying the E_uc-myc gene displayed a prelymphomagenic state characterized by the polyclonal expansion of pre-B cells, at expense of B cells in the bone marrow and also in the spleen; both pre-B and B cells were also larger than normal (reviewed by Cory and Adams, 1988). Some E_uN-myc mice display strikingly similar patterns while others do not even though they develop inheritable B-lymphoid malignancies. Currently it is not clear whether variations in these aberrations of normal B cell development in some E_uN-myc mice is due to specific constructs, different endogenous genetic constitutions among various E_uN-*myc* mice, or other factors. However, it is clear that the E_uN-*myc* mice appear to generate mature B cells and an immune system that functions reasonably normally. Whether this implies that the decline in N-*myc* expression at the pre-B/B cell stage is of no consequence with respect to normal differentiation in this pathway obviously will require many additional analyses. For example, we have not yet determined the expression levels of the transgenic N-*myc* or of the normal c-*myc* gene in normal, transgenic B lineage cells; it is possible that N-*myc* expression is extinguished in normal cells that progress beyond the pre-B stage or that perturbations in c-*myc* expression levels (see below) allow apparently normal differentiation to occur.

CROSS-REGULATION OF *MYC*-GENE EXPRESSION.

We have found that the relative level of c-*myc* expression is inversely correlated with relative N-*myc* expression levels in neuroblastomas and Wilm's tumors (Alt et al., 1986; Nisen et al., 1986). Results from other laboratories demonstrated an inverse relationship between high level expression of one *myc*-family member (c, N, or L) and expression of other family members in small cell lung carcinomas (Nau et al., 1986). To further address this potential inverse relationship, we assayed c-*myc* expression in rat embryo fibroblasts (REFs) that were neoplastically transformed by introductions of an activated Ha-*ras* gene plus an N-*myc* expression vector (Yancopoulos et al., 1985). Although normal rat embryo fibroblasts express substantial levels of c-*myc*, the REFs transformed with LTR-N-*myc* vectors plus an activated Ha-*ras* oncogene did not express significant levels of c-*myc* (Alt et al., 1986). Together, these findings led us to propose that high level N-*myc* expression in certain contexts may be able to "cross-regulate" c-*myc* expression (Alt et al., 1986; Nisen et al., 1986).

We have observed coordinate *baseline* expression levels (for N-*myc*, defined as the expression level in a neuroblastoma with a single copy number of the N-*myc* gene) of two or in some cases all three *myc* genes in normal and transformed cells (eg. Zimmerman et al., 1986; Nisen et al., 1986). One possibility consistent with the findings of cross-regulation in some cells and coordinate expression is others is that high level expression of N-*myc* (i.e., greater than 20-fold the baseline levels seen in unamplified neuroblastomas or pre-B cells) can lead to down-regulation (cross-regulation) of c-*myc* expression. Support for this idea comes from findings that N-*myc* expression above a similar threshold level is necessary to down-regulate class I gene expression in neuroblastomas (Bernards et al., 1986). However, numerous 3T3 cell clones which expressed greater than 20-fold baseline levels of N-*myc* from a construct co-amplified with a dihydrofolate reductase vector all expressed endogenous c-*myc* (Zimmerman and Alt, unpublished). Therefore, the exact nature of this putative cross-regulatory phenomenon and the requirements to produce it remain to be determined. The E_uN-*myc* cell lines offer an ideal system to further examine these questions.

All transformed cell lines generated from the E_uN-*myc* transgenics expressed *very* high levels of the N-*myc* transgene (Figure 4). Expression in E_uN-*myc* B cell neoplasias (including pre-B tumors) are at least 50-fold greater than the expression levels in the A-MuLV transformed pre-B lines and more similar to those of neuroblastomas in which the N-*myc* gene is amplified 50-100 fold (Figure 4; N-*myc* expression in the non-transgenic pre-B line is barely detectable in this exposure but is still more than 20-fold greater than the undetectable levels of more mature B lymphoid neoplasias from non-transgenic mice). Thus, E_uN-*myc* expression levels were much higher than endogenous expression levels found in normal pre-B cells or A-MuLV transformed pre-B cell lines, not to mention B or plasma cell stage tumors or normal cells where N-*myc*

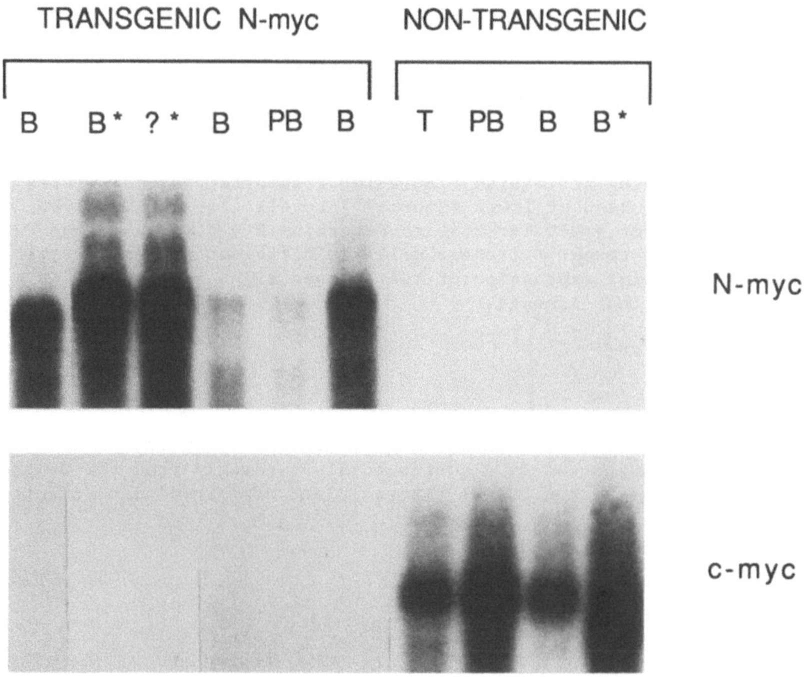

Figure 4. Cross-Regulation of c-myc Expression in Tumors That Express High E_uN-myc Levels. 10ug of total RNA from B cell (B), pre-B cell (PB) potential plasma cell (?) and T cell (T) tumors were assayed by Northen blotting procedures for hybridization to N- and C-myc-specific probes.

expression is undetectable (Zimmerman et al., 1986; Figure 4). Strikingly, none of the E_uN-myc cell lines representing any B cell differentiation stage expressed endogenous c-myc at detectable levels; the corresponding types of tumors from normal mice expressed significant levels of c-myc whether or not they expressed endogenous N-myc (Fig. 4). Thus, as observed in naturally occurring neuroblastomas and derived neoplastic fibroblast lines whose transformation was associated with high-level N-myc expression, the E_uN-myc transformed B lineage tumor cell lines totally lacked detectable c-myc expression.

Previous studies have suggested that deregulated c-myc expression may downregulate c-myc expression from normal c-myc alleles; in support of this possibility B cell tumor lines generated from E_uc-myc transgenics did not express the endogenous c-myc gene (reviewed by Cory and Adams, 1988). It should be emphasized that neither the putative auto- or cross-regulatory activities of myc genes has been shown to be a direct effect. Decreased normal gene expression (myc or class I) associated with high level myc gene expression may be secondary to other (perhaps differentiative) events associated with high-level expression of these genes. For example, a possible explanation for our apparently discrepant results with the 3T3 cells and REFs is that activities of N-myc and activated Ha-ras genes plus other potential alterations in endogenous gene expression were selected by the neoplastic transformation assay. The permanent cell lines may have already had different secondary alterations which in combination with high level N-myc expression did not lead to cross-regulation. Also notable is the observation that auto-regulation of c-myc expression by the E_uc-myc B cell tumors occurs at E_uc-myc expression levels similar to those of endogenous c-myc genes in normal cells (Cory and

Adams, 1988). N-myc cross-regulation, so far, has only been observed in transformed cells that express dramatically higher N-myc levels than their normal counterparts. This difference may offer a clue as to why c-myc but apparently not N-myc deregulation frequently is involved in naturally occurring B cell neoplasias; both deregulated expression as well as increased expression may be required for N-myc but not c-myc transformation in this lineage. In this regard, it remains possible that when expressed at high levels, N-myc shares more overlapping activities (including tranformatting activities) with c-myc than when expressed at lower ("normal") levels (Alt et al., 1986). Finally, it will be of great interest to determine N-myc versus c-myc expression levels in normal B-lineage cells at different stages to begin to assess how differential expression of these genes might be involved in progression through that lineage.

ACKNOWLEDGEMENTS.

This work was supported by the Howard Hughes Medical Organization and NIH grant CA42335. R.D. was supported by a postdoctoral fellowship from the Deutsche Forschungsgemeinschaft. R.A.D. was the recipient of a Physician Scientist Award, AI-00602.

REFERENCES

Adams, J., Harris, A., Pinkert, C., Corcoran, L., Alexander, W., Cory, S., Palmiter, R., Brinster, R. (1985). The c-myc oncogene driven by Ig enhancers induces lymphoid malignancy in transgenic mice. Nature 318, 533-538.

Alt, F.W., DePinho, R.A., Zimmerman, K., Legouy, E., Hatton, K., Ferrier, P., Tesfaye, A., Yancopoulos, G.D., Nisen, P. (1986). The human myc-gene family. Cold Spring Harbor Quant. Biol. 51, 931-941.

Bernards, R., Dessain, S.K. Weinberg, R.A. (1986). N-myc amplification causes down-modulation of MHC Class I antigen expression in neuroblastoma. Cell 47, 667-674.

Cory, S., and Adams, J. (1988) Transgenic Mice and Oncogenesis. Ann. Rev. Immunol. 6, 25-48.

DePinho, R.A., Legouy, E., Feldman, L.B., Kohl, N.E., Yancopoulos, G.D., Alt, F.W. (1986). Structure and expression of the murine N-myc gene. Proc. Nat'l. Acad. Sci. USA 83, 1827-1831.

DePinho, R., Hatton, K., Tesfaye, A., Yancopoulos, G., and Alt, F.W. (1987) The Human Myc Gene Family: Structure and Activity of L-myc and an L-myc Pseudogene. Genes and Development I: 1311-1326.

Kaye, F., Battey, J., Nau, M. Brooks, D., Seifter, E., DeGreve, J., Birrer, M., Sausville, E., and Minna, J. (1988) Structure and Expression of the Human L-myc Gene Reveals a Complex Pattern of Alternative mRNA Processing. Mol. Cell. Biol. 8, 186-195.

Klein, G. and Klein, E. (1985) Myc-Ig Juxtoposition by Chromosomal Translocations: Some New Insights, Puzzles, and Paradoxes. Immunol. Today, 6, 208-215.

Kohl, N.E., Leguoy, E., DePinho, R.A., Smith, R., Gee, C., Alt, F.W. (1986). Human N-myc is closely related in organization and nucleotide sequence to c-myc. Nature 319, 73-77.

Legouy, E., DePinho, R., Zimmerman, K., Collum R., Yancopoulos, G., Mitsock, L., Kriz, R. and Alt, F. (1987a) Structure and Expression of the Murine L-myc Gene. EMBO J. 6, 3359-3366.

Legouy, E., DePinho, R., Zimmerman, K., Ferrier, P., Collum, R., and Alt, F.W. (1987b) Structure and Expression of Myc-Family Genes, Nuclear Oncogenes, Alt, F.W., Harlow, E., and Ziff, E., Editors, Current Communications in Molecular Biology, Cold Spring Harbor Press, 1987.

Marcu, K.B. (1987) Regulation of expression of the c-myc protooncogene. Bioessays, 6:28-34.

Nau, M., Brooks, B., Battey, J., Sausville, E., Gasdar, A., Kirsh, I., McBride, O., Bertness, V., Hollis, G., Minna, J. (1985). L-myc, a new myc-related gene amplified and expressed in human small cell lung cancer. Nature 318, 69-73.

Nisen, P.D., Zimmerman, K.A., Cotter, S.V., Gilbert, F., Alt, f.W. (1986). Enhanced expression of the N-myc gene in Wilms tumors. Cancer Res. 46, 6217-6222.

Stanton, L., Schwab, M., Bishop, J.M. (1986). Nucleotide sequence of the human N-myc gene. Proc. Nat'l. Acad. Sci. USA 83, 1772-1776.

Yancopoulos, G.D., Nisen, P.D., Tesfaye, A., Kohl, N.E., Goldfarb, M.P., Alt, F.W. (1985). N-myc can cooperate with ras to transform normal cells in culture. Proc. Nat'l. Acad. Sci. USA 82, 5455-5459.

Zimmerman, K.A., Yancopoulos, G.D., Collum, R.G., Smith, R.K., Kohl, N.E., Denis, K.A., Nau, M.M., Witte, O.N., Toran-Allerand, D., Gee, C.E., Minna, J.D., Alt, F.W. (1986). Differential expression of myc family genes during murine development. Nature 319, 780-783.

Plasmacytoma Induction by J Series of v-*myc* Recombinant Retroviruses: Evidence for the Requirement of Two (*raf* and *myc*) Oncogenes for Transformation

J. TROPPMAIR[1], M. HULEIHEL[1], J. CLEVELAND[1], J.F. MUSHINSKI[2], J. KURIE[2], H.C. MORSE, III[3], J.S. WAX[4], M. POTTER[2], and U.R. RAPP[1]

Introduction

In genetically susceptible BALB/cAn mice intraperitoneal injection of pristane induces plasmacytomas. More than 95% of these plasmacytomas show chromosome (6;15) or (12;15) translocations, which result in continuous expression of c-myc (Ohno et al.1984). Direct evidence for a causative role of myc in tumor induction came from experiments using the J series of recombinant retroviruses (Rapp et al. 1985, Potter et al. 1987). The J2 and J3 viruses carry an avian MH2/MC29 hybrid v-myc and a mouse v-raf oncogene. However, in the J3 virus gag-raf gene is taken out of frame by a 256 bp deletion spanning the gag/raf border. The removal of v-raf specific sequences by this deletion leaves the minimal transforming sequence of v-raf intact (Heidecker et al., in preparation). The J5 virus carries a v-myc oncogene derived from MC29. Injection of pristane-primed mice with the J2 virus induced myeloid tumors and lymphocytic neoplasms, whereas only myeloid tumors were seen in animals inoculated with the J5 virus. Importantly, plasmacytoma development was accelerated upon infection with J3 virus. This occurred in the absence of c-myc activation by chromosomal translocation providing direct evidence for a role of the translocation-activated c-myc in pristane-induced plasmacytomagenesis since abundant expression of v-myc appears to have replaced a requirement for translocation-activation of c-myc. The lack of plasmacytomas in J2 inoculated animals was thought to be due to the preponderance of other faster developing myeloid and lymphoid tumors. In the absence of pristane only lymphocytic neoplasms were induced by J2 (Table 1).

To determine the basis for the observed differences between the J-viruses in their ability to accelerate plasmacytomagenesis further experiments addressed the following questions: (i) Are qualitative differences in the v-myc gene (entirely MC29 derived v-myc in J5 versus a MC29/MH2 hybrid v-myc in J3) responsible for the inability of J5 to induce plasma cell transformation? or (ii) Is the presence of the gag-myc fusion protein incompatible with B cell transformation ? The possibility that gag-fusion affects target cell specificity had to be considered since in the case of v-abl a gag-

[1] Laboratory of Viral Carcinogenesis, National Cancer Institute, Frederick, MD 21701, [2] Laboratory of Genetics, National Cancer Institute, National Institute of Health, Bethesda, MD 20892, [3] Laboratory of Immunopathology, National Institute of Allergy and Infectious Diseases, National Institute of Health, Bethesda, MD 20892, [4] Hazelton Laboratories, Rockville, MD 20850

Current Topics in Microbiology and Immunology, Vol. 141

Table 1: Cell lineage specificity of tumors induced in weanling BALB/cAn mice after intraperitoneal inoculation with J2, J3 and J5 with or without preceding pristane priming (0.5 ml pristane 7 to 25 days before virus)

VIRUS	GENE	PRISTANE	NO.MICE	INCIDENCE	CELL LINEAGE* MY	PC	TL
J2	raf/myc	-	15	2/15	0	0	2
		+	30	21/30	18	0	3
J3	Δraf/myc	-	45	0/45	0	0	0
		+	114	35/114	9	25	1
J5	myc	-	14	0/14	0	0	0
		+	15	14/15	14	0	0

* MY=myeloid tumors, PC=plasmacytomas, TL=thymic lymphocytic neoplasms

Table 2: Effects of J viruses on growth pattern of NIH 3T3 cells

VIRUS	FOCI OF MORPHOLOGICALLY TRANSFORMED CELLS (1)	FOCAL OVERGROWTH (2)
J2	+	N.A. (3)
J3	-	+
J3*	+	N.A.
J5	-	+
Y7	+	N.A.

(1) distinct areas of densely growing cells resulting from a block to post confluence inhibition of cell division in NIH 3T3 cells. Foci appear 10 to 12 days after virus infection.
(2) distinct areas of criss crossing refractile cells appearing 4 to 6 days after infection.
(3) N.A.: not applicable due to early foci of morphologically transformed cells

abl fusion protein is essential for lymphoid cell but not for fibroblast transformation (Prywes et al. 1983). We therefore introduced mutations in J5 v-myc, which restricted expression of a myc protein to the gag-free form. (iii) Will another dual oncogene virus similar to J2 but carrying a less transforming raf gene allow for plasmacytoma induction? Truncation-activated A-raf is less transforming than v-raf in 3611 MSV. It was therefore expressed in a retrovirus in conjunction with MC29 v-myc in sense (Y7) or antisense (Y5) orientation. (iv) Does the ability of J3 to accelerate plasmacytomagenesis result from a potential selection of mutant virus with altered properties? Therefore virus was recovered from J3 induced plasmascytomas and examined for transforming activity.

RESULTS AND DISCUSSION

The series of myc-viruses examined is summarized in Fig.1. In the J5D virus a portion of MH2 derived v-myc containing the majority of mutations in MH2 versus MC29 v-myc was transferred onto the J5 background. In the J5B virus the gag-myc fusion protein (p107) is taken out of frame, but the MC29 derived v-myc protein (p58) is still made from a subgenomic RNA. In Y7 MC29 v-myc was combined with a truncation-activated A-raf.

The predicted expression of the mutant genes in all of these virus constructs was confirmed by Northern hybridization and immunoprecipitation (data not shown). Several of these viruses have been examined for plasmacytoma acceleration in pristane-primed BALB/cAn mice. The data (Fig.2) can be summarized as follows: First, the virus J5B which lacks specifically gag-v-myc behaves like J5 and only induces myeloid tumors, (ii) Second, combining MC29 v-myc and a truncation-activated A-raf (as done in Y7) leads to plasmacytoma acceleration comparable to J3, (iii) Finally, altered virus was recovered from the ascites of 3 out of 3 successive J3 induced plasmacytomas. Infection of NIH 3T3 cells with this virus, which we termed J3*, rapidly (within 4 to 6 days) gives rise to foci of criss crossing refractile cells as is typical for v-raf containing viruses. In contrast viruses carrying only v-myc (like J3 and J5) induce focal overgrowth foci, which develop after confluency with cell lines normally exhibiting postconfluence inhibition of cell division (Table 2).

These results show that (i) the target cell specificity of J5 is not affected by the absence of the gag/myc fusion protein and (ii) adding an activated A-raf oncogene to MC29 v-myc (Y7) broadens the target cell spectrum and allows for plasmacytoma induction, which is not seen with MC29 myc alone. Thus MC29 myc can contribute to plasmacytoma induction. Moreover preliminary data demonstrate that transfer of J3 type myc into the J5 background (virus J5D) does not confer plasmacytoma inducing ability to J5, and (iii) initial characterization of J3* as a focus forming virus strongly suggests selection of altered virus in the course of plasmacytoma development accelerated by J3. The basis for the altered biological properties of recovered J3 virus is currently being examined. Considering the fact that the minimal transforming sequence of v-raf is left intact by the deletion-induced frameshift mutation used to inactivate v-raf in J3, it appears likely that v-raf expression was restored in J3* by a second frameshift mutation.

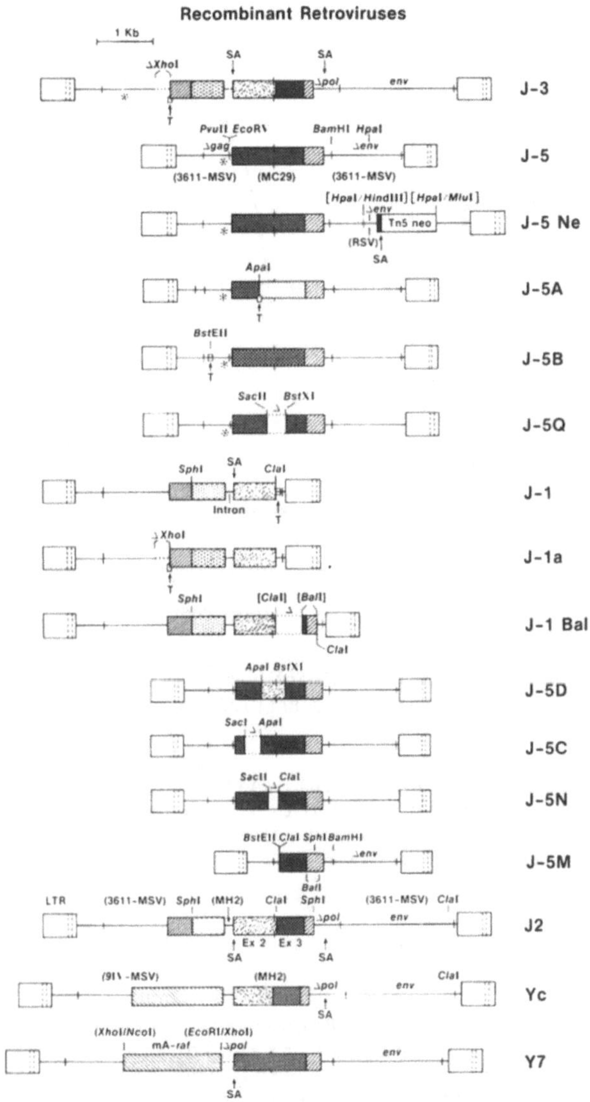

Fig.1: Genomic organization of the J series of recombinant retroviruses. The origin of specific v-onc sequences in the constructs is indicated by: MC29 v-myc, MH2 v-myc, 3611 MSV v-raf, MH2 v-mil

In conclusion several lines of evidence now indicate a requirement for both, raf and myc oncogenes, in J viruses for plasmacytoma acceleration: (i) J viruses carrying only v-myc genes and no functional or truncated v-raf gene do not accelerate plasmacytomas; (ii) Combining v-myc with a truncation activated A-raf gene in the Y7 virus accelerates plasmacytomas. The v-myc negative, A-raf positive Y5 virus control was inactive; and (iii) J3 virus recovered from plasmacytomas has

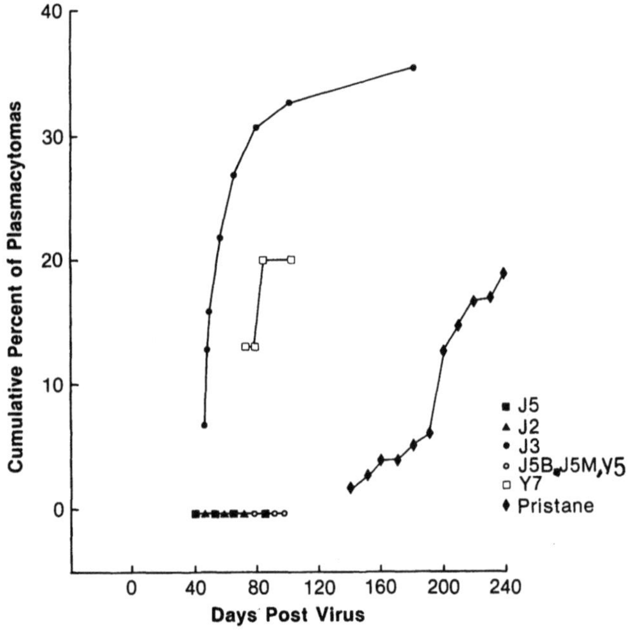

Fig.2:Effects of J series of recombinant v-myc retroviruses on plasmacytoma acceleration in pristane-primed BALB/cAn mice

biological properties consistent with recovery of active v-raf. These results strongly support the view that plasmacytoma induction seen with J3 results from the synergistic action of myc and reactivated raf oncogenes.

REFERENCES

Ohno S, Migita S, Wiener F, Babonits M, Klein G, Mushinski JF, Potter M (1984) Chromosomal translocations activating myc sequences and transduction of v-abl are critical events in the rapid induction of plasmacytomas by pristane and abelson virus. J Exp Med 159:1762-1777

Potter M, Mushinski JF, Mushinski EB, Brust S, Wax JS, Wiener F, Babonits M, Rapp UR, Morse HC III (1987) Avian v-myc replaces chromosomal translocation in murine plasmacytomagenesis. Science 235:787-789

Prywes R, Foulkes JG, Rosenberg N, Baltimore D (1983) Sequences of A-MuLV protein needed for fibroblast and lymphoid cell transformation. Cell 34:569-579

Rapp UR, Cleveland JL, Frederickson TN, Holmes KL, Morse HC III, Jansen HW, Pastschinsky T, Bister K (1985) Rapid induction of hemopoietic neoplasms in newborn mice by a raf(mil)/myc recombinant murine retrovirus. J Virol 55, 23-33

Synergy of an IgH Promoter-Enhancer-Driven c-*myc*/v-Ha-*ras* Retrovirus and Pristane in the Induction of Murine Plasmacytomas

R. Clynes[1], L. W. Stanton[1,4], J. Wax[2], S. Smith-Gill[3], M. Potter[3], and K. B. Marcu[1]

INTRODUCTION

Intraperitoneal injection of BALBc/An mice with a single 0.5 ml dose of the mineral oil pristane leads to the appearance of malignant Ig-secreting (usually IgA) plasma cells in 25% of mice after minimal latent periods of 120 days and mean latent periods between 210-220 days (Potter and Wax 1983). Over 95% of these plasmacytomas (PCTs) contain rcpt(12;15) or rcpt(6;15) chromosomal translocations in which the c-myc gene on chromosome 15 is juxtaposed with immunoglobulin heavy or light chain loci on chromosomes 12 and 6 respectively . These structural alterations disrupt the normal regulation of the c-myc gene which is thought to be causally related to the development of malignancy (reviewed in Potter 1984). Indeed infection of pristane primed mice with wild-type Moloney (Mo-MuLV) helper virus and the replication defective J-3 retrovirus, which contains a partially deleted raf oncogene and a hybrid MH2/MC29 avian v-myc oncogene induced shortened latency plasmacytomas which lacked c-myc translocations (Potter et. al. 1987). On the other hand experiments which used murine retroviruses harboring an LTR driven murine c-myc gene only induced myeloid malignancies in pristane primed mice (Wolff et. al. 1986). A role for second oncogenes in plasmacytogenesis was shown by the induction of short latency PCTs harboring c-myc translocations in pristane primed mice infected with Abelson virus and Mo-MuLV helper virus (Potter et. al. 1973; Ohno et. al. 1984). Contributions of other activated oncogenes in plasmacytogenesis have not been reported.

In the present study, we describe the properties of a highly efficient PCT inducing retrovirus, RIM, that contains the coding exons of a normal murine c-myc cDNA linked to the IgH enhancer and an IgV_H promoter in addition to an LTR driven v-Ha-ras gene. RIM virus consistently induced a high incidence of IgM secreting PCTs with short latent periods in pristane treated mice. Another retroviral construct harboring only v-Ha-ras induced only a few PCTs while another only containing the IgH promoter-enhancer driven myc cassette did not induce PCTs.

1 Department of Biochemistry, SUNY at Stonybrook, NY 11790
2 Hazelton Laboratories of America Inc., Rockville, MD 20850
3 Laboratory of Genetics, National Cancer Institute,Bethesda, MD 20895
4 Present Address: Department of Microbiology and Immunology University of California Medical Center, San Francisco CA 94143

Current Topics in Microbiology and Immunology, Vol. 141

RESULTS

Production of Retroviruses

Retroviral vectors were constructed in which v-Ha-ras and c-myc were placed under the transcriptional control of an Mo-MuLV LTR and IgH transcriptional enhancer and promoter sequences respectively (Fig. 1). Following transfection of these retroviral constructs along with pZAP, a wild-type Mo-MuLV proviral clone (Hoffman et. al. 1982), by the calcium phosphate technique (Graham and van der Eb 1973) into NIH3T3 cells, high titer retroviral stocks were obtained from neo resistent NIH3T3 clones. Viruses were designated R (LTR driven v-Ha-ras), IM (IgH promoter-enhancer driven c-myc) and RIM (combined version of R and IM).

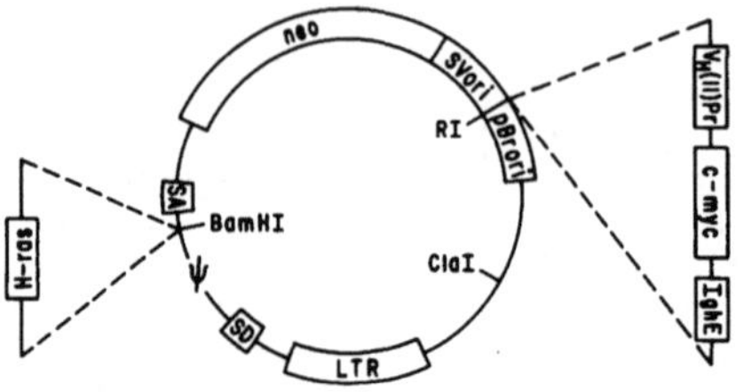

Fig. 1. Retroviral constructs. Details of the vector constructions are given elsewhere (Clynes et. al. submitted). The R retrovector contains the v-Ha-ras gene of the Ha-MuSV rat retrovirus (Gruss et. al. 1984) on a 700 bp SacI-PstI fragment inserted into the BamHI site of a derivitive of ZIPNeoSVX(1) (Cepko et. al. 1984) by linker ligation. The IM and RIM retrovector constructions contain the coding exons of a normal murine c-myc cDNA (Stanton et. al. 1983) on a 1.4 Kb XhoI fragment flanked by IgH gene transcriptional control elements inserted into the EcoR1 site of the modified ZIPNeoSVX(1) and R retrovectors respectively. The subgenomic 2.3 Kb myc expression cassette has the MPC-11 V_H gene's promoter on a 204 bp EcoR1-NcoI fragment (Gilmore et. al. 1987) at its 5' end and the IgH enhancer on a ~700 bp XbaI-EcoRI fragment (Gillies et.al. 1983) on its 3' end

IgM Secreting Plasmacytoma Induction in Pristane-Primed Mice

In the absence of pristane priming RIM virus was injected into fifteen 3 week old and ten 3 month old BALBc/An mice. Although no PCTs were induced three percent of adults and twenty percent of 3 week old mice developed thymic lymphocytic neoplasms (Table 1). In contrast in single dose pristane conditioned BALB/cAn mice infected with RIM virus, PCTs were induced with a mean latency of 60-70 days in 28 and 83 % of 2-4 month and 3 week old mice respectively. Observations of sacrificed animals indicated that RIM/pristane inoculated mice had evidence of plasma cell ascites as early as 40 days post pristane. This requirement for pristane was flexible in that pristane could be administered either 15 days preceding or up to 60 days following viral infection but the optimal time was 1 to 3 days before viral infection. RIM virus also effectively induced PCTs in CDF_1 mice (Table 1) which are resistent to PCT induction by pristane alone (Potter 1984a).

A few PCTs were induced in pristane primed BALBc/An mice infected with R virus and no PCTs were induced with IM virus (Table 2). Neither sequential nor simultaneous infection with the R and IM viruses gave rise to PCTs although 22% of the latter mice developed lymphocytic tumors due to their higher Mo-MuLV dosage.

In contrast to previously studied plasmacytomas which predominantly secrete IgA, (Potter et. al. 1972) 79 out of 94 RIM induced PCTs (84%) secreted IgM. One secreted IgG and two animals had both IgG and IgM components while only 9 (<10%) secreted IgA.

Table 1: Induction of Neoplasias by RIM Virus: frequency and latency of PCTs depends on pristane, animal age and genetic background

Mouse Strain	Pristane (ml)	Virus (day)	Mice No.	Mice Age	Tumors (%) Myeloid	PCT	L.N.[a]	PCT Latency Period mean range in days post-pristane
BALB/cAn	-	0	10	2M	0	0	4	-
"	-	0	15	3W	0	0	20	-
"	0.5	3-15	122	2-4M	1	28	1	76 (60-118)
"	0.2-0.3	0-1	76	3-4W	0	83	0	60 (49-91)
"	0.5	(-10)[b]	23	3M	0	22	0	77 (72-111)
"	0.5	(-20)[b]	23	3M	0	22	0	62 (62)
CDF_1	-	0	14	3W	0	0	7	-
"	0.5	4	38	2M	5	74	0	81 (63-118)
"	0.2	0	50	3W	3	63	0	80(55-119)

M = months; W = weeks; CDF_1 = (BALB/c x DBA/2)F_1

a = lymphocytic neoplasm characterized by enlarged thymus glands, spleen, mesenteric and peripheral lymph nodes.

b = virus was inoculated 10 or 20 days prior to pristane

Table 2: R and IM viruses fail to induce PCTs in Pristane Treated BALB/cAN Mice

Viruses	Pristane (ml)	Virus (days post-pristane)	Mice No.	Mice Age	Tumors (%) Myeloid	PCT	LN
IM	0	-	10	2M	0	0	0
IM	0.5	7-14	75	2M	1.3	1.3	1.3
IM	0.3	0	30	3W	0	0	26.6
R	0	-	10	2M	0	0	0
R	0.3	0	15	3W	0	0	20.0
R	0.5	10-41	75	2M	2.6	2.6	0
IM->R	0.5	7->10	40	2M	2.0	0	5.0
R->IM	0.5	7->10	40	2M	2.0	0	7.0
R+IM	0.5	7	40	2M	5.0	0	22.0

Structure and Expression of Proviral Oncogenes and Endogenous c-Myc Genes

Southern analysis of DNA extracted from frozen tumor samples was performed to investigate the proviral structures present in RIM induced PCTs (Fig. 2). As expected the 2.3 Kb EcoRI fragment containing the viral c-*myc* expression cassette was present in all RIM induced PCTS and the endogenous c-*myc* alleles remained unrearranged as a 21 kb EcoRI fragment. All tumors in Fig. 2 contain a v-Ha-*ras* 700 bp BamHI fragment in addition to endogenous BALBc/An *ras* hybridizing sequences. Southern hybridization of SstI digestions to a *neo* probe revealed the expected size of the entire recombinant provirus since SstI only cleaves the integrated provirus once in each LTR. Nine RIM induced PCTs in Fig. 2 contain an intact 7.4 kb provirus while two, 2177 and 2274 show altered sizes. However, our other Southern data indicate that the rearrangements in the latter two recombinant proviruses occur outside of the *ras* and *myc* genes. PC 2143 was obtained with R virus and contains the expected 5.1 kb provirus as well as a typical endogenous c-*myc* rearrangement caused by a chromosomal translocation.

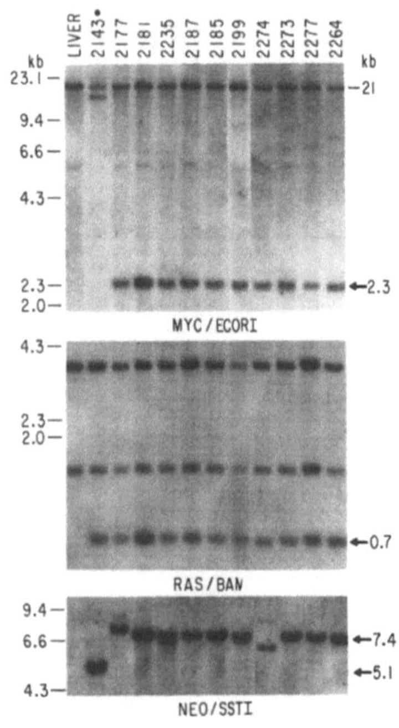

Fig. 2. Contexts of retroviral *myc*, *ras* and *neo* genes in PCTs. Genomic DNAs of RIM and R virus induced PCTs were restricted by the indicated enzymes and submitted to Southern hybridization with random primer labeled restriction fragments. The c-*myc* probe was a 1.4 Kb XhoI fragment derived from the pMc-myc54 cDNA clone (Stanton et al. 1983). The v-Ha-*ras* probe was a 700 bp BamHI fragment and the *neo* probe was a 1.3 Kb HindIII-SmaI segment both isolated from the R retrovector in Figure 1. Liver DNA was of BALB/c origin. A lane denoted by a filled in circle is an R virus induced PCT. Arrows indicate retroviral specific bands

Nuclease S1 mapping analysis was employed to investigate the transcription of proviral *ras* and *myc* genes as well as the endogenous c-*myc* genes in RIM induced PCTs. Using a 5′ end-labeled S1 probe derived from the c-*myc* expression cassette there were at least 6 fragments sized between 210 and 410 nt protected from S1 digestion. This indicates that there are multiple transcriptional initiation sites within the V_H promoter (Fig. 3 top). The full length protected band of 410 nt is a result of read through transcription of *myc* presumed to initiate within the 5′ LTR. A second c-*myc* probe (320-4) was employed to confirm that the viral c-*myc* gene but not the endogenous c-*myc* genes are transcriptionally active in RIM induced PCTs (Fig. 3 bottom). As a consequence of the construction of the RIM vector only 46 nt of exon 1 sequences are present in viral c-*myc* transcripts whereas endogenous c-myc transcripts contain a greater portion of exon 1 sequences including all those present in the 320-4 probe. Transcripts from the normal c-myc alleles would protect a 298 nucleotide segment from the 320-4 probe as seen for the RNA of 230-23-8 cells, a pre-B cell line with normal c-myc alleles. However, RNA from six RIM induced tumors protected only a 183 base fragment as expected for viral c-myc transcripts. Hence the endogenous c-myc alleles are transcriptionally silent as well as unrearranged, while the RIM proviral V_H promoter efficiently transcribes viral c-myc in these plasmacytomas.

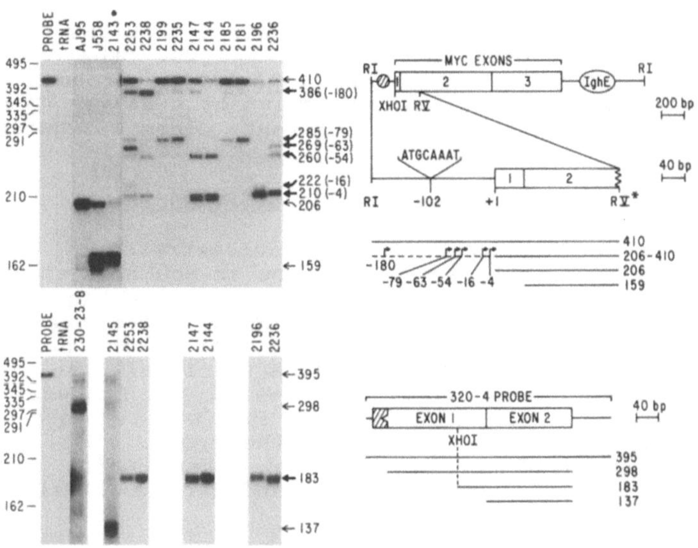

Fig. 3. Nuclease S1 analysis of proviral and endogenous c-myc RNA transcription. Top: A 5′ end-labeled S1 mapping probe was derived from the 2.3 Kb EcoRI myc expression cassette as indicated. Samples include: AJ95 (mature B cell line with an amplified c-myc locus) (Citri et. al. 1987), J558 (pristane induced PCT with a translocated myc gene) (Stanton et. al. 1983), 2143 (R virus induced PCT with a rearranged cellular myc gene, see Figure 1), all other samples are RIM induced PCTs. Bold arrows denote V_H promoter driven myc RNAs and light arrows are myc RNAs which initiated upstream of the expression cassette (410 nt), within exon 1 (206 nt) or intron 1 (160 nt) of the cellular myc gene. Hatched circle represents the ATGCAAAT octamer motif (Shreidereit et.al. 1987). Bottom: S1 mapping with 320-4 probe (derived from an MPC-11 c-myc cDNA clone) (Julius et. al. in press). Samples include: 230-23-8 (an MuLV pre B cell line with normal myc alleles (Yang et.al. 1985; Alt et. al. 1982), 2145 (RIM induced PCT lacking the viral myc gene but possessing an endogenous c-myc rearrangement), all other samples are RIM induced PCTs. Myc transcripts originating from the RIM provirus would all migrate at 183 nt

In order to detect v-Ha-ras transcription in RIM induced PCTS a 5′ end-labeled rat v-Ha-ras probe was used in nuclease S1 analysis. As seen in Figure 4 this probe detected comparable levels of v-Ha-ras transcripts in all the tumor samples. Transcription of these v-Ha-ras sequences is presumed to initiate in the proviral 5′ LTR. Levels of ras expression were similar in RIM and R virus induced PCTs. Protection of endogenous murine c-Ha-ras transcripts was not observed by this rat v-Ha-ras probe.

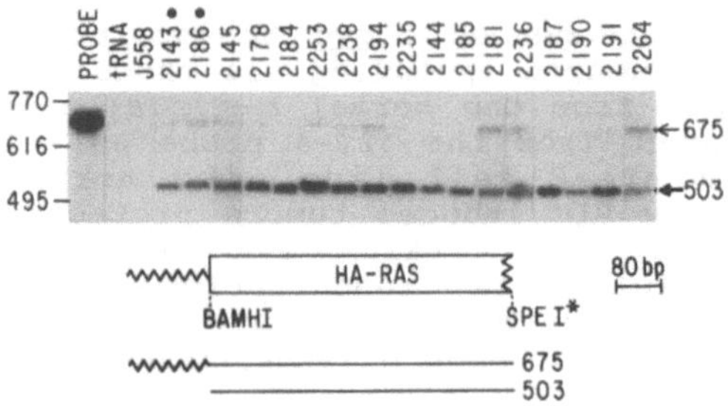

Fig. 4. Nuclease S1 analysis of proviral v-Ha-*ras* transcription. A 5′ end labeled v-Ha-*ras* probe was obtained from the R retrovector. 2143 and 2186 are R virus induced PCTs and the other lanes are RIM induced PCTs. Arrows indicate the full-length probe (675 nt) and bands protected from S1 nuclease digestion by v-Ha-*ras* RNAs (503 nt). The jagged line in the probe map represents plasmid sequences

RIM induced PCTs are monoclonal B cell malignancies

Clonality of the RIM induced PCTs was demonstrated by two different methods. First using a J_H specific probe in Southern analysis it was seen that RIM induced PCTs had at most two rearranged J_H alleles (Fig. 5 right). The appearance of the germline 6.2 Kb EcoRI J_H fragment in all the samples is due to contaminating normal cells in these tissue samples. Another band observed in all the tumor samples is a 2.3 Kb band corresponding to the viral *myc* expression cassette and is revealed in all RIM induced PCT samples because of the presence of IgH enhancer sequences in the probe. Besides these two ubiquitous bands there are at most two others which correspond to rearranged J_H loci. These findings indicate a clonal origin for these B cell malignancies.

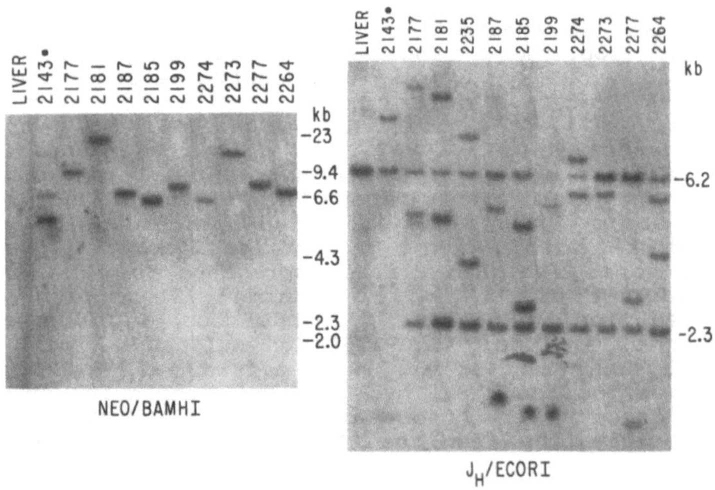

Fig. 5. Southern analysis indicates that RIM virus induced PCTs are monoclonal malignancies. DNAs of RIM induced PCTs were restricted by the indicated enzymes and hybridized to ^{32}P labeled DNA fragments: a 1.3 Kb HindIII-SmaI segment of the pN(S) *neo* gene and a 1.6 Kb BamHI-EcoRI fragment of the pj11 plasmid (Marcu et. al. 1980) which contains Ig J_H segments 3 and 4 and the 3′ flanking IgH enhancer. Liver DNA is BALB/c. A filled circle denotes an R virus induced PCT

A viral specific neo probe was used in Southern analysis of BamHI digested DNA as a second demonstration of clonality (Fig. 5 left). The 3' BamHI site of the proviral neo hybridizing fragment is donated by flanking cellular DNA and is unique to each proviral insertion. The presence of a unique neo hybridizing fragment in each of the tumor samples in Figure 5 indicates that each tumor contains a single proviral integration site. The submolar band present in PCT 2143 is due to a partial digestion.

Moloney Integration is not Necessary for Malignant Transformation

To investigate the possibility that the wild-type Moloney helper virus present in the viral inoculum contributed to tumorigenesis we screened for Moloney proviral integration using an AKR derived ecotropic specific viral envelope probe (Chattopadhyay et. al. 1980). This probe cross-hybridizes with the single endogenous ecotropic locus in Balbc/An present on a 4.3 kb PvuII fragment (Potter et. al. 1984b) as well as to a 3.0 kb fragment of Mo-MuLV. The 3.0 kb band was present in two thymocytic and two myeloid neoplasms obtained from pristane-primed mice infected with retroviral stocks as well as in M15T, a T cell tumor induced in an infected neonatal BALB/cAn mouse (Primi et. al. in press). The transformed phenotype of these tumors which lacked integrated recombinant viruses (data not shown) may in part be a result of Moloney proviral insertional activation of cellular proto-oncogenes. In contrast 8 of 9 RIM induced PCs in Fig. 6 lacked the 3.0 Kb band and hence we conclude that the Mo-MuLV helper virus does not significantly contribute to the transformed phenotype of RIM induced PCTs.

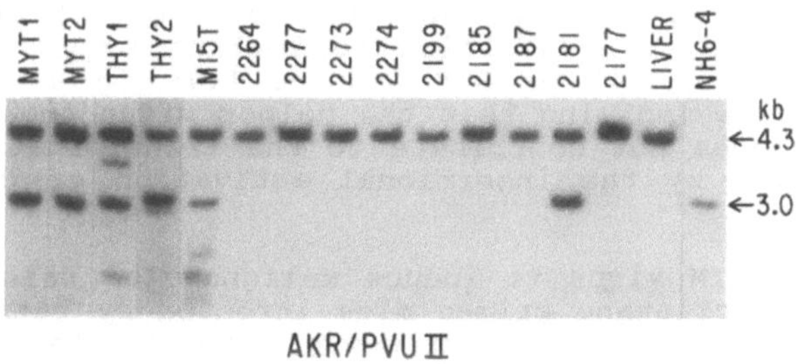

Fig. 6. Southern analysis for Mo-MuLV envelope sequences in RIM induced PCTs and other non-PCT retrovirally induced myeloid and T lymphoid malignancies. DNAs were restricted with PvuII and analyzed by Southern hybridization with a 400 bp DNA probe derrived from the ecotropic envelope gene of an Akv-1 provirus (Chattopadhyay et. al. 1980). MYT1, MYT2, THY1, and THY2 are myeloid tumors and thymic lymphomas obtained from RIM virus infected pristane primed BALB/cAn mice. M15T is an immature T cell neoplasm obtained from IM virus infected neonatal BALB/c mice (Primi et. al. in press). All other lanes are RIM virus induced PCTs and the liver DNA is of BALB/c origin. NH6-4 is an NIH3T3 line producing RIM and Mo-MuLV which is not of BALB/c origin and therefore lacks an endogenous ecotropic locus

DISCUSSION

The RIM virus has been shown to induce short latency, monoclonal, IgM secreting PCTs in pristane primed BALBc/An mice with incidences of 28% in adult mice and 83% in 3 week old mice. CDF_1 mice, which are genetically resistant to PCT induction by pristane alone were found to be susceptible to PCT induction by pristane plus RIM virus, indicating that RIM virus circumvented normal resistence mechanisms.

The variable incidence (20 to 90%) and the clonality of the tumors both suggest that other genetic events in addition to RIM virus infection may be necessary for the malignant transformation of the target B cells. However the very high incidence of tumors in 3 week old mice and our recent observations of plasma cell ascites as early as 40 days post pristane make a requirement for additional genetic mutations in the target cell less appealing. Indeed much of the time needed for the development of these tumors may simply reflect the time required to allow for both infection of a vulnerable target cell and its subsequent clonal expansion. The latter process may require the appearance of antigenic and/or other proliferative stimuli within the oil granuloma.

Two events which are not required for the malignant transformation of these tumors are c-myc translocations and Moloney virus integration. We find, with only a few minor exceptions, that RIM induced tumors contain transcriptionally silent, unrearranged endogenous c-myc alleles. This observation extends the results of a previous study involving a v-myc retrovirus (Potter et. al. 1987) by showing that the presence of a retrovirally transduced normal c-myc gene can substitute for the requirement for a c-myc translocation in plasmacytomagenesis. In addition to lacking c-myc rearrangements, the majority (18 of 24) of RIM induced PCTs in adult mice lacked Moloney proviruses implying that the helper virus present in the RIM viral inoculum does not contribute to the transformation of these B cell malignancies by the insertional activation of cellular proto-oncogenes.

The inability of IM virus to induce malignancies raises two issues. In another study pristane primed mice infected by retroviruses which contained an Mo-MuLV LTR-driven c-myc gene developed only myeloid malignancies (Wolff et. al. 1986). However in the present study c-myc expression is targeted to the B cell compartment by flanking Igh transcriptional elements. Indeed preliminary results of immunopreciptations of c-myc from cellular extracts from RIM infected cell lines have indicated that high levels of c-myc protein are present in RIM infected B cell lines but not in RIM infected myeloid lines. Secondly, while the IM virus was unable to induce short latency PCTs, we have shown that the additional presence of the v-Ha-ras oncogene in the RIM virus imparts a strong plasmocytogenic effect suggesting a role for activation of ras family genes in plasmacytogenesis.

REFERENCES

Alt FW, Rosenberg N, Enea V, Siden E and Baltimore D (1982) Multiple immunoglobulin heavy-chain gene transcripts in Abelson murine leukemia virus-transformed lymphoid cell lines. Mol.& Cell Biol. 2: 386-390

Cepko CL, Roberts BE, and Mulligan RC (1984) Construction and applications of a highly transmissable murine retrovirus shuttle vector. Cell 37: 1053-1062

Chattopadhyay S, Lander M, Rands E and Lowy R (1980) Structure of endogenous murine leukemia virus DNA in mouse genomes. Proc. Natl. Acad. Sci. USA 77: 5774-5778

Citri Y, Braun J and Baltimore D (1987) Elevated myc expression and c-myc amplification in spontaneously occuring B lymphoid cell lines. J. Exp. Med. 165: 1188-1194

Clynes R, Wax J, Stanton LW, Smith-Gill S, Potter M and Marcu KB (submitted) Rapid induction of IgM secreting murine plasmacytomas by pristane and an IgH promoter-enhancer driven c-myc/v-Ha-ras retrovirus. Proc. Natl. Acad. Sci. USA

Gilmore GL, Yang JQ, Marcu KB, and Birshtein BK (1987) Absence of somatic mutation in the variable region of MPC 11 variants expressing a different heavy chain isotype. J. Immunol. 139:619-624

Gillies SD, Morrison SL, Oi VT and Tonegawa S (1983) A tissue specific transcriptional enhancer is located in the major intron of a rearranged immunoglobulin heavy chain gene. Cell 33: 717-728

Graham FL and van der Eb AJ (1973) Transformation of rat cells by DNA of human adenovirus 5. Virology 52: 456-467

Gruss P, Ellis TY, Shih TY, Noenig M, Scolnick M and Khoury G (1981) SV40 recombinant molecules express the gene encoding p21 transforming protein of Harvey murine sarcome virus. Nature (London) 293: 486-488

Hoffman JW, Steffen D, Gusella J, Tabin C, Bird S, Cowing D and Weinberg RA (1982) DNA methylation effecting the expression of murine leukemia proviruses. J. Virol. 44: 144-157

Julius MA, Street AJ, Fahrlander PD, Yang JQ, Eisenmann RN and Marcu KB (in press) Translocated c-myc genes produce chimeric transcripts containing antisense sequences of the immunoglobulin heavy chain locus in mouse plasmacytomas. Oncogene

Marcu KB, Banerji J, Penncavage NA, Lang R and Arnheim N (1980) 5' flanking region of immunoglobulin heavy chain locus displays length heterogeneity in germlines of inbred mouse strains. Cell 22:187-196

Ohno S, Migita S, Wiener F, Babonits M, Klein G, Mushinski JF, and Potter M (1984) Chromosomal translocations activating myc sequences and transduction of v-abl are critical events in the rapid induction of plasmacytomas by pristane and Abelson virus. J Exp Med 159: 1762-1777

Potter M (1972) Immunoglobulin producing tumors and myeloma proteins in mice. Physiol. Rev. 52: 631-719

Potter M, Sklar MD and Rowe WP (1973) Rapid viral induction of plasmacytomas in pristane primed BALB/c mice. Science 182: 592-594

Potter M and Wax JS (1983) Peritoneal plasmacytomagenesis in mice: comparison of pristane dose regimens. J. Natl. Cancer Inst. 71: 391-395

Potter M (1984a) Genetics of susceptibility to plasmacytoma development in BALB/c mice. Cancer Surveys 3: 249-264

Potter M, Hartley JW, Wax JS and Gallahan D (1984b) Effect of MuLV-related genes on plasmacytogenesis in BALB/c mice. J.Exp. Med. 160: 435-440

Potter M, Mushinski JF, Mushinski EB, Brust S, Wax JS, Wiener F, Babonits M, Rapp UR and Morse HC (1987) Avian v-myc replaces chromosomal translocation in murine plasmacytogenesis. Science 235: 787-789

Primi D, Jouvin-Marche E, Clynes RA, Marolleau J-P, Gris C, Marcu KB and Cazenave P-A (in press) Dynamics of immunoglobulin and T-cell receptor genes recombinations during lymphocyte development. CRC Rev. in Immunol.

Sheidereit C, Heguy A, and Roeder RG (1987) Identification and purification of a human lymphoid-specific octamer-binding protein (OTF-2) that activates transcription of an immunoglobulin promoter in vitro. Cell 51: 783-793

Stanton LW, Watt R and Marcu KB (1983) Translocation, breakage and truncated transcripts of c-myc oncogene in murine plasmacytomas. Nature (London) 303: 401-406

Wolff L, Mushinski JF, Gilboa E and Morse HC III (1986) Induction of hematopoietic tumors using a viral construct containing c-myc cDNA from normal mouse spleen. Curr. Top. in Microbiol. and Immunol. 132: 33-37

Yang JQ, Bauer SR, Mushinski JF and Marcu KB (1985) Chromosome translocations clustered 5′ of the murine c-myc gene qualitatively affect promoter usage: implications for the site of normal c-myc regulation. EMBO J. 4: 1441-1447

The Genetics of Susceptibility to RIM-Induced Plasmacytomagenesis

B. Mock, J. Wax[1], R. Clynes[2], K.B. Marcu[2], and M. Potter

INTRODUCTION

BALB/cAn mice are susceptible to the induction of plasmacytomas (PCTs) by the intraperitoneal injection of pristane. An average of 60% of the mice inoculated with pristane develop tumors with a mean latent period of 220 days (Potter et al. 1984). DBA/2 mice are resistant to PCT formation (0% incidence by Day 220). Resistance in DBA/2 is a dominant trait as (BALB/cAnPt x DBA/2N) F_1 mice are solidly resistant to the plasmacytomagenic process. The incidence of PCTs in F2 (<2%; unpublished data) and backcross progeny (11%; Potter et al. 1984) between BALB/cAnPt (PCT^S) and DBA/2N (PCT^R) progenitors is suggestive of a model for 3 recessive genes for resistance to PCT formation. Over 95% of the tumors which develop in BALB/cAn mice carry either rcpt (6;15) or rcpt (12;15) translocations involving an Ig locus and c-myc directly or indirectly. Attempts have been made to bypass some of the earlier genetic events in the pathway to PCT formation through the introduction of specific oncogenes via retroviral vectors. Vectors carrying v-abl (Potter et al. 1973) or combinations of v-raf and v-myc (Potter et al. 1987) have produced PCTs in at least 20% of the pristane-primed BALB/cAn mice inoculated with these retroviral constructs by as early as day 80 post pristane. More recently, a new construct containing an IgH promoter-c-myc-IgH enhancer cassette and a MoMuLV-LTR driven Ha-ras oncogene has been examined for its effect on the development of PCTs in BALB/c mice (Clynes et al., 1988; this vol.) and in this report, the genetics of susceptibility to PCT formation in resistant and susceptible mice.

MATERIALS AND METHODS

BALB/cAnPt, DBA/2N, (BALB/cAnPt x DBA/2N)F_1 and (CxD)F_1 X DBA/2N backcross progeny were generated and maintained in a closed conventional colony at Hazleton Labs under contract number N01-CB-71085. Three week old mice were inoculated i.p. with 0.2 ml of pristane one day prior to receiving a wild type Moloney (Mo-MuLV) helper virus and a replication defective retrovirus (referred to as RIM) that contains the coding exons of a normal murine c-myc cDNA linked to the IgH enhancer and an IgV_H promoter in addition to an LTR driven v-Ha-ras gene (Clynes et al. 1988; this volume). Mice were followed for the development of plasmacytomagenesis by examination of ascites smears.

Laboratory of Genetics, National Cancer Institute, National Institutes of Health, Bethesda, MD 20892 USA, [1]Hazleton Laboratories, Rockville, MD, [2]Biochemistry Department, SUNY at Stony Brook, Stony Brook, NY 11794-5215

Current Topics in Microbiology and Immunology, Vol. 141

RESULTS AND DISCUSSION

By day 120 of this PCT induction study with RIM, 100% of the BALB/cAnPt and 0% of the DBA/2N mice had developed PCTs. Susceptibility is dominant with this induction scheme; 78% of the $(CxD)F_1$ mice developed PCTs. Interestingly, 45% of the backcross progeny were susceptible to PCT formation (Fig. 1), indicating that a single gene is likely to control the development of RIM-induced PCTs in BALB/cAn mice ($\chi^2_1 = 0.51$, $p = 0.5$, i.e. no significant deviation from 50% susceptible: 50% resistant expected for a single gene model).

These tumor induction patterns are in contrast to those seen in mice given only pristane in which three or more recessive genes in the DBA/2 produce what appears to be a dominant resistant phenotype in F1 mice. The scheme of PCT-induction via pristane plus the RIM retroviral construct provides a unique system to examine the effect of a single dominant gene involved in the susceptibility of BALB/cAn mice to PCT formation. The introduction of immunoglobulin and _myc_ sequences through the retroviral vector appears to have bypassed some of the early genetic events which precede tumor development and may allow one to examine the effect of a single gene further along the pathway. Our current efforts are focused on a linkage analysis of restriction fragment length polymorphisms (RFLPs) associated with PCT^S backcross progeny.

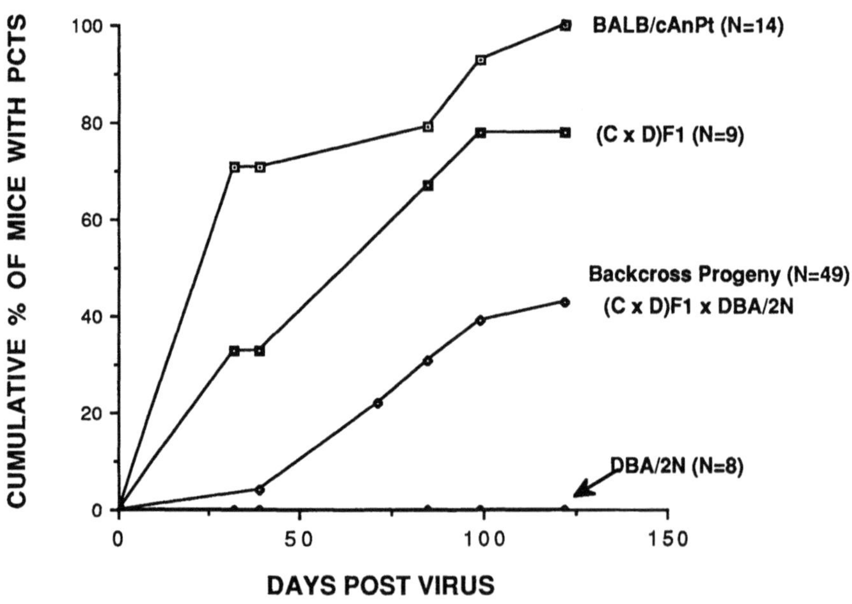

Fig. 1 The incidence of plasmacytomas in a set of backcross progeny generated from a cross between $(C \times D)F_1$ females (PCT^S) and DBA/2N males (PCT^R)

Literature

Clynes, R., Stanton, L.W., Wax, J., Potter, M., Marcu, K.B. 1988. Rapid induction of IgM secreting murine plasmacytomas by pristane and an IgH-promoter-enhancer driven c-myc/v-Ha-ras retrovirus. Proc. Natl. Acad. Sci., in press.

Clynes, R., Stanton, L.W., Wax, J., Smith-Gill, S., Potter, M. and Marcu, K.B. 1988. Synergy of an IgH promoter-enhancer driven c-myc/v-Ha-ras retrovirus and pristane in the induction of murine plasmacytomas. Current Topics in Microbiol. and Immunol., this volume.

Potter, M., Sklar, M.D., Rowe, W.P. 1973. Rapid viral induction of plasmacytomas in pristane-primed BALB/c mice. Science 182: 592-594.

Potter, M., Hartley, J.W., Wax, J.S., and Gallahan, D. 1984. Effect of MuLV-related genes on plasmacytomagenesis in BALB/c mice. J. Exp. Med. 160: 435-440.

Potter, M., Mushinski, J.F., Mushinski, E.B., Brust, S., Wax, J.S., Wiener, F., Babonits, M., Rapp, U.R. and Morse, H.C., III. 1987. Avian v-myc replaces chromosomal translocation in murine plasmacytomagenesis. Science 235: 787-789.

Significance of Chromosome 8 Breakpoint Location in Burkitt's Lymphoma: Correlation with Geographical Origin and Association with Epstein-Barr Virus

F. BARRIGA, J. KIWANUKA, M. ALVAREZ-MON, B. SHIRAMIZU, B. HUBER, P. LEVINE, and I. MAGRATH

INTRODUCTION

Burkitt's lymphoma (BL) is a B cell neoplasm, which occurs throughout the world, predominantly in children and young adults [Magrath 1983]. Its relatively high incidence in equatorial Africa has led to the African form of the disease being designated as "endemic" (eBL), while cases outside equatorial Africa are referred to as "sporadic" (sBL) Burkitt's lymphoma. A number of differences between eBL and sBL have been described [Magrath 1986] and are summarized in table 1.

The Epstein-Barr virus (EBV), which was discovered in a BL cell line derived from an African patient [Epstein 1964], is present in a latent phase in about 95% of eBLs and some 10% to 15% of sBLs. These observations make it highly unlikely that EBV can have a role in the pathogenesis of all cases of Burkitt's lymphoma, and have served to shed doubt on a causal role for EBV even in EBV positive cases. The discovery of non-random, reciprocal chromosomal translocations in all cases of BL [Zech 1976] has focused attention on a genetic event as the critical pathogenetic step, and the potential role of EBV has remained entirely speculative. The translocations invariably include chromosome 8 as one partner and one of the immunoglobulin chain loci on chromosomes 14 (heavy chain), 2 (kappa chain) or 22 (lambda) as the other. The 8;14 translocation is the most frequent, occurring in some 80% or more of tumors, regardless of geographic origin. At a molecular level, the translocations result in the juxtaposition of the c-myc oncogene to one of the immunoglobulin loci, an event that has suggested involvement of cis-acting transcriptional enhancing elements present in the immunoglobulin loci in the deregulation of c-myc associated with the translocations.

Recently, a difference between sBL and eBL with regard to the location of the breakpoints in chromosome 8 has been described [Pellici 1986]. In sBL the breakpoint lies within c-myc or in the immediate 5' flanking sequences in the majority of tumors, while in eBL, the breakpoint is usually far upstream of c-myc, although most, if not all tumors with a far 5' breakpoint have mutations in the first exon of c-myc. This observation implies that the pathogenetic events and the mechanism of deregulation of c-myc differ in sBL from eBL, a finding

Pediatric and Epidemiology Branches, National Cancer Institute, Bethesda, Maryland,
and Burroughs Wellcome, Research Triangle, North Carolina.

Current Topics in Microbiology and Immunology, Vol. 141

Table 1. DIFFERENCES BETWEEN ENDEMIC AND SPORADIC BURKITT'S LYMPHOMA

	eBL	sBL
Average annual incidence children < 15 yrs.)	1Ø per 1ØØ,ØØØ	Ø.2 per 1ØØ,ØØØ
Occurrence	Climatically determined	Not climatically determined
Association with EBV	95%	15%
Chromosome 8 breakpoints	Upstream of c-myc	Within c-myc
Immunological features	CALLA - T 1 +, B2 +	CALLA + T 1 -, B2 -
Common sites of tumor	Jaw Abdomen Orbit Paraspinal	Abdomen Bone marrow Nasopharynx Lymph nodes
Response of recurrent tumor (including CNS)	Good	Poor

which is consistent with the clinical and biological differences referred to. Moreover, genetic differences between sBL and eBL leave open the possibility that EBV has a causal role in the majority of cases of eBL. Indeed, a correlation between EBV association and specific structural changes in c-myc would strongly support this possibility. To examine this issue further, we have determined the breakpoint locations on chromosome 8 in 56 cases of BL and related them to the presence of the EBV genome in the tumor cell DNA.

SAMPLES, ORIGIN AND DIAGNOSTIC CRITERIA.

Table 2. shows the type and origin of the 56 BLs analyzed. Samples of eBL, obtained at the time of tumor biopsy or resection and kept frozen, were obtained from the Burkitt's Lymphoma Project in Ghana. Samples of sBL were obtained from patients treated at NCI (1Ø) and other institutions in the U.S. (1Ø).
Most of the sBL cell lines were derived from patients seen at NCI; most of them and all 6 eBL have been extensively studied. All cell lines had a t(8;14) karyotype and so did all the fresh tumors in which cytogenetic data was available (1Ø sBL, Ø eBL). Histological diagnosis was made at the referring institution, immunophenotyping being done in a few patients. All the tumors included in the analysis had rearrangements of the immunoglobulin heavy chain locus detected by Southern blot utilizing J_H and C probes.

Table 2.ORIGIN AND TYPE OF 56 BLs STUDIED

	Endemic	Sporadic	Total
Fresh tumors	19	20	39
Cell lines	6	11	17
Total	25	31	56

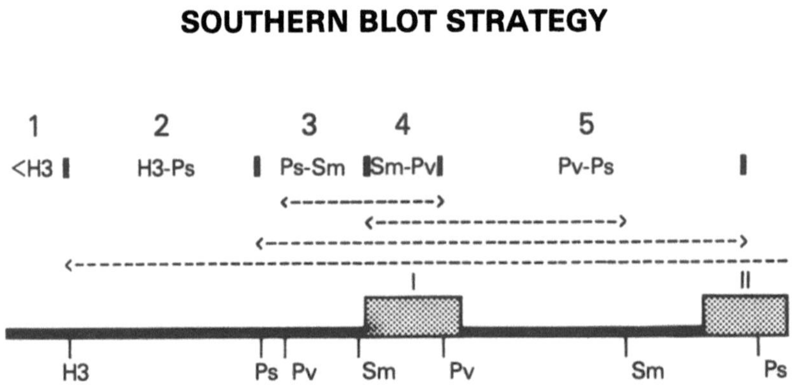

Figure 1.

CHROMOSOME 8 BREAKPOINT ANALYSIS

Southern Blot Strategy.

Tumor cell DNA was digested with Hind III, EcoRI, PstI, PvuII and SmaI and Southern blot analysis was done using probes for the first exon(PvuII-PvuII) [Dalla Favera 1983] and the third exon (ClaI-EcoRI) [Dalla Favera 1984] of the c-myc gene. Chromosome 8 breakpoints were inferred from the analysis of c-myc rearrangements and were arbitrarily classified into the 5 regions shown figure 1:

1 - Breakpoints in 5' regions distant from the gene:
No rearranged fragments were detected by Southern blotting of DNA digested EcoRl, Hind III Pst I or Sma I and hybridized with appropriate probes. Some tumors with breakpoints outside the 5' EcoRI restriction site showed a rearrangement of c-myc using the first exon probe in Pvu II blots as previously described [Pellici 1986]. The reárranged band in such tumor was usually approximately 1.8 Kb in size - the

size expected by the loss of the PvuII site at the 3' end of the first exon.

2,3- Breakpoints in the 5' flanking sequences:
Rearrangements of c-myc were seen with EcoRI, and Hind III; Sma I digests revealed no rearrangements and the first and third exon probes detected a rearranged fragment of the same size in both Hind III and EcoRI blots, demonstrating continuity between these exons. Breakpoints in region 3 were distinguished from breakpoints in region 2 by the presence of rearrangements in Pst I or Pst I and Pvu II digests.

4 - Breakpoints within the first exon:
Rearrangements were detected in Hind III, EcoRI, Pst I, Sma I and PvuII digests.

5 - Breakpoints within the first intron:
Rearrangements of c-myc were detected with all enzymes except PvuII. In addition the rearranged fragments detected by 1st and 3rd exon probes in Hind III and EcoRI digests were of different sizes, indicating discontinuity between these two exons.

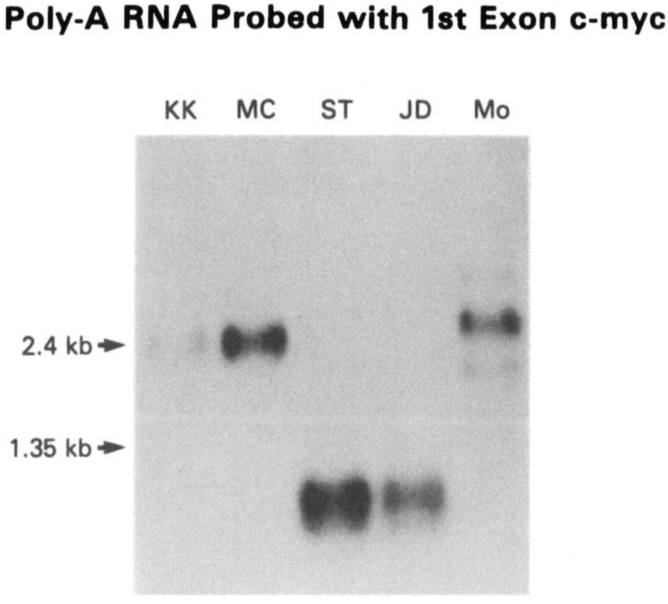

Figure 2.

Confirmation of First Intron Breakpoints by Northern Blot Analysis and S1 Nuclease Protection Assays.

Northern blot analysis of RNA from some of the cell lines used in this study confirmed a breakpoint location in the first intron since 1st and 3rd exon probes hybridized with different sized transcripts. The 3rd exon probe detected a 2.4 kb message, while the 1st exon probe failed to hybridize with this message, but instead detected a 900 bp transcript (fig 2.)

An S1 nuclease protection assay utilizing an 860n single stranded probe spanning the first exon (PvuII-PvuII) and encompassing the two c-myc promoters, P1 and P2 provided further confirmation of the location of some of the breakpoints (fig 3). When the c-myc gene is not broken 3'of the promoters, transcripts are initiated from both P1 and P2, as shown by protection of the expected lengths of the probe (510n and 349n). In normal cells transcripts are initiated predominantly from P2, (4 or 5 to 1) whereas both promoters are equally active in Burkitt's lymphoma and hepatoma cells). In the Burkitt's lines examined, Daudi has a breakpoint distant from c-myc and transcripts are initiated from both promoters. Namalwa and ST486 have breakpoints in the first intron by Southern blot analysis. In the first case no RNA fragments are protected by the probe, indicating that transcription is not initiated from P1 or P2, and first exon sequences are absent from the transcript. In ST486, however, in which Southern blot analysis showed a breakpoint in the intron and northern analysis showed transcription of the first exon separately from the third (fig 3), the protection of

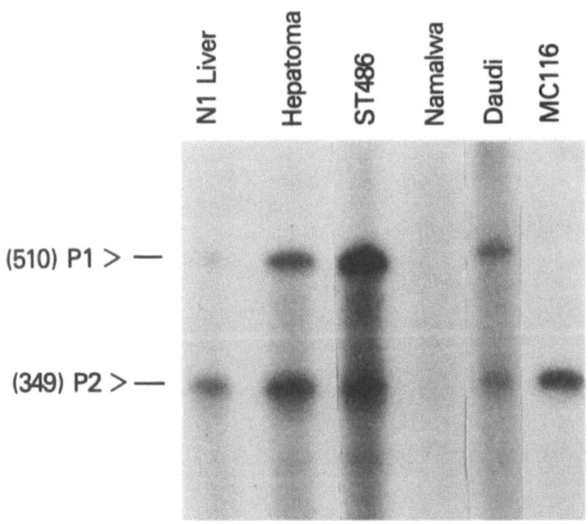

Figure 3.

the probe lengths which indicate transcription from both P1 and P2 is clearly accomplished by transcripts originating from the separate 1st exon remaining on chromosome 8.

In one tumor derived cell line (MC116), with a breakpoint very close to the 5' PvuII site, S1 protection experiments revealed only transcripts initiated from P2. The significance of this is discussed below.

BREAKPOINT LOCATIONS

Fig. 4 shows the distribution of chromosome 8 breakpoints in the sample analyzed. The tumors have been divided into eBL and sBL and the distribution of breakpoints among the 5 designated regions is shown. The presence of the EBV genome in the tumors analyzed was detected by Southern blot analysis using a probe derived from the Bam HI "K" fragment of EBV. All eBLs analyzed were EBV positive. Nine of 31 of the sBL were positive. Table 3 shows the correlation between breakpoints, geographic origin of the tumor and EBV association in the tumors and cell lines studied. There was segregation of EBV association with far 5' breakpoints. Other breakpoint regions included both positive and negative tumors.

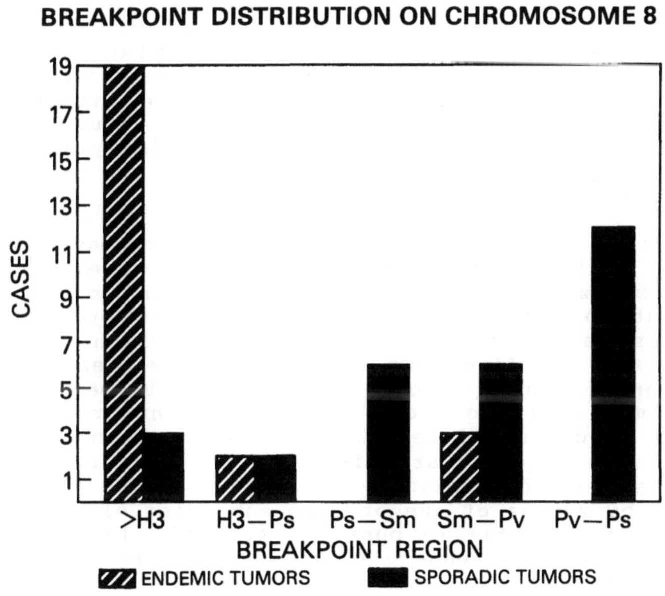

Figure 4.

Table 3. CORRELATION BETWEEN BREAKPOINTS, GEOGRAPHIC ORIGIN AND EBV ASSOCIATION

BREAKPOINT	ENDEMIC	SPORADIC	EBV +	EBV -
Within the HindIII fragment	6	27	12	21
Outside HindIII	19	4	22	1

DISCUSSION

Although it is widely believed that the chromosomal translocations characteristic of Burkitt's lymphoma result in the deregulation of c-myc, and that this constitutes a critical event in the malignant transformation, the actual mechanisms by which the gene is deregulated are not clearly understood. There is much new information, however, regarding the regulation of the expression of c-myc, and increasingly the evidence points towards multiple mechanisms of pathological deregulation. Since regulatory regions are present both within the gene and in its 5' flanking sequences, it is apparent that the position of the breakpoints themselves may well be critical, by virtue of separating regulatory regions from the gene or causing structural and functional damage to the regulatory elements themselves. In this regard, the analysis of chromosomal breakpoint locations in c-myc in relation to the different regulatory regions may provide useful preliminary indications as to the mechanism of deregulation which applies in a particular tumor. In addition, the selective association of EBV with a specific breakpoint location, or subset of breakpoint locations, would strongly suggest a specific role for the virus in the production or maintanance of the transformed phenotype, implying that EBV may be an essential component of neoplastic growth in the presence of some structural changes in c-myc, but not others.

The present analysis considerably extends previously available data [Pellici 1986] both with regard to the number of tumors examined, and by more precisely localizing breakpoints. We have also demonstrated an apparent correlation between EBV association and breakpoint location. Our analysis raises a number of questions which are experimentally approachable. The first set of questions relates to the mechanism of deregulation of c-myc. Table 4 lists possible mechanisms of deregulation which may apply to each major breakpoint group, the rational basis of which is discussed below. It must be born in mind that in each case, additional factors are likely to be operative. One of these probably results from the proximity of c-myc to immunoglobulin (Ig) sequences - perhaps by virtue of one or more enhancing elements in the Ig region, although it is entirely possible that the Ig sequence has a more general role, e.g. maintenance of an appropriate chromatin environment for c-myc expression. It is also possible that the necessity for immunoglobulin gene rearrangements during normal B cell differentiation facilitates or predisposes to chromosomal translocations involving the 14q region, and that immunoglobulin sequences have no direct role in the altered c-myc expression. Finally, the abnormal expression of other oncogenes, perhaps brought about by additional chromosomal abnormalities may be a requirement for neoplastic transformation.

Table 4. MECHANISMS LEADING TO CONSTITUTIVE EXPRESSION OF C-MYC IN BURKITT'S LYMPHOMA

DISTANT BREAKPOINT WITH MUTATIONS IN 1ST EXON (mostly endemic)

Disabling of elongation block
EBV gene product(s) necessary e.g act via c-myc regulatory elements ?

IMMEDIATE 5' BREAKPOINTS

Imbalance of normal regulatory elements

1ST EXON OR INTRON BREAKPOINTS

Bypass of elongation block - new transcription start site in intron - and deletion of regulatory elements

Tumors with breakpoints far distant to the gene are found predominantly in patients with eBL. These tumors are almost invariably EBV positive (Table 3). Only one sBL with a breakpoint distant to c-myc that was EBV negative was found. Another 3 sBL with this type of breakpoint were EBV positive. Although not statistically significant at this point, because sBL with distant breakpoints are so few in number, if more sBL with far 5' breakpoints and EBV positivity are found, then the probability that EBV is a necessary component of pathogenesis in this subset of BL would be highly likely. That this subset is common in endemic regions, and rare elsewhere suggests that the environment, and possibly differences in the age of acquisition of EBV, are major predisposing factors.

Although the c-myc gene is intact in tumors with breakpoints distant to the gene, sequence analysis of c-myc in some of these tumors, and Southern blot analysis in others has shown the presence of mutations and other genetic alterations, especially in the first exon/first intron region. A particularly frequent mutation that can be detected by Southern blot results in the deletion of the PvuII site in the 3' end of the first exon and a rearrangement of the PvuII-PvuII fragment encompassing the first exon. This mutation was found in 12/23 tumors with breakpoints distant to the gene in our series. The significance of this mutation is not known, but one possible consequence is abrogation of the physiologic block to elongation of the message that has been described in this region of c-myc [Bentley 1986].

Recently, various cis elements that regulate the transcription of c-myc promoters have been described 5' of the gene, and within the Hind III fragment [Hay 1987]. Breakpoints in this region result in removal of some elements and could abrogate the function of others, resulting in deregulation of the gene. Analysis of one of our tumor cell lines (MC116) confirmed this possibility. We found, by S1 protection analysis, that c-myc transcription in MC116 was only detectable from the P2 promoter. This cell line had a breakpoint very close to, or possibly within, a region known to contain both a negative regulatory element for P1 and P2, and an element necessary for P1 function in the absence of 5' enhancer regions. The pattern of transcription in this tumor was identical to that seen in experiments performed by Hay, Bishop and Levens [Hay 1987] with c-myc constructs which included a similar amount of the 5' region of c-myc to that translocated with the c-myc gene in MC116, and is entirely consistent with elimination of the function of both of these elements. In similar fashion it may be possible to predict the pattern of transcription from P1 and P2 in tumors with breakpoints in the region immediately upstream of c-myc, since the positions of the positive and negative regulatory elements have been defined.

Finally, in tumors with breakpoints inside the first exon or intron, both the 5' regulatory elements and the normal c-myc promoters are separated from the gene and transcription starts from a cryptic promotor in the first intron [Cory 1986]. Constitutive expression of the gene in this setting may result from the lack of the 5' regulatory regions, perhaps coupled to enhancer function provided by the Ig region juxtaposed to c-myc by virtue of the translocation. Two additional points of interest arise from this subset of tumors. Firstly it was exclusively found in sBL in this series. Thus it is presumably at least rare in endemic regions. Secondly, it was associated with EBV in approximately half of the cases. It is entirely possible that regulatory elements for c-myc, capable of acting on the cryptic promotor, are present in the intron. The precise position of the breakpoint in relation to such elements may determine the necessity of additional transcriptional enhancing factors provided by EBV.

In summary, our data indicate that the breakpoint location on chromosome 8 in BL may itself be an integral factor in the deregulation of c-myc. Moreover, the likelihood that EBV is associated with some breakpoint locations and not others suggests that in such cases viral genes may provide necessary factors for transformation or maintenance of the transformed state.

REFERENCES

Bentley DL and Groudine M: A block to elongation is largely responsible for decreased transcription of c-myc in differentiated HL-60 cells. Nature 1986; 321: 702-706.

Cory S: Activation of cellular oncogenes in hemopoietic cells by chromosomal translocation. Advances in Cancer Research 1986; 47: 189.

Dalla Favera R, Gellman EP, et. al.: Cloning and characterization of different human sequences related to the onc gene (V-myc) of avian myelocytomatosis virus (MC-29). Proc. Nat. Acad. Sci. USA 1982; 79: 6497-6501.

Dalla Favera R, Martinotti S, et. al.: Translocation and rearrangements of the c-myc oncogene locus in human undifferentiated B-cell lymphomas. Science 1983; 219: 963-967.

Epstein MA, Achong BG, Barr YM: Virus particles in cultured lympho blasts from Burkitt's lymphoma. Lancet 1964; 1:702-703

Hay N, Bishop MJ, Levens D.: Regulatory elements that modulate expression of human c-myc. Genes and Development 1987; 1: 659-671.

Magrath IT: Burkitt's lymphoma: clinical aspects and treatment. In Wolander DW (ed.) Diseases of the lymphatic system. 1983 Springer Verlag, p.103.

Magrath IT: Burkitt's lymphoma as a human tumor model: New concepts in etiology and pathogenesis. In Pediatric Hematology Oncology Reviews. Pochedly C (ed.) Praeger, Westport,1986.pp 1-51

Pellici PG, Knowles DM, et. al.: Chromosomal breakpoints and structural alterations of the c-myc locus differ in endemic and sporadic forms of Burkitt's lymphoma. Proc. Natl. Acad. Sci. USA 1986; 83: 2984-2988.

Zech L, Hoagland U, et al.:Characteristic chromosomal abnormalities in biopsies and lymphoid cell lines from patients with Burkitt's and non-Burkitt's lymphoma. Int. J. Canc. 1976; 17:47-56.

B Cell Proliferation in Follicles, Germinal Centre Formation and the Site of Neoplastic Transformation in Burkitt's Lymphoma

I. C. M. MacLennan, Y. L. Liu, and N. R. Ling

INTRODUCTION

Both the nature of germinal centres and the origin of Burkitt lymphoma have given rise to an exceptional level of divergent speculation. The histological similarities between this tumor and germinal centres was recognised as long ago as 1961 by O'Conor. Mann et al (1976) noted that several American patients with Burkitt lymphoma had tumors which appeared to arise in germinal centres in lymph nodes or Peyer's patches. In a careful review of the histopathological evidence for this relationship Lennert (1978) concluded that on morphological grounds the Burkitt lymphoma cell most closely resembled an intermediate between a true centroblast and a B blast cell. This paper reports the findings of a study of synchronised germinal centre formation in rats. It shows that germinal centre formation is most apparent in primary B cell responses; secondary responses being associated with massive extrafollicular B cell responses. Finally the phenotype of germinal centre cells is compared with that of cells derived from Burkitt lymphoma. It is concluded that there is an extraordinary concordance between the cells found in normal germinal centres and the neoplastic cells of Burkitt lymphoma.

METHODS

Cell Lines

The following cell lines derived from patients with Burkitt lymphoma were studied BL2 (Derived by Dr G Lenoir, International Agency for Research on Cancer, Lyon); Daudi (Klein et al 1968); EB2 (Epstein et al 1965); EB4 (Epstein et al 1966); Kieti, Obagi and WW2BL (kindly provided by Professor A B Rickinson, the Department of Cancer Studies, Birmingham; Namalwa (Klein et al 1972); Raji and Jijoye (Pulvertaft 1965); Y4 (Lowe et al 1985).

Monoclonal Antibodies

The following monoclonal antibodies used in this study were included in the Third International Workshop on Leukocyte Typing (Ling et al 1987) BU-16 (CD9), J5 (CD10), BU-12 (CD19), L27 (CD20), BL-13 (CD21), MHM6 (CD23), BA-1 (CD24), WR-17 (CD37), KiB3, NU-B1 and KB61. OKT10 (CD38) was derived from the US type culture collection. AC2 (CD39) was kindly provided by Dr M Rowe (Rowe et al 1982).

Department of Immunology, University of Birmingham, Birmingham B15 2TJ, U.K.

Current Topics in Microbiology and Immunology, Vol. 141

HIS 14 an antibody which identifies an antigen found on all B cells and HIS 22 which binds to an antigen found on follicular mantle B cells but not marginal zone or germinal centre B cells (Kroese et al 1985) were a kind gift from Dr Kroese and Professor Nieuwenhuis.

Quantitative Analysis of Antigen Expression on Cell Lines

This was carried out using an indirect immunofluorescence technique with a FACS IV (Becton Dickinson). The results were expressed as the 'mean gating value' after correction for background, which gives an approximate measure of antibody bound and hence amount of surface antigen per cell (Ling et al 1987).

Animals

Wistar rats were used which were bred under clean conditions in the animal house of the Department of Immunology, University of Birmingham. These animals were used because of the very low background level of germinal centres in their spleens.

Antigens and Immunization Procedures, Immunohistological Techniques and Quantitative Microscopy Methods

These are described in detail in Liu et al (1988) for the rat studies and by Hoffman-Fezer et al (1976) for human lymph nodes, spleens and tonsils.

RESULTS

Synchronised Germinal Centre Reaction

Experimental design: Conditions were found in which synchronised germinal centre formation could be studied in Wistar rats. In order to do this T cell help was provided by priming rats by intraperitoneal injection of alum-precipitated spider crab hemocyanin (MSH) mixed with 5 x 10^9 chemically killed Bordetella pertussis organisms. One month later the primary B cell response to two haptens 2.4-dinitrophenyl (DNP) and phenyloxazalone (OX) was studied following simultaneous i.v. injection with 50 μg DNP-MSH and 50μg Ox-MSH. The reason for using a two-hapten system was to allow simultaneous identification of Ox-binding cells and DNP-binding cells in tissue sections by immuno-enzymatic methods using Ox-horse radish peroxidase and DNP-alkaline phosphatase (Liu et al 1988). This allowed endogenously-produced hapten-specific surface immunoglobulin to be distinguished with confidence from passively bound antibody (Fig. 3b). The cell cycle status of cells during the response was determined either by injecting 5 mg Bromodeoxyuridine (BrdUrd) 2 hours before spleens were taken for analysis or by administration of BrdUrd at 0.8 mg/ml in rats' drinking water for 48 hrs prior to sacrifice. Cells which had taken up the label were identified by immunohistology (Liu et al 1988).

The follicular response to DNP-MSH and OX-MSH in the spleen: This will be described in relation to the time from immunization. The results are summarised in Figs. 1a, 2, and are illustrated by immunohistology in Figs. 3a,b,c,d and e.

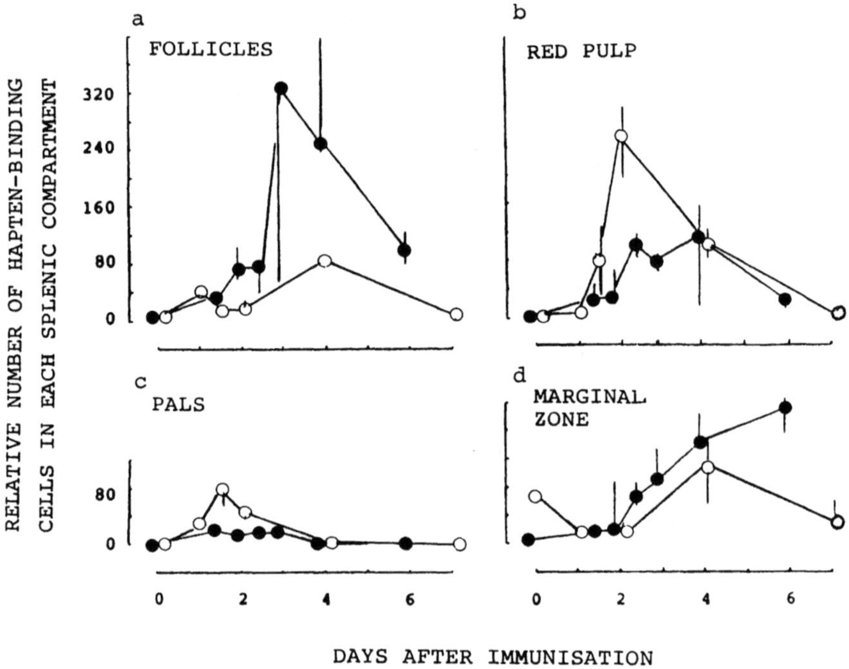

Fig. 1. The relative number of DNP-binding cells in each splenic compartment during primary ●—● and secondary ○—○ responses to DNP-MSH. The relative number of cells in a compartment was determined by dividing the number of hapten-binding cells in a given compartment by the total area of spleen examined using the Weibel point counting technique (1963). Vertical bars represent the range of observations for different rats. The lines are drawn through medians

At the time of immunization with the hapten-protein conjugates: Almost all the follicles in the spleen were primary follicles. i.e. they appeared as a uniform area in sections almost entirely filled with small B cells with the following phenotype: Surface (S) IgM+ve SIgD+ve, HIS 22+ve, HIS 14+ve. Very few cells were labeled with a 2 hr BrdUrd pulse (Fig. 3a).

0-24 hr: Hapten-binding B cells were seen to have accumulated in small numbers in follicles. Most of these cells were not in cell cycle and were still HIS 22+ve.

24-60 hr: Over this period the number of hapten-binding cells in follicles increased dramatically. A high proportion of these cells were in cell cycle as judged by BrdUrd uptake. These cells were seen to have lost the antigen recognised by the monoclonal antibody HIS 22.

72 hr: At this time the B blasts formed a confluent mass in the centre of the follicle. These were sIg +ve in that many bound one or other of the haptens (Fig. 3b) and in common with all B cells were HIS 14+ve but HIS 22-ve. Small HIS 22+ve HIS 14+ve SIgD +ve population of recirculating B cells were now displaced to form a follicular mantle. At 3 days a significant level of apoptosis was seen to be taking place in the B blast population. From this stage a significant number of hapten-binding B cells became apparent in the marginal zone (Fig. 2d)

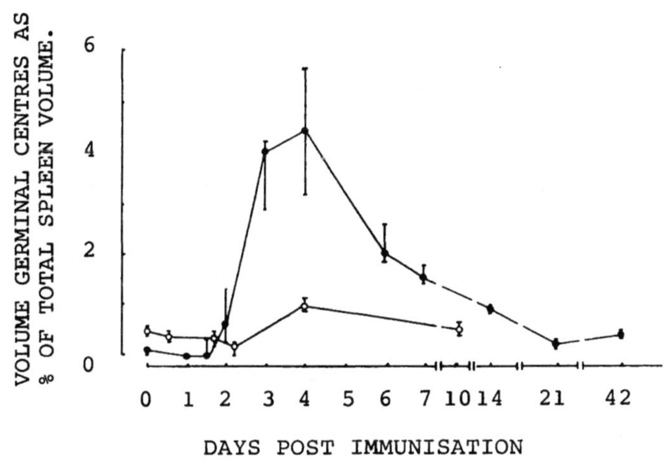

Fig. 2. The volume of HIS 22-ve areas within follicles expressed as a per cent of total spleen volume during primary ●—● and secondary ○—○ responses to DNP-MSH

96 hr: This is the height of the follicular reaction (Figs. 1, 2). The classical appearance of a germinal centre was now apparent (Fig. 3c) SIg-ve HIS 22-ve HIS 14+ve centroblasts had appeared in the part of the follicle adjacent to the T zone. Most of these centroblasts were labeled by a 2 hour pulse of BrdUrd. Surrounding the centroblasts were centrocytes which were SIg+ve, as assessed by hapten-binding, HIS 14+ve and HIS 22-ve. Very few of these cells were labeled by the 2 hr BrdUrd pulse. However, after 48 hr BrdUrd administration in drinking water all of these cells were labeled (Fig. 3c). There was a high rate of apoptosis in the centrocyte-rich area; tingible body macrophages being laden with BrdUrd +ve nuclear debris.

If rats given BrdUrd between 48 and 96 hr were kept for a further 2 days without BrdUrd, the germinal centres were largely unlabeled; suggesting rapid dilution of label by cells in active cell cycle. By contrast, if rats were kept for 3 weeks after labelling between 48 and 96 hr some of the hapten-binding cells in the marginal zone remained labeled (Fig. 3f). This finding implies that they had been in cycle during the BrdUrd administration but over the next 21 days were not in cell cycle (Liu et al 1988).

After 96 hr: By 7 days, the volume of germinal centres had fallen to one third of that at 4 days (Fig. 2). This process of germinal centre involution continued progressively so that very little germinal centre reaction was apparent at 3 weeks from immunization. However, in a proportion of follicles small foci of hapten-binding B blasts, which were labeled by 2 hr BrdUrd pulse, could be identified for several weeks (Fig. 3e). These seem likely to be the source of B cell activation which maintains T cell-dependent responses.

Plasma cells: Between 3 days and 4 days substantial numbers of plasma cells which bound a large amount of one or other of the haptens were found in red pulp of the spleen (Figs 1d & 3e). By 7 days these were no longer apparent. Antibody production by this stage had largely switched to the bone marrow (Chapple et al 1988).

Hapten-binding memory B cells not in cell cycle: During the response to the two hapten-protein conjugates substantial numbers of hapten-binding cells appear in the marginal zone of the spleen (Figs. 1d & 3e). Kinetic studies show that these are derived from cells in active cell cycle. However, once in the marginal zone they are not in cell cycle. These memory B cells appear in the marginal zones from as early as 60 hours and persist in this site for many weeks.

The Main Proliferative Response to Reexposure to DNP-MSH is Outside Follicles

Experimental design: Groups of animals which had been primed with MSH and boosted with both DNP-MSH and OX-MSH, as described in section 1 were re-immunised 6 weeks after the second immunization with 50μg DNP-MSH i.v. The response was followed as described in section 1.1 (Figs. 1, 2). The immunohistological appearance at the height of the reaction is illustrated in Fig. 3d.

Follicular response: The follicular response was considerably less marked than that observed after the rats were first exposed to the hapten-protein conjugates. The peak of the follicular response was again at 4 days but the volume of splenic germinal centres was only one third of that seen at the same stage in the primary anti-hapten response. As in the primary response persistent small foci of proliferating hapten-specific B blasts were seen in a proportion of follicles for several weeks from immunization (Fig. 3e)

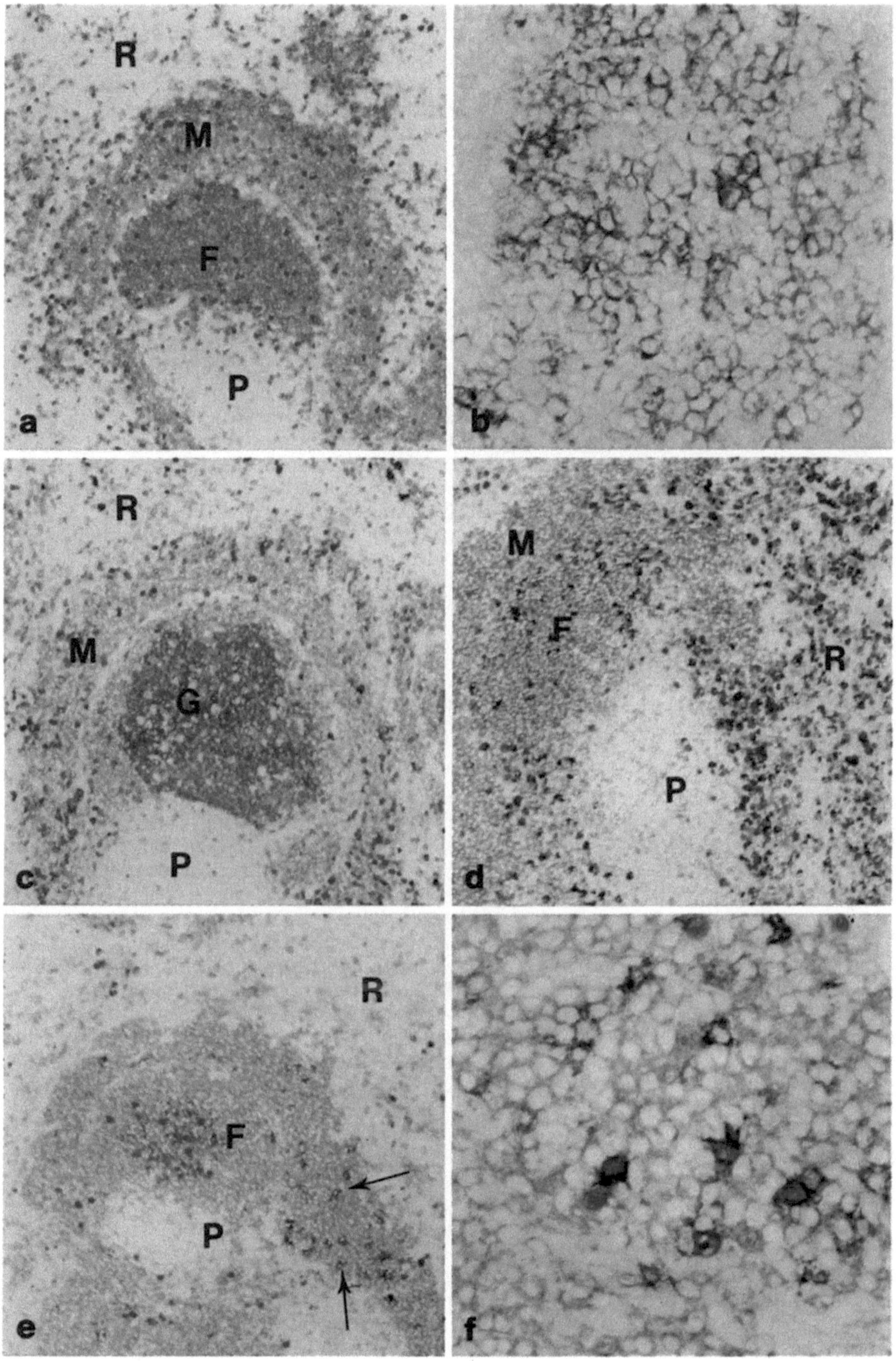
R
M
F
P
a
b
R
M
G
P
c
M
F
R
P
d
R
F
P
e
f

Fig. 3. Photomicrographs of rat spleen sections. M = marginal zone F = follicle G = germinal centre R = red pulp. P = T cell-rich periarteriolar region. Cells with nuclei stained red have taken up BrdUrd in vivo. Cells stained gold have bound the pan B antibody HIS 14. Cells stained blue have bound DNP-alkaline phosphatase. In Fig. 3b cells with a red stained membrane have bound Ox-peroxidase

Fig. 3a. Section of spleen from a rat one month after priming with MSH. This rat was subjected to 48 hrs labeling with BrdUrd before sacrifice, indicating the small proportion of cells (stained red) which had been through cell cycle in this period (microscope magnification x 60)

Fig. 3b. DNP-binding (blue) and Ox-binding (red) blasts in a follicle 72 hrs after immunizing MSH-primed rats with DNP-MSH and Ox-MSH (microscope magnification x 400)

Fig. 3c. A germinal centre (microscope magnification x 60) in a follicle 96 hrs after immunizing MSH-primed rats with DNP-MSH and Ox-MSH. The germinal centre is stained uniform red indicating its cells had taken up BrdUrd over the previous 48 hrs. White areas in the germinal centre are tingible body macrophages. The small B cells have been displaced to the outside of the follicle to form a follicular mantle

Fig. 3d. Extrafollicular and red pulp blast reaction 48 hrs following reimmunization of a rat with DNP-MSH. The rat was primed with MSH, one month later boosted with DNP-MSH and Ox-MSH and reimmunized with DNP-MSH 6 weeks after this. DNP-binding cells are stained blue. Cells which had taken up BrdUrd in a 2 hr pulse are stained red (microscope magnification x 60)

Fig. 3e. The spleen of a rat 6 weeks after immunizing an MSH-primed rat with DNP-MSH and Ox-MSH (microscope magnification x 80). Small numbers of B blasts labeled by a 2hr BrdUrd pulse are seen in the follicle. Some of these are DNP-binding. Note DNP-binding memory B cells in the marginal zone (arrowed). These have not taken up BrdUrd

Fig. 3f. DNP-binding (blue). Marginal zone memory B cells from an MSH-primed rat three weeks after it had been boosted with DNP-MSH and Ox-MSH. This rat had been given BrdUrd between 48 and 96 hrs following boosting. A significant proportion of the marginal zone hapten-binding cells are BrdUrd labeled (red). This indicates they had been in cell cycle during the pulse label but had not been in cell cycle in the 17 days since then. B cells are stained gold with HIS 14. Microscope magnification x 400

Extrafollicular response: This was by far the most impressive component of the response.

0-24 hrs: The immediate effect of administration of DNP-MSH was the selective loss of DNP-binding cells from the marginal zone (Fig. 1d). The Ox-binding cells, however, remained at that site (Liu et al 1988). At the same time DNP-binding cells were seen to accumulate in the outer aspect of the T cell- and interdigitating cell-rich periarteriolar lymphocytic sheath (PALS), but at this stage few of these cells took up BrdUrd in a 2 hr pulse.

24 hrs - 48 hrs: In this period there was massive proliferation of blasts in the outer PALS. Also blasts appeared in marginal zone bridging channels and the red pulp of the spleen (Fig. 3d). The number of blasts in these three sites at 48 hrs was of the same order as that found in follicles at the peak of the primary response. The median proportion of hapten-binding cells which were labeled by the 2 hr BrdUrd pulse in these sites was 48% in the red pulp and 64% in the PALS.

After 48 hrs: The extrafollicular response while being dramatic is short-lived. The red pulp and PALS become virtually devoid of hapten-binding blasts by 4 days, although substantial numbers of non-cycling plasma cells remained in the red pulp at this time. Even these were mostly gone by 7 days when the only residual activity was in follicles (Fig. 1b & c).

Marginal zones: The selective depletion of DNP-binding cells from the marginal zones had completely reversed by 96 hrs. At this stage there were increased numbers of DNP-binding memory cells in that site. As in the case of the primary response, substantial numbers of marginal zone memory cells could be identified for several weeks.

The Phenotype of Cell Derived from Burkitt Lymphoma Closely Resembles that Found in Human Germinal Centres

In this part of the study the phenotype of: (a) Eleven Burkitt lymphoma cell lines and (b) germinal centre cells in frozen sections from five human tonsils and six reactive lymph nodes 3 human spleens. The results of this analysis are shown in Fig. 4. They show a good correlation between the median level of expression of 12 distinct antigens, identified by monoclonal antibodies, on the Burkitt lymphoma lines and the expression of these antigens on centroblasts and centrocytes in normal secondary lymphoid tissue. Some cell lines showed antigen expression which deviates markedly from the median. Gregory et al (1987) have compared the expression of CD39, BLA, CD10 and CD23 on freshly isolated Burkitt lymphoma cells with that of cell lines established from these cells. They found that the freshly explanted tissue uniformly had a level of antigen expression which was close to the medians we describe for Burkitt lymphoma lines in this study. The established cell lines tended to show some deviation from this phenotype. Complete phenotypic analysis was carried out on several different cell types (data not shown): Follicular mantle B cells (Ling et al 1987), splenic marginal zone cells (Ling et al 1987), bone marrow B cells (Chapple et al 1988), CD19+ve SIg-ve bone marrow B cell progenitors (Chapple et al 1988), EBV-lymphoblastoid cell lines and pre-B cell lines (Ling et al 1987). None of these cells have a phenotype which resembles that of Burkitt lymphoma lines or germinal centre cells.

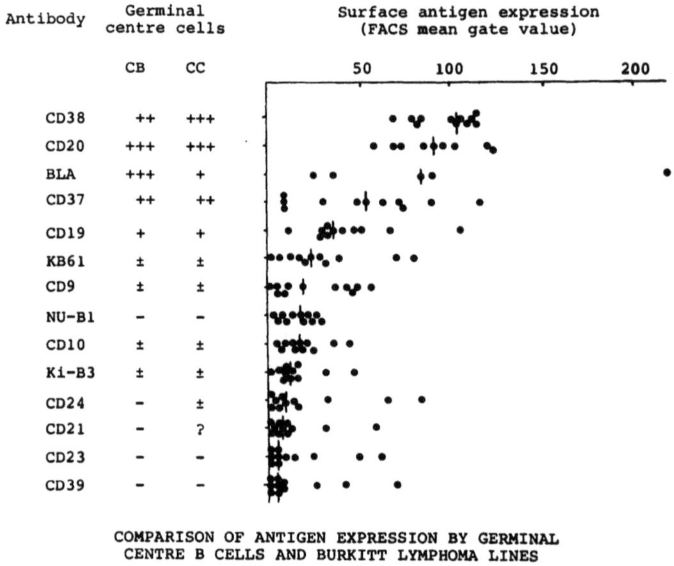

Fig. 4. The relationship between the phenotype of cell lines derived from Burkitt lymphoma and the phenotype of normal germinal centre cells CB = centroblasts, CC = centrocytes. The + and - scale represents intensity of staining by the monoclonal antibodies indicated using indirect immunoperoxidase. The results for the Burkitt lymphoma lines are shown on the right, each point in each row represents the result with a single cell line, vertical lines are drawn through the median points

DISCUSSION

The studies in rats described in this paper indicate that germinal centre formation is more characteristic of primary B cell responses than secondary responses. On the other hand the cells giving rise to centroblasts have probably been in active cell cycle for 2 to 3 days before losing their surface immunoglobulin. The follicular B blasts, which proliferate in follicles in the first 3 days of the primary anti-hapten response described, already have some characteristics of centroblasts i.e. Rapid proliferation, loss of CD22 expression and loss of SIgD. However, unlike centroblasts, but like Burkitt lymphoma cells, they still express surface immunoglobulin as assessed by hapten-binding. Once these B blasts become confluent in the centre of follicles they show a significant rate of apoptosis. In this respect they resemble Burkitt lymphoma cells. It is unclear if the phenotype of human follicular B blasts resembles Burkitt lymphoma cells. Although there are considerable difficulties in studying these cells from or in human tissues there is clearly a need for their phenotype to be determined.

The results of the present study do not conflict with the hypothesis that germinal centres are the site of somatic hypermutation in rearranged Ig V-region genes (MacLennan and Gray 1986). The careful study by Siekevitz et al (1987) has recently shown that extensive proliferation can occur in memory B cells in mice without further somatic mutation in their Ig V-region-genes. If the sites of B cell proliferation in secondary responses in their study were equivalent to those described here, then it is likely that the proliferation was occurring outside germinal centres. A recent report from Dalla-Favera's group (Cesarman et al 1987) has indicated that point mutations characteristically occur in c-myc genes translocated to Ig J segment genes in Burkitt lymphoma. This somatic mutation is seen to occur within the 70 base pairs at the 3' border of the first exon of c-myc. The possibility that this base pair substitution is switched on in centroblasts as part of a process normally affecting Ig V-region genes is intriguing. This substitution was found in endemic, sporadic and AIDS-associated Burkitt lymphoma (Neri et al 1988). The possible role of somatic mutation in translocated c-myc in the neoplastic process remains to be determined.

REFERENCES

Cesarman E, Dalla-Farera R, Bently D, Grondine M (1987) Mutations in the first exon are associated with altered transcription of c-myc in Burkitt Lymphoma. Science 238: 1272

Chapple MR, Johnson GD, MacLennan ICM (1988) IgM and IgD positive lymphocyte subpopulations in normal human bone marrow: a phenotypic study. In preparation.

Epstein MA, Barr YM Achong BG (1965) Studies with Burkitt's lymphoma. Wistar Inst Monogr Philadelphia 4: 69

Epstein MA, Barr YM Achong BG (1966) Preliminary observations on new lymphoblast strains (EB4, EB5) from Burkitt tumours in a British and a Ugandan patient. Brit J Cancer 20: 475

Gregory CD, Rowe M, Rickinson AB, Hardie D, Ling NR (1987) Lymphoblastoid progression of Burkitt's lymphoma cells in vitro. In: McMichael AJ et al (eds) Leucocyte Typing III, Oxford University Press, Oxford, p 352

Hoffman-Fezer G, Rodt H, Eulitz M, Thierfelder S (1976) Immunohistochemical identification of T- and B-lymphocytes delineated by the unlabeled antibody enzyme method. I. Anatomical distribution of theta-positive and Ig-positive cells in lymphoid organs in mice. J Immunol Methods 13: 261

Klein E, Klein G, Nadkarni JS, Nadkarni JJ, Wigzell H, Clifford P (1968) Surface IgM-kappa specificity on a Burkitt lymphoma cell in vitro and in derived culture lines. Cancer Res 28: 1300

Klein G, Dombos L, Gothoskar B (1972) Sensitivity of Epstein-Barr virus (EBV) producer and non-producer human lymphoblastoid cell lines to superinfection with EB virus. Int J Cancer 10: 44

Kroese FGM, Opstelten D, Wubbena AS, Deenen GJ, Alten J, Schwander EH, De Leij L, Nieuwenhuis P (1985) Monoclonal antibodies to Rat B lymphocyte (sub)-populations. Adv Exp Med Biol 186: 81

Lennert K (1978) Malignant lymphomas other than Hodgkin's Disease. Springer-Verlag, Berlin

Ling NR, MacLennan ICM, Mason DY (1987) B cell antigens: new and previously defined clusters. In: McMichael AJ et al (eds), Leucocyte Typing III. Oxford University Press, Oxford, p 302

Liu Y-J, Oldfield S, MacLennan ICM (1988) Memory B cells in T cell-dependent antibody responses colonize the splenic marginal zones. Eur J Immunol 18: in press

Lowe JA, Ling NR, Forrester JA, Cumber AJ, Rose WCJ (1985) Variants of lymphoid lines produced with ricin-A chain-monoclonal antibody conjugates. J Immunol Methods 76: 93

MacLennan ICM, Gray D (1986) Antigen-driven selection of virgin and memory B cells. Immunol Rev 91: 61

Mann RB, Jaffe ES, Braylan RC, Nanba K, Frank MM, Ziegler JL, Berard CW (1976) Non-endemic Burkitt's lymphoma. A new B cell tumour related to germinal centres. New Engl J Med 295: 685

Neri A, Barriga F, Knowles DM, Magrath IT, Dalla-Favera R (1988) Different regions of the immunoglobulin heavy chain locus are involved in chromosomal translocations in distinct pathogenetic forms of Burkitt lymphoma. Proc. Natl. Acad. Sci. In press

O'Conor GT (1961) Malignant lymphoma in African children. II. A pathological entity. Cancer (Philad.) 14: 270

Pulvertaft RJV (1965) A study of malignant tumours in Nigeria by short-term tissue culture. J Clin Path 18: 261

Rowe M, Hildreth JEK, Rickinson AB, Epstein MA (1982) Monoclonal antibodies to Epstein-Barr virus-induced transformation-associated cell surface antigens: binding patterns and effect upon virus-specific T-cell cytotoxicity. Int J Cancer 29: 373

Siekevitz M, Kocks C, Rajewsky K, Dildrop R (1987) Analysis of somatic mutation and class switching in naive and memory B cells generating adoptive primary and secondary responses. Cell 48: 757

Weibel ER (1963) Principles and methods for the morphometric study of the lung and other organs. Lab Invest 12:131

Altered Growth Phenotype of a Burkitt's Lymphoma Line Following the Introduction and Stable Expression of the EBNA 2A Gene

J. GORDON[1], M.J. MILLSUM[1], M. FINNEY[1], J.A. CAIRNS[1], G.R. GUY[2], C.D. GREGORY[3], S.D. ABBOT[3], A.B. RICKINSON[3], F. WANG[4], and E. KIEFF[4]

SUMMARY

Using recombinant retrovirus-mediated transfer, the coding region for the Epstein-Barr virus-encoded EBNA 2A gene was selectively introduced and stably expressed in the Louckes Burkitt lymphoma cell line. Transfected cells displayed the following altered growth characteristics when compared with the parental line: (i) maintenance of proliferation at reduced serum concentrations, (ii) improved seeding efficiency, (iii) production of a unique growth-enhancing factor, (iv) increased sensitivity to the unique factor. These growth-related changes were accompanied by the appearance of large amounts of soluble CD23 fragments in medium conditioned by Louckes cells expressing EBNA 2A. Tumor-promoting phorbol esters were found to induce similar phenotypic change in the Louckes parental line as seen on the introduction of the EBNA 2A gene. Possible mechanisms by which EBNA 2A provides a growth advantage to B lymphoma cells are discussed.

INTRODUCTION

The almost invariable association between the Epstein-Barr virus (EBV) and Burkitt lymphoma (BL) within endemic areas has led to the nomination of the former as a causative agent for the latter (Epstein and Achong, 1983). Possibly reflecting a potential role in oncogenic promotion, EBV can transform resting B lymphocytes into permanently growing, but non-tumorigenic lymphoblastoid cell lines (Gordon et al, 1984). A contribution from the EBV nuclear antigen EBNA 2 to the immortalisation process has been indicated from the observation that mutant virus containing a deletion in the coding region for this protein is non-transforming (Gordon et al, 1986) while B-lymphoblastoid cells carrying the EBNA 2A gene display improved growth characteristics over those expressing the EBNA 2B allele(Rickinson, Young and Rowe, 1987). Furthermore, the specific introduction and expression of the EBNA 2A gene sequence into rodent fibroblasts has been shown to reduce the serum requirements of these cells (Dambaugh et al, 1986). Recently, it was found that, following transfection by the EBNA 2A gene on a retrovirus promoter, the EBV-genome negative Louckes BL cells dramatically and, apparently, specifically up-regulated their expression of the 45 kDa membrane-bound CD23 molecule (Wang et al, 1987). When released by proteolytic cleavage to its 25-35 kDa soluble form, this glycoprotein has identity with a B-cell-derived

Departments of [1]Immunology, [2]Biochemistry and [3]Cancer Studies, University of Birmingham, Birmingham, U.K. and [4] 12th Floor, Thorn Building, Brigham Hospital, MA, USA.

Current Topics in Microbiology and Immunology, Vol. 141

B-cell growth factor (Swendeman and Thorley-Lawson, 1987). In this study, we have explored the possibility that the introduction of the EBNA 2A gene into Louckes cells might modify their growth characteristics and have explored a contribution from the CD23 molecule in this regard.

MATERIALS AND METHODS

Retrovirus Infection and Selection of Transfected Cells

The construction of recombinant murine leukemia virus-derived expression vectors containing EBV DNA and their introduction into EBV-genome negative Louckes cells using Polybrene have been described in detail elsewhere (Wang et al, 1987). The presence of the neomycin phosphotransferase gene in this vector system allowed selection of successfully transfected cells by virtue of their resistance to the neomycin analog G418. Transfected cells were maintained in medium containing G418 up until one week prior to the experiments described when they were then washed and transferred into regular culture medium (RPMI 1640) supplemented with 10% fetal calf serum (FCS). Louckes cells transfected with retrovirus constructs containing the EBNA 2A gene (but not other gene sequences) were found to maintain stable expression over periods of 2-3 months following their removal from the selection medium (unpublished observations).

Growth Characteristics

Transfected and parental Louckes cells were grown to logarithmic phase in RPMI 1640 containing 10% FCS, washed extensively, reseeded at a concentration of 3 x 10^5 cells per ml and then, after 24hr, washed again before replating in 96 micro-well plates at cell numbers and FCS concentrations indicated in the text. Cells grown for 72hr received a [^{3}H]dThd pulse (1μCi) for the final 16hr and the amount of radioactivity incorporated into DNA was determined in triplicate. All experiments described were performed a minimum of four times with the same qualitative outcome and representative results are shown.

Measurements of the CD23 Molecule

Soluble CD23 fragments were assayed in culture supernatants using a recently described ELISA-based system (Cairns et al, 1988). CD23 expression at cell surfaces was determined by indirect immunofluorescence using MHM6 as the first layer and a fluoresceinated sheep anti-mouse immunoglobulin antibody at the second step. Intensity of stain was measured on a FACS IV flow cytometer (Becton-Dickinson, CA). Other surface antigens were similarly assessed using the following monoclonal antibodies at the first step: Jo5 (University of Birmingham), specificity for polymorphic determinants of MHC Class II DR molecules; AC2 (M. Rowe, Birmingham), reactivity within the CD39 cluster; OKT9 (Ortho-Diagnostics), specificity for the transferrin receptor.

RESULTS

EBNA 2A Expression Results in Reduced Serum Requirement and Improved Plating Efficiency

Louckes cells expressing the EBNA 2A gene were compared with non-transfected counterparts for their ability to establish growth over a range of cell densities in varying FCS concentrations. It is clear from Fig. 1 that cells containing the EBNA 2A gene demonstrated a significantly reduced serum requirement by comparison with the parental cell line. Furthermore, high proliferative indices were achieved from lower initial seeding numbers as a consequence of introducing the EBNA 2A gene (Fig. 1).

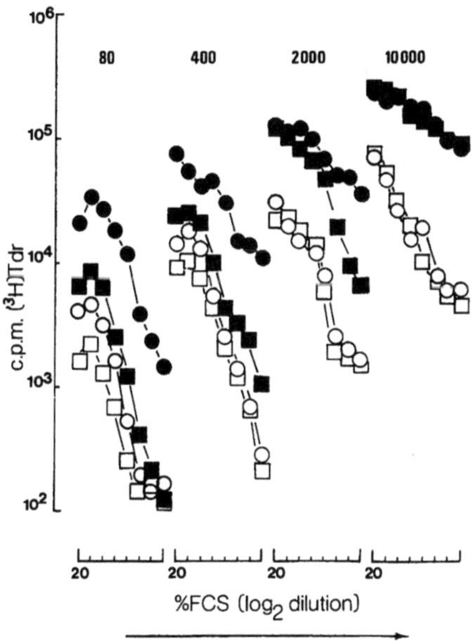

Fig. 1 Growth Phenotype of Louckes cells. Cells were plated at numbers indicated (top) and cultured in serum concentrations shown for 72hr, receiving a [^{3}H]dThd. pulse for the final 16hr. Louckes Parent cells (○,□), Louckes EBNA 2A cells (●,■) cultured in round - (○,●) or flat-bottom wells (□,■).

These alterations in growth phenotype were most pronounced under conditions which encouraged cell intimacy such as seen when cells were seeded into round-bottomed wells or when cultures were initiated at relatively high seeding densities. In other experiments (not detailed), it was shown that the altered growth phenotype was specific to Louckes cells expressing EBNA 2A inasmuch as cells transfected with retrovirus constructs containing either EBNA 1, EBNA 3 or the vector alone displayed growth characteristics indistinguishable from the parental line.

<u>EBNA 2A Expression Results in Production and Response to Unique Growth-Enhancing Factor</u>

In order to explore a possible contribution from a soluble factor to the improved growth characteristics observed on introducing the EBNA 2A gene, parental and transfected cells were used to condition medium for 24hr prior to its harvesting and assay for growth-enhancing activity. First, it can be seen from Fig. 2 that both the parental and EBNA 2A-containing cell lines produced an autostimulatory component that was equally effective at supporting the growth of both cell-types seeded at limiting numbers in low serum.

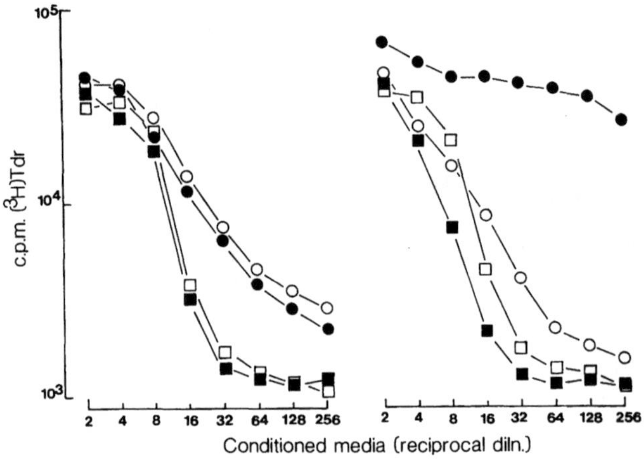

Fig. 2 Influence of conditioned medium on cell growth. Cells were plated at 2000 per well in 0.5% FCS and cultured as for Fig. 1 Left hand panel, Louckes Parent cells; right hand panel, Louckes EBNA 2A cells. The influence of medium conditioned by 3 x 10^5 Louckes Parent cells (○,□) or Louckes EBNA 2A cells (●,■) over a 24hr period is shown on cells cultured in round- (○,●) or flat-bottomed wells (□,■).

However, Louckes cells expressing EBNA 2A were found to condition their medium with an additional activity which did not follow the dilution curve for the common autostimulatory component and which, furthermore, selectively augmented the proliferation of the autologous cell line (Fig. 2). Again, such change was most evident when cells were cultured under conditions which encouraged cell intimacy.

Using a recently developed ELISA-based assay (Cairns et al, 1988), the conditioned media from the two lines were measured for their content of soluble CD23 molecules. In four experiments, medium conditioned by EBNA 2A-expressing cells contained 75-190ng per ml of CD23 molecules (mean = 142ng per ml) whereas supernatants from parental cells, invariably, carried less than 10ng per ml of CD23 fragments. Thus, it is possible that the unique factor present in medium conditoned by EBNA 2A-containing cells had identity with soluble CD23.

EBNA 2A Expression Results in Responsivity to Growth Promotion via a CD23 Antibody

Certain CD23 antibodies have been found to trigger cell-cycle progression in tonsillar B lymphocytes which have been activated with the tumor-promoting phorbol myristic acetate (PMA) (Gordon et al, 1987). In the present study, it was shown that Louckes cells containing the EBNA 2A gene, but not the parental clone, were provided with a growth stimulus upon the introduction into culture of the stimulating CD23 antibody EBVCS4 (Table 1). This demonstrates that not only does EBNA 2A expression result in a dramatic up-regulation of CD23 but also that the induced CD23 molecule is functionally active. Interestingly, parental and transfected cells were equally responsive to a commercial growth factor preparation which contains a cocktail of growth-promoting activities (Table 1).

Table 1. Triggering of Louckes Cells via CD23[a]

	Control	EBVCS4(20μg/ml)	EBVCS4(2μg/ml)	BCGF[b]
Louckes Parent	295	255	329	27,346
Louckes EBNA 2A	341	21,846	6,091	32,767

a Cells plated at 2000 per flat-bottom microwell and grown for 72hr in medium containing 0.5% FCS. Results are expressed as incorporation of [^{3}H] dThd. (c.p.m.) following a 16hr pulse at termination of culture.

b EBVCS4 antibody kindly provided by B. Sugden (McArdle Institute, WI); BCGF is a commercial preparation purchased from Cellular Products Inc. (Buffalo, N.Y.).

PMA Partially Mimics the EBNA 2A Gene Product

In an attempt to gain insight into the biochemical signalling pathways which might be used by the EBNA 2A gene product, Louckes parental cells were treated with PMA, an activator of Protein Kinase-C, and then examined for any alteration of phenotype. It can be seen from Fig.3 that CD23 expression was significantly up-regulated by PMA treatment. This was specific inasmuch as another activation marker CD39, described by the monoclonal antibody AC2, was unaffected and the level of transferrin receptors remained unchanged. However, MHC Class II expression was invariably increased on PMA-treated Louckes cells (Fig.3). It was subsequently found that introducing the EBNA 2A gene into Louckes cells also resulted in a similar up-regulation of Class II expression (data not shown). Although identical antigenic changes were occurring on PMA treatment as with EBNA 2A expression, the former resulted in an inhibition of growth by contrast with the improvement seen with the latter (experiments not detailed). The augmentation of CD23 expression by PMA was tightly dependent on the density of Louckes cells in culture. It can be seen from Fig. 3 that there was a close linear correlation (r=0.9955) between

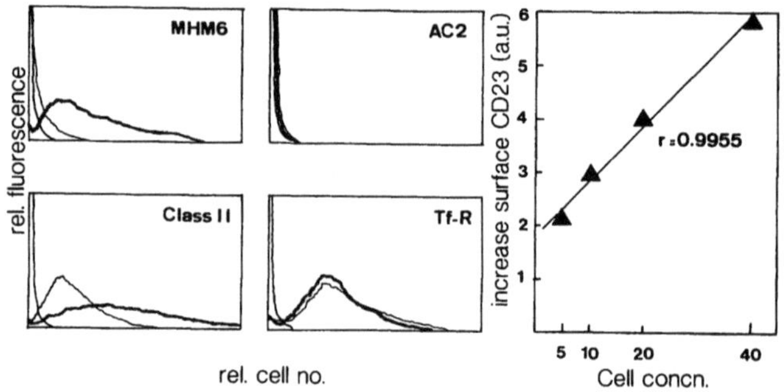

Fig. 3 Influence of PMA on surface phenotype. Louckes Parent cells were cultured for 24hr with PMA (1ng/ml) and assessed for surface antigen expression on a FACS IV as indicated: heavy line, PMA-induced; medium line, expression on control cells; faint line, control (background) staining. To the right is shown the increase in CD23 expression (MHM6) in relation to the number of cells (x 10^4) cultured with PMA.

the number of cells seeded in PMA-containing medium and the level of induced CD23 24hr later. This raises the possibility that PMA-mediated augmentation of CD23 expression may have occurred indirectly via an induced soluble factor or cell surface component.

ACKNOWLEDGEMENTS

This work was supported with funding from the M.R.C. (U.K.), the Leukaemia Research Fund and the U.S. National Institutes of Health.

DISCUSSION

The results outlined in this paper demonstrate that the specific introduction and stable expression of the EBNA 2A gene confers on EBV-genome negative BL cells a clear growth advantage which is mediated, at least in part, through the induced CD23 molecule. These observations would suggest an involvment of EBNA 2 in the pathogenesis of endemic BL. It is then initially surprising that both EBNA 2 and the EBV-encoded latent membrane protein (LMP) should both be down-regulated in freshly-isolated BL cells (Rowe et al, 1987). However, this suppressed phenotype has now been related to a correspondingly diminished expression of adhesion molecules which serve to promote T-cell mediated cytolysis (Gregory et al, 1988). These observations, taken together, would argue in favour of the scenario whereby EBV promotes oncogenic change by bestowing autonomy of growth on proliferating clones and thereby increasing the risk of chromosomal error rather than one where the virus was simply serving as a growth-enhancing carrier to an already malignant cell. It would make little sense that endemic BL develops first outside the influence of EBV only to acquire an association with it later, one which would provide the potential means for tumor elimination (through up-regulation of adhesion molecules) in the absence of any discernible benefit. Thus, the transposition of the myc-gene to the immunoglobulin locus can best be seen as REPLACING the growth autonomy initially conferred on cells by EBNA 2 (and possibly LMP) through CD23 induction thereby allowing the subsequent down-regulation of those elements and escape from immune surveillance.

A novel observation of the present study was the indication that EBNA 2A might be providing cells with the means to respond to the soluble CD23 molecule in addition to bringing about a dramatically increased production of this factor. The basis of this selective response is currently being investigated. Whatever the mechanism responsible, it is of interest that a single transforming gene appears to provide both the components necessary for establishing an autocrine loop.

Finally, the ability of PMA to induce in Louckes cells all of the antigenic changes seen on introducing the EBNA 2A gene suggests that these two transforming agents might be using a common biochemical signalling pathway. In this regard, it is of interest that, albeit a relatively crude preparation of EBNA, has been described as possessing cAMP-independent Protein Kinase activity (Kamata et al 1981). This possibility now warrants further investigation.

REFERENCES

Cairns J A, Flores-Romo L, Millsum M J, Guy G R, Ledbetter J A, Gillis S, Gordon J (1988) Soluble CD23 is released by B lymphocytes cycling in response to interleukin 4 and anti-Bp50 (CDw40) Eur. J. Immunol. (In press).

Dambaugh T, Wang F, Hennessy K, Woodland E, Rickinson A, Kieff E (1986) Expression of Epstein-Barr virus nuclear protein 2 in rodent cells. J. Virol. 59:453-462.

Gordon J, Ley S C, Melamed M, English L S, Hughes-Jones N C (1984) Immortalised B lymphocytres produce B cell growth factor. Nature (London) 310: 145-147.

Gordon J, Walker L, Guy G, Brown G, Rowe M, Rickinson A (1986) Control of human B-lymphocyte replication II. Transforming Epstein-Barr virus exploits three distinct viral signals to undermine three separate control points in B-cell growth. Immunology 58: 591-595.

Gordon J, Webb A J, Guy G R, Walker L (1987) Triggering of B lymphocytes through CD23: epitope mapping and studies using antibody derivatives indicate and allosteric mechanism of signalling. Immunology 60: 517-521.

Gregory C D, Murray R J, Edwards C J, Rickinson A B (1988) Down-regulation of cell adhesion molecules LFA-3 and ICAM-1 in Epstein-Barr virus-positive Burkitt's lymphoma underlines tumour cell escape from virus-specific T cell surveillance. J. Exp. Med. (In press).

Kamata J, Takaki K, Hinuma T, Watanabe Y (1981) Protein kinase activity associated with Epstein-Barr virus-determined nuclear antigen. Virology. 113: 512-520.

Rickinson A B, Young L S, Rowe M (1987) Influence of the Epstein-Barr virus nuclear antigen EBNA 2 on the growth phenotype of virus-transformed B cells. J. Virol. 61:1310-1317.

Rowe M, Rowe D T, Gregory C D, Young L S, Farrell P J, Rupani H, Rickinson A B (1987) Differences in B cell growth phenotype reflect novel patterns of Epstein-Barr virus latent gene expression in Burkitts lymphoma cells. EMBO J. 6:2743-2751.

Swendeman S and Thorley-Lawson D A (1987) The activation antigen BLAST-2, when shed, is an autocrine BCGF for normal and transformed B cells. EMBO J. 6:1637-1642.

Wang F, Gregory C D, Rowe M, Rickinson A B, Wang D, Birkenbach M, Kikutani H, Kishimoto T, Kieff E (1987) Epstein-Barr virus nuclear antigen 2 specifically induces expression of the B-cell activation antigen CD23. Proc.Natl.Acad. Sci. USA 84:3452-3456.

Soluble CD23/BLAST-2 (S-CD23/Blast-2) and Its Role in B Cell Proliferation

S. L. SWENDEMAN and D. THORLEY-LAWSON

SUMMARY

The two chains (25 and 12 Kd) of the S-CD23/Blast-2-complex both have activity as autocrine growth factors (AGF) for EBV-LCL however, the bulk of the activity is found in the 25 Kd chain. Several lines of evidence indicate that the 25 and 12 Kd chains are complexed and only the 25 Kd chain is related to the membrane bound CD23 molecule. Lastly, both EBV-LCL and BL use an AGF system, however, the BL factor lacks CD23 determinants and does not appear to require cell-to-cell contact whereas the EBV-LCL-AGF has CD23 determinants and the cells require cell-cell contact for growth. A model to account for these observations is presented.

INTRODUCTION

CD23 (Blast-2) is a 47,000 mol. wt. glycoprotein found on the surface of B cells activated by mitogens, anti-receptor antibody, antigens or by infection with the human herpesvirus Epstein-Barr Virus (EBV) (1). Using EBV infected B cells it was shown previously that CD23 expression was required for transformation (2) and that CD23 is rapidly shed into the culture supernatant. This shed material (S-CD23/Blast-2) could be purified by immunoaffinity chromatography. The purified material consisted of two polypeptides of 25,000 and 12,000 mol. wt. Functionally, the S-CD23 acted as an autocrine growth factor (AGF) for EBV lymphoblastoid cells (EBV-LCL) and normal B blasts and had weak activity as a comitogen with PHA stimulated T cells (IL-1-like) (3). The molecule had no effect on resting cells. Lastly it was shown that S-CD23 was the major component responsible for the AGF activity detected in the supernatants of EBV-LCL.

In the present study we have addressed the questions as to the origins of the 25 Kd and 12 Kd polypeptides, which polypeptide expresses the biological activity and propose a possible model to explain the role of CD23 and S-CD23 in B cell proliferation.

MATERIALS AND METHODS

Maintenance of Cell Line, Preparation of Conditioned Media, Purification of S-CD23/Blast 2 and the Autocrine Growth Factor (AGF) Assay

These were performed as described previously (3,4).

Department of Pathology, Tufts University School of Medicine, 136 Harrison Avenue, Boston, MA 02111 USA

Current Topics in Microbiology and Immunology, Vol. 141

Biochemical and Immunochemical Analysis

IEF was performed in glass tubes (.50 mm x 6 cm) in 5% polyacrylamide. Denaturing gels and samples contained 8M Urea and 2% NP-40, non-denaturing gels and samples 10% glycerol. Gels were prefocused for 30 minutes at 300 volts and then run for 2 hours at 600 volts. The noodles were then extruded, soaked for 30 min. in SDS-PAGE sample buffer and overlaid onto a 12% SDS-PAGE gel. For preparative IEF, 200 mg of S-CD23/Blast were applied. The gel noodle was sliced into 4 mm pieces which were soaked in 1 ml of H_2O for 6 hours to remove excess Urea and ampholines. The pH of the water was measured and the water replaced with 1 ml of serum free media and incubated for 48 hours at 4°C. Partial cleavage of protein was performed as described by Cleveland (1977) (5). Pulse chase analysis, immunoprecipitation, Western blotting and radiolabeling were as described previously (6).

RESULTS

The 25 Kd but not 12 Kd Chain of S-CD23/Blast-2 is Derived from CD23

A variety of immunochemical and biochemical approaches were taken to characterize the origin of the 25 Kd and 12 Kd chain of S-CD23/Blast-2. The results are summarized in Table 1. When immunoprecipitation was performed with a panel of anti-CD23 monoclonal antibodies [EBVCS #1, 2, 4, 5 (7) and MHM6 (8)] both chains were always coprecipitated, however, in Western blot analysis the antibody MHM6 only bound to the 25 Kd chain. This suggests that the 25 Kd chain is derived from CD23 and the 12 Kd chain coprecipitates due to association with the 25 Kd chain. Pulse chase analysis confirms that the 25 Kd chain is derived from CD23, however, it was not possible to chase the label in the 25 Kd chain into the 12 Kd chain. Lastly, partial cleavage maps demonstrated that CD23 and the 25 Kd protein shared peptides in common, but the 25 Kd and 12 Kd chain did not. Taken together this data suggests that the membrane bound form of CD23 gives rise to the 25 Kd chain, and the 12 Kd chain is a separate gene product that associates with the 25 Kd chain.

To further demonstrate this point S-CD23/Blast-2 was subjected to isoelectric focusing in the presence and absence of Urea followed by second dimension SDS-PAGE analysis. As may be seen in Fig. 1a in the presence of Urea the 25 Kd chain focuses as a series of spots around pH 5.4 whereas the 12 Kd chain focuses at pH 4.3. In the absence of Urea, however, half of the 25 Kd molecules and all of the 12 Kd molecules focus at an intermediate pH of 4.9 the remainder of the 25 Kd molecule focusing at pH 5.4. This result indicates that under non-denaturing conditions the 25 Kd and 12 Kd chain are associated and comigrate.

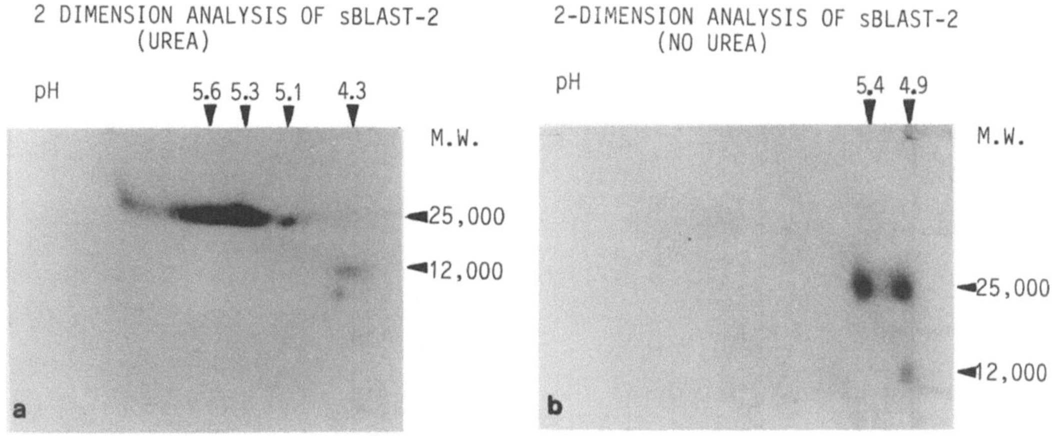

TABLE 1: Relationship of the 25 Kd and 12 Kd Chains of S-CD23 to Membrane CD23 - 12 Kd is Distinct from 25 Kd

	Relationship to CD23	
	25 Kd[d]	12 Kd
Western Blot[a]	+	-
Pulse Chase[b]	+	-
Limited Proteolysis[c]	+	-
Immunoprecipitation[a]	+	+

a Western blot and immunoprecipitation studies were performed with CD23 monoclonal antibodies. The antibodies only reacted with the 25 Kd chain on Western blot although they precipitated both chain.

b Radiolabeled CD23 was shed and recovered as 25 Kd chain no label was detected at 12 Kd.

c Limited proteolysis revealed identifical polypeptides between CD23 and 25 Kd but none between 25 Kd and 12 Kd.

d The 25 Kd chain has been shown to be derived from CD23 since its partial amino-acid sequence matches that of a CD23 cDNA (9,10).

Isolated 25 Kd Chain has Biological Activity

The 25 and 12 Kd chains were separated on a preparative scale using IEF. The gel was then sliced and individual slices assayed for biological activity in the autocrine growth factor assay. As shown in Fig. 2 the bulk of the activity recovered from the gel migrated with the 25 Kd chain although a small but significant amount of activity was also detected with the 12 Kd chain.

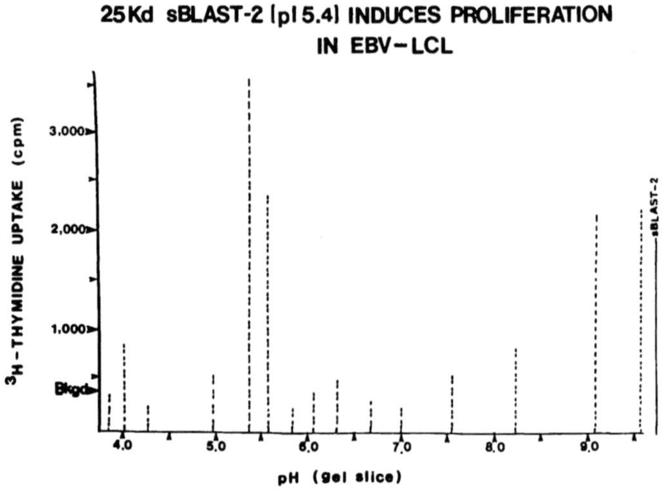

Figure 2

Table 2 summarizes three experiments with different preparations of S-CD23/Blast-2 and demonstrates that the same observations were made in each case.

TABLE 2: Both chains of the S-Blast-2 complex have autocrine growth factor activity.

		Proliferation[a]					
		1		2		3	
pI of sample	M.W. of chain	cpm(SD)[b]	SI[c]	cpm(SD)	SI	cpm(SD)	SI
4.1 - 4.3	12 Kd	708(1.2)	2.5	463(1.4)	4.8	1039(1.2)	6.1
5.2 - 5.5	25 Kd	3453(1.3)	12.5	912(1.6)	9.4	1224(1.3)	7.2
Background	(gel slice from elsewhere)	278(1.2)	1.0	97(1.2)	1.0	170(1.4)	1.0
S-BLAST-2[d]		1870(1.3)	6.7	406(1.2)	4.1	509(1.2)	3.0

a - results of three separate experiments; b - SD = geometric standard deviation; c - SI = cpm in experimental/cpm in background; d - unseparated S-BLAST-2 equivalent to the amount loaded on the IEF.
Note: Weak but significant proliferation is seen with the 12 Kd protein whereas stronger proliferation is seen with the 25 Kd protein.

A Burkitt's Lymphoma Responds to and Produces Growth Factors

We have been able to demonstrate previously that both EBV-LCL and Burkitt's lymphoma (BL) lines respond to purified S-CD23/Blast-2 and that most of the AGF activity from EBV-LCL conditioned media (EBV-LCL-CM) can be removed by anti-CD23 monoclonal antibodies (3). As shown in Table 3, media conditioned by the BL^- line RAMOS also contain growth factors that stimulate both an EBV-LCL and the BL lines. However, in this case the AGF activity could not be removed by anti-CD23 antibodies (not shown).

TABLE 3: Autocrine growth factor from EBV^- BL

Source of and % of Conditioned Medium		Proliferation			
		ER[a]		RAMOS[b]	
		cpm(SD)	SI	cpm(SD)	SI
ER[c]	50	552 (1.1)	12	895 (1.3)	17
	37.5	228 (1.1)	5	689 (1.3)	9
	25	---		512 (1.2)	9
	12.5	96 (1.8)	2	192 (1.2)	4
	--	47 (1.6)	1	54 (2)	1
RAMOS	50	870 (1.2)	19	613 (1.1)	11
	37.5	380 (1.1)	8	333 (1.6)	6
	25	246 (1.2)	5	277 (1.2)	5
	12.5	119 (1.2)	3	152 (1.8)	3
	--	47 (1.6)	1	56 (2)	1

a - ER is an EBV LCL

b - RAMOS is an EBV^- BL

c - similar results are obtained with purified S-BLAST-2

DISCUSSION

In this study we have demonstrated that the 25 Kd chain of S-CD23/Blast-2 is derived from the membrane form of CD23. This result was expected as amino acid sequence from the 25 Kd chain has been used to clone a cDNA for CD23 (9,10). More interestingly the isolated 25 Kd chain possesses biological activity in the AGF assay on EBV-LCL. The 25 Kd chain is non-covalently associated with a 12 Kd chain that appears unrelated to CD23. The isolated 12 Kd chain also has weak but significant AGF activity. Whether this represents inherently weak activity or is a consequence of the denaturing conditions used during the separation procedure is unclear. In Table 4 is presented a summary of the biological activities we have found associated with S-CD23/Blast-2. We have also assayed isolated 25 Kd and 12 Kd protein in the T cell proliferation assay and found a similar distribution of activity (high with 25 Kd, low with 12 Kd) to the AGF activity.

TABLE 4: Biological Activities of S-CD23/Blast-2

Autocrine Growth Factor	
EBV-LCL[a]	++
Normal B Lymphoblasts	++
Resting Lymphocytes	-
IL-1 Like Activity:	
Comitogen for PHA Activated T Cells[a]	+
Proliferation of D10 Cells	+/-
Proliferation of Fibroblasts	-

a Also tested on isolated 25 Kd and 12 Kd chain
++ Positive activity
+ Weak but significant
+/- Weak, borderline significance
- No activity

Although the BL-RAMOS also produces and responds to AGF, the activity in the BL-CM is distinct from that in EBV-LCL-CM since it cannot be removed with anti-CD23 antibodies. A summary of the properties of the EBV-LCL and BL systems is presented in Table 5.

TABLE 5: Properties of the Growth of EBV-LCL and the BL-RAMOS

		EBV-LCL	BL
Growth Factor Activity on:	LCL	+	+
	BL	+	+
Removed by Anti-CD23		+	-
CD23 on Cells		+++++	+/-
Cell Contact		+	-
Cloning Efficiency		1-2%	>80%

It is worthy of note that the BL cells express very low levels of CD23, grow as single cells and clone with high efficiency. On the other hand EBV-LCL grow as clumps and clone very inefficiently. Thus, the BL appear to be using a soluble growth factor whereas the EBV-LCL require cell-to-cell contact to generate growth stimulation. This has led us to propose a model for the mode of action of CD23 in EBV-LCL growth. This model, shown diagrammatically in Fig. 3, suggests that growth stimulation is generated by the interaction of CD23 with a ligand/ receptor on an adjacent cell. This interaction is relatively weak since EBV-LCL require the interaction of the adhesion molecules I-CAM and LFA-1 (12) in order to remain aggregated. Once a signal is transduced the two molecules are shed as a complex with the 25 Kd chain derived from CD23 and the 12 Kd chain derived from the other ligand/ receptor. In normal B cells this would result in a down regulation of proliferation since no further molecules are synthesized. In the case of EBV-LCL both receptor/ligands are constitutively synthesized resulting in repeated growth stimulation, shedding and accumulation in the culture supernatant. This model leads to the conclusion that the AGF activity from EBV-LCL is really a tissue culture artifact due to the accumulation of shed complexes, some of which dissociates and can re-bind to the cell surface. Thus, when we purify material on a CD23 antibody column we would isolate dissociated 25 Kd and 25 Kd complexed with 12 Kd, leaving dissociated 12 Kd behind. This is, indeed, the result we see in Fig. 1b. Furthermore, the model predicts that the 25 Kd plus 12 Kd complex will be inert biologically since the receptor/ligand site on each molecule is occupied.

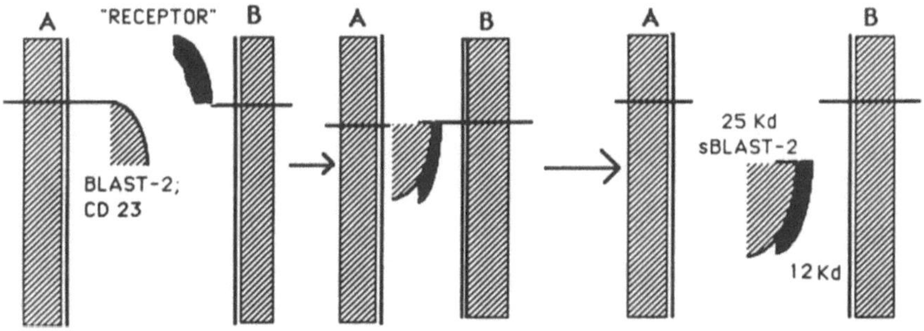

Figure 3.

REFERENCES

1. Thorley-Lawson DA, Nadler LM, Bhan AK, Schooley RT (1985) Blast-2 (EBVCS). An early cell surface marker of human B cell activation, is superinduced by Epstein Barr Virus. J Immunol 136:3007-3012.
2. Thorley-Lawson DA, Mann KP (1985) The early events in EBV infection provide a model for B cell activation. J Exp Med 162:45-49.
3. Swendeman S, Thorley-Lawson DA (1987) The activation antigen Blast-2, when shed, is an autocrine BCGF for normal and transformed B cells. The EMBO J 6:1637-1642.
4. Gordon J, Ley SC, Melamed MD, Aman P, Hughes-Jones NC (1984) Soluble factor requirements for the autostimulatory growth of B lymphocytes immortalized by Epstein-Barr virus. J Exp Med 159:1554-1559.
5. Cleveland DW, Fisher SG, Kirschner MW, Laemmli UK (1977) Peptide mapping by limited proteolysis in sodium dodecyl sulfate and analysis by gel electrophoresis. J Biol Chem 252:1102-1106.
6. Thorley-Lawson DA, Swendeman SL, Edson C (1986) Biochemical analysis suggests distinct functional roles for the BLAST-1 and BLAST-2 antigens. J Immunol 136:1745-1751.
7. Kinter C, Sugden B (1981) Identification of antigenic determinants unique to the surfaces of cells transformed by Epstein-Barr virus. Nature 294:458-460.

8. Rowe M, Hildreth JEK, Rickinson AB, Epstein MA (1982) Monoclonal antibodies to Epstein-Barr virus-induced, transformation-associated cell surface antigens: binding patterns and effect upon virus-specific T-cell cytotoxicity. Int J Cancer 29:373.
9. Ikuta K, Takami M, Kim CW, Honjo T, Miyoshi T, Tangaya Y, Kawabe T, Yodoi J (1987) Cloning of cDNA for human IGe receptor: homology with animal lectins. Proc Natl Acad Sci (USA) 84:819.
10. Kikutani H, Inui S, Sato R, Barsumian EL, Owaki H, Yamasaki K, Kaisho T, Uchibayashi N, Hardy RR, Hirano T, Tsunasawa S, Sakiyama F, Suemura M, Kishimoto T (1986) Molecular structure of the human lymphocyte receptor for immunoglobulin E. Cell 47:657.
11. Rothlein R, Dustin ML, Marlin SD, Springer TA (1988) A human intercellular adhesion molecule (ICAM-1) distinct from LFA-1. J Immunol 137:1270-1276.

Burkitt's Lymphoma Variant Translocations: Distribution of Chromosomal Breakpoints and Perturbated Regulation of a Mutated c-*myc* Gene

B. Henglein[1], M. Lipp[2], P. Hartl[2], S. Adolph[3], H. Hameister[3], D. Eick[1], A. Polack[1], S. Joos[1], F. Baas[4], G.M. Lenoir[5], and G.W. Bornkamm[1]

INTRODUCTION

Variant translocations in Burkitt's lymphoma have the breakpoint of the chromosomal translocation 3' of c-myc on chromosome 8 and 5' of immunoglobulin kappa or lambda light chain constant genes on chromosome 2 or 22. The distance of the breakpoints to c-myc is variable. In a few cases the breakpoints are localized within 10 kb 3' of c-myc (BL37, KK124, and BL2) (Hollis et al. 1984, Denny et al. 1985, Showe et al. 1987). In the majority of cases the breakpoints are, however, at an unknown large distance from c-myc. Major structural alterations in c-myc were observed only in two cases with variant translocations. These represent duplications of the first exon (JBL2) (Taub et al. 1984) and of the coding part of c-myc (BL64) (Hartl and Lipp 1987). The duplication turned out to be constitutional in BL64, whereas in the case of JBL2 no normal tissue was available for analysis. Minor structural changes in the c-myc gene were found in all cases of variant translocations studied at the sequence level (LY47, KK124, BL2, BL64, BL60) (Rabbitts et al. 1984, Denny et al. 1985, Showe et al. 1987, Hartl and Lipp 1987, Strobl and Polack unpublished) and in a large panel of cell lines studied for restriction enzyme polymorphisms (Pelicci et al. 1986, Szajnert et al. 1987).

The c-myc gene is assumed to be deregulated in Burkitt's lymphoma cells with variant translocations: the transcripts are derived only from the c-myc allele on the $8q^+$ chromosome with preferential usage of the first promoter P1 whereas the normal allele is silent (for review see Bornkamm et al. 1988). The mutations in and around exon 1 might interfere with the regulation of transcript elongation, recently shown to be an important mechanism of c-myc regulation (Bentley and Groudine 1986, Nepveu et al. 1986, Eick and Bornkamm 1986). It is not known whether somatic mutations are sufficient to deregulate c-myc expression or whether incoming regulatory elements located in the immunoglobulin locus dominate c-myc regulation in cis. Another important question is whether a second gene in addition to c-myc might be affected by the variant chromosomal translocations.

The work on the structure of the breakpoints in variant translocations presented in the first part ot this paper may help to address these questions. In the second part we provide evidence that regulation at the level of RNA elongation is perturbated in a Burkitt's lymphoma line with variant translocation and mutations in the first exon and intron of c-myc.

(1) Institut für Med. Mikrobiologie, Abt.Virologie, D-7800 Freiburg, (2) Inst.f.Biochemie der Universität, D-8000 München, (3) Inst.f. Humangenetik der Universität, D-7900 Ulm, (4) Dept.Pediatrics, Acad. Medical Centre, NL-1105 AZ-Amsterdam, (5) IARC, F-69732 Lyon

Current Topics in Microbiology and Immunology, Vol. 141

RESULTS AND DISCUSSION

Breakpoints of t(2;8) Translocations

The breakpoints of the variant translocations, although dispersed over a large distance on chromosome 8, are always located close to immunoglobulin constant regions and are therefore accessible to molecular cloning using immunoglobulin gene probes. As shown previously, the breakpoint of LY91 is located either within or close to kappa joining sequences (Rappold et al. 1984). Using a kappa constant probe we have isolated a phage containing a 9.5 kb BamHI fragment which carries the breakpoint of the chromosomal translocation. From this phage a 1.1 kb SacI fragment with unique sequences was subcloned which is derived from chromosome 8 (probe 8q305). The breakpoint of the cell line BL64 had been cloned previously (Hartl and Lipp 1987). Probe 8q262 represents a 700 bp XhoI-BamHI fragment derived from chromosome 8 close to the BL64 breakpoint. A 2.4 kb EcoRI-BamHI fragment derived from chromosome 8 proximal to the breakpoint of JBL2 was kindly provided by P. Leder (probe 8q373). Two other probes derived from the region 48 kb (8q264) and 75 kb (8q267) downstream of c-myc have been obtained by chromosomal walking by two of us (P.H. and M.L.). Only the breakpoint of the cell line JI turned out to be located within the 75 kb 3′ of c-myc as shown by Sun et al. (1986) and Klobeck et al. (1987). The breakpoints of BL21, BL64, LY66, LY91 and JBL2 are located further downstream towards the telomere. By chromosomal walking using the breakpoints of BL64, LY91 and JBL2 as probes additional large pieces of chromosome 8 band q24 were cloned from a phage library.

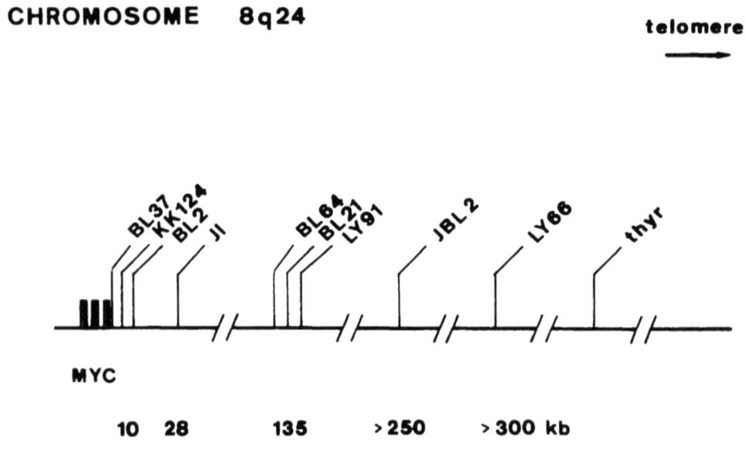

Fig. 1. Map of the region of chromosome 8q24 downstream of c-myc

Mapping of the various phage inserts revealed that (i) the breakpoints of BL64 and LY91 are located close to each other, (ii)

that phages from this region do not overlap with those obtained by cloning with the JBL2 breakpoint as a probe, and (iii) that neither region is contiguous to the cloned 75 kb region downstream of c-myc. Analysis of the other cell lines revealed that the breakpoint of BL21 is located between those of BL64 and LY91, whereas that of LY66 is outside the cloned region. To establish the order of breakpoints relative to each other and to c-myc, and to see whether the thyroglobulin gene is without or within the region affected by the variant chromosomal t(2;8) translocations (Baas et al. 1985), in situ hybridization to metaphase chromosomes were performed with the probes described above. The data revealed the following order: centromere - c-myc - JI - BL64 - BL21 - LY91 - JBL2 - LY66 - thyroglobulin - telomere (Fig. 1).

Of course, knowing the order of breakpoints relative to each other, does not tell us anything about distances. To get an idea of the distances involved we started to establish a physical map of the region downstream of c-myc using pulsed field gel electrophoresis. The results of this analysis are summarized in Fig. 1 and allowed us to draw the following conclusions: (i) the cluster of breakpoints of the cell lines BL64, BL21 and LY91 is located at a distance 135 kb downstream of c-myc, (ii) the human pvt-1 locus encompassing the region around the JBL2 breakpoint (Graham and Adams 1986, Mengle-Gaw and Rabbitts 1987) is located at least 250 kb 3′ of c-myc, (iii) the breakpoint of LY66 is located at least another 40 kb further downstream, and (iv) the thyroglobulin gene is located at an unknown distance further towards the telomere. Using Northern blotting and the probes described above, we failed to detect any transcriptional activity in the cloned region 3′of c-myc.

Perturbation of Regulation in a Mutated c-myc Gene

The c-myc gene located on the $8q^+$ chromosome of variant translocations appears invariably to be mutated in or around exon 1. The hypothesis that these mutations interfere with the regulation at the level of RNA elongation is supported by the fact that the density of polymerases on c-myc exon 2 and 3 is higher in Burkitt's lymphoma cells compared to that in EBV immortalized normal cells (Cesarman at al. 1987). We have attempted to establish a functional assay for the elongation block in human B cells in order to test whether Burkitt's lymphoma cells can respond in such an assay. Extrapolating from our work on HL60 cells (Eick and Bornkamm 1986) we have asked the question whether DMSO induces a block of RNA elongation also in BJAB cells. BJAB is an EBV negative human B cell line of uncertain disease origin. With respect to c-myc, BJAB cells were regarded as normal controls because the cells do not carry one of the chromosomal translocations characteristic for Burkitt's lymphoma cells and present no evidence for structural alteration of the c-myc gene.

Treatment of BJAB cells with DMSO induced a rapid decrease of c-myc RNA. Nuclear run-on assay revealed that transcription of the first exon remained unaltered whereas the density of polymerases decreased substantially on the distal part of the gene (Fig. 2). We therefore conclude that DMSO induces a c-myc RNA elongation block also in B cells. The respective experiment in which a Burkitt's lymphoma line with variant translocation was treated with DMSO and analysed for the transcriptional activity along the c-myc gene, is shown in Figure 2. Remarkably, even without DMSO treatment, the transcriptional pattern along the c-myc gene was quite different in BL2 cells than that in BJAB or HL60 cells. The polymerases were found to be not equally distributed along the c-myc gene. The density of polymerases for instance was surprisingly low on the first exon of BL2 cells and substantially higher upon other parts of the gene. The relative ratios of signals observed on various fragments of c-myc were not altered by DMSO induction. The same was true when other cell lines with variant translocations (JI, BL37, L660) were studied (data not shown). This demonstrates that the RNA elongation block at the end of exon 1, although inducible in BJAB cells, cannot be induced by DMSO in these Burkitt's lymphoma cell lines with variant translocations.

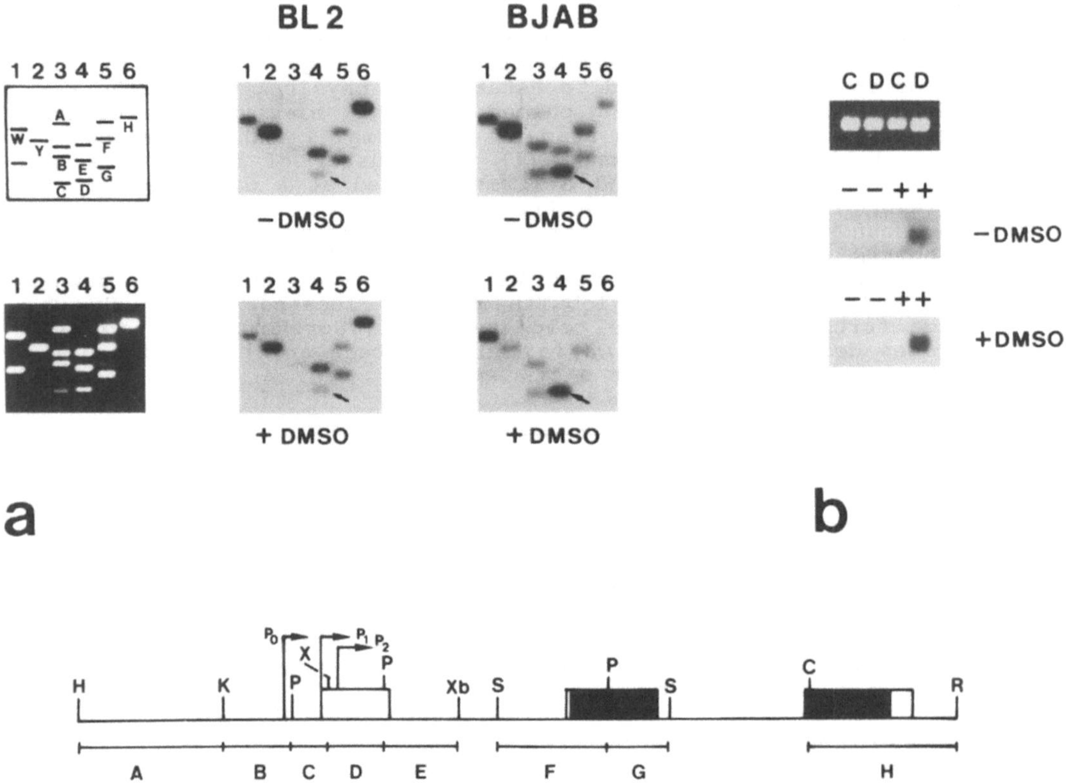

Fig. 2. Transcriptional activity of the c-myc gene in BJAB and BL2 cells in response to DMSO. Miniblots of a gel containing the various parts of the c-myc gene (left part of [a]) were hybridized with ^{32}P-labelled RNA synthesized in isolated nuclei of BL2 and BJAB cells. Nuclei were prepared from cells treated with or without 1.25% DMSO (v/v) for one hour (a). The arrows point to fragment D representing exon 1. Lanes 1 and 2 contain the human bcr gene (W) and immunoglobulin heavy chain μ gene (Y), respectively. The strand specificity of transcription is illustrated by hybridization of the ^{32}P-labelled nuclear run-on RNA of BJAB cells to single stranded fragments cloned in phage M13 (b)

Even though DMSO is incapable to induce the RNA elongation block, it nevertheless exhibited a pleiotropic action on c-myc expression in Burkitt's lymphoma lines. Shortly after exposure of BL2 cells to DMSO a transient decrease of c-myc RNA was observed (Fig. 3).
The rapid decrease of c-myc RNA after one hour has a strong posttranscriptional component, since the run-on signals did not change significantly within this time period. Notably, the reincrease in c-myc RNA revealed an altered ratio of promoter usage of c-myc. RNA initiated at the second promoter P2 increased whereas

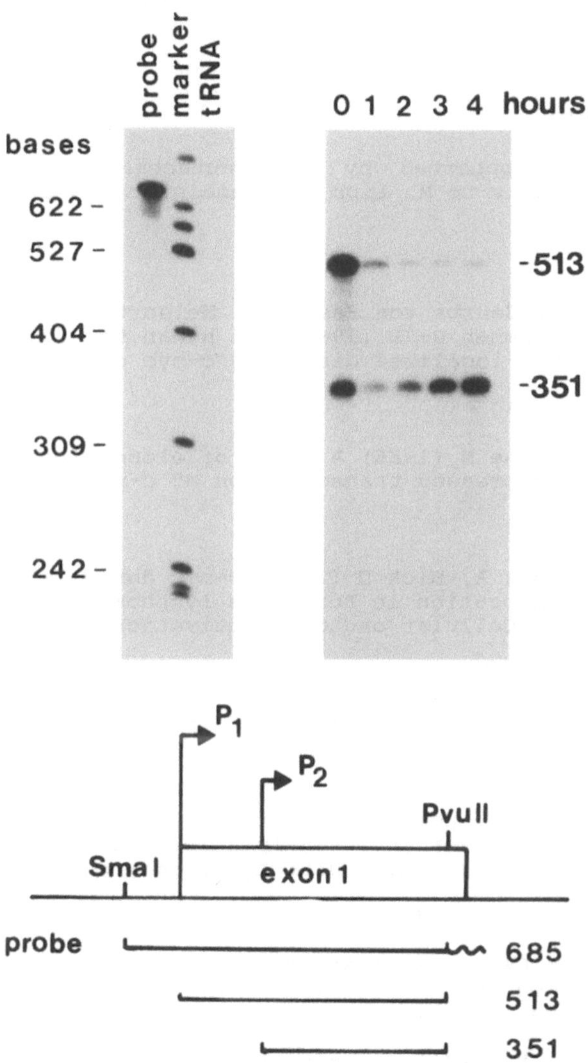

Fig. 3. Altered c-myc promoter usage in BL2 cells in response to 1.25% DMSO (v/v) as shown in an S1 protection experiment using a single stranded SmaI-PvuII fragment as a probe spanning most of c-myc exon 1

RNA initiated at P1 remained invariably low. The question whether this reincrease of c-myc RNA is due to activation of the normal allele cannot be definitely answered from this experiment. It appears unlikely in the light of the fact that in BL18 and BL67 cells with t(8;14) translocations and truncated c-myc genes DMSO caused also an intermittent decrease of c-myc RNA without the normal c-myc allele being induced (data not shown). Preferential usage of promoter P1 is in the case of a nontruncated c-myc gene a consistent feature of Burkitt's lymphoma (and mouse plasmacytoma) cells. The shift in promoter usage observed in BL2 cells after DMSO treatment provides evidence that somatic mutations in and around exon 1 cannot be solely responsible for the altered promoter usage in Burkitt's lymphoma cells.

ACKNOWLEDGEMENTS

This work was suppported by die Deutsche Forschungsgemeinschaft (DFG) through grants to M. Lipp, H. Hameister and G.W. Bornkamm.

REFERENCES

Baas F, Bikker H, Geurts von Kessel A, Melsert R, Pearson PL, de Vijlder JJM, van Ommen G-JB (1985) The human thyroglobulin gene: A polymorphic marker localized distal to c-myc on chromosome 8 band q24.
Hum Genet 69:138-143

Bentley DL, Groudine M (1986) A block of elongation is largely responsible for decreased transcription of c-myc in differentiated HL60 cells.
Nature 321:702-706

Bornkamm GW, Polack A, Eick D (1988) c-myc deregulation by chromosomal translocation in Burkitt's lymphoma.
In: Klein G, (ed) Cellular oncogene activation, Marcel Dekker, New York, p 223

Cesarman E, Dalla-Favera R, Bentley D, Groudine M (1987) Mutations in the first exon are associated with altered transcription of c-myc in Burkitt lymphoma.
Science 238:1272-1275

Denny CT, Magrath GF, Hollis IT, Kirsch IR (1985) Burkitt lymphoma cell line carrying a variant translocation creates new DNA at the breakpoint and violates the hierarchy of immunoglobulin gene rearrangement.
Mol Cell Biol 5:3199-3207

Eick D, Bornkamm GW (1986) Transcriptional arrest within the first exon is a fast control mechanism in c-myc expression.
Nucleic Acids Res 14:8331-8346

Graham M, Adams JM (1986) Chromosome 8 breakpoint far 3' of the c-myc oncogene in a Burkitt's lymphoma 2;8 variant translocation is equivalent to the murine pvt-1 locus.
EMBO J 5:2845-2851

Hartl P, Lipp M (1987) Translocation of Burkitt's lymphoma by site-specific recombination via the kappa light-chain joining signals. Mol Cell Biol 7:2037-2045

Hollis GF, Mitchell KF, Battey J, Potter H, Taub R, Lenoir GM, Leder P (1984) A variant translocation places the lambda immunoglobulin genes 3' to the c-myc oncogene in Burkitt's lymphoma. Nature 307:752-755

Klobeck G, Combriato G, Zachau HG (1987) N segment insertion and region-directed somatic hypermutation in a Kappa gene of a t(2;8) chromosomal translocation. Nucleic Acids Res. 15:4877-4888

Mengle-Gaw L, Rabbitts TH (1987) A human chromosome 8 region with abnormalities in B cell, HTLV-I+ T cell and c-myc amplified tumours. EMBO J 6:1959-1965

Nėpveu A, Marcu KB (1986) Intragenic pausing and antisense transcription within the murine c-myc locus. EMBO J 5:2859-2865

Pelicci PG, Knowles DM, Magrath I, Dalla-Favera R, (1986) Chromosomal breakpoints and structural alterations of the c-myc locus differ in endemic and sporadic forms of Burkitt lymphoma. Proc. Natl. Acad. Sci. USA 83:2984-2988

Rabbitts TH, Forster A, Hamlyn PH, Baer R (1984) Effect of somatic mutation within translocated c-myc gene in Burkitt's lymphoma. Nature 309:592-597

Rappold GA, Hameister H, Cremer T, Adolph S, Henglein B, Freese UK, Lenoir GM, Bornkamm GW (1984) c-myc and immunoglobulin kappa light chain constant genes are on the $8q^+$ chromosome of three Burkitt lymphoma lines with t(2;8) translocations. EMBO J 12:2951-2955

Showe LC, Moore RCA, Erikson J, Croce CM (1987) MYC oncogene involved in a t(8;22) chromosome translocation is not altered in its putative regulatory regions. Proc Natl Acad Sci USA 84:2824-2828

Sun LK, Showe LC, Croce CM (1986) Analysis of the 3' flanking region of the human c-myc gene in lymphomas with the t(8;22) and t(2;8) chromosomal translocations. Nucleic Acids Res 14:4037-4050

Szajnert MF, Saule S, Bornkamm GW, Wajcman H, Lenoir GM, Kaplan JC (1987) Clustered somatic mutations in and around first exon of non-rearranged c-myc in Burkitt lymphoma with t(8;22) translocation. Nucleic Acids Res. 15:4553-4565.→ic Acids Res., 15, 4877-4888

Taub R, Kelly K, Battey J, Latt S, Lenoir GM, Tantravahi U, Tu Z, Leder P (1984) A novel alteration in the structure of an activated c-myc gene in a variant t(2;8) Burkitt lymphoma. Cell 37:511-520

Differentiation Associated c-*myc* Expression in Phorbol Ester and Lymphokine Stimulated B-Type Chronic Lymphocytic Leukemia Cells

K. NILSSON[1], A. BJÖRKLAND[1], M. CARLSSON[1], L.-G. LARSSON[1], K. FUNA[2], and T. TÖTTERMAN[1]

INTRODUCTION

Detailed phenotypic studies of human leukemia/lymphoma cells over the last decade have demonstrated that most hematopoietic tumors represent clonal expansions of malignant cells arrested at a pre-terminal stage of differentiation (Greaves 1979, Nilsson et al 1985a). The importance of this blocked differentiation for the aberrant growth behaviour of the tumors is further suggested by in vitro experiments with established leukemia/lymphoma cell lines (e.g. HL-60 and U-937). These demonstrate that induction of differentiation will lead to maturation and irreversible growth arrest, similar to what occurs in normal hematopoietic cells undergoing terminal differentiation (Nilsson et al 1985b).

The genetic changes causing the deregulation of the gene programs for growth and differentiation frequently involve translocations (e.g. in Burkitt's lymphoma (BL), follicular lymphoma (FL) and chronic myeloid leukemia), trisomies and deletions. In a few B-cell tumors the nature of some frequently translocated genes is known, i.e. the protoonco-gene c-myc in BL and the putative oncogene bcl-2 in FL. In BL the c-myc appears to be deregulated by a translocation event in all cases. The hypothesis has therefore been put forward that the myc translocation affects the possibility of the tumor cells to differentiate terminally and become quiescent in the G_0 phase of the cell cycle (Klein and Klein 1986). In murine and rat myelomas the presence of equivalent c-myc translocation suggests a similar pathogenetic role of c-myc in the development of these B-cell tumors.

The studies on the HL-60 and U-937 cell lines in vitro, correlating c-myc expression with the process of differentiation, supported this view on the importance of c-myc in the oncogenesis of B-cell tumors (Westin et al 1982, Reitsma et al 1983, Einat et al 1985). The recent results from analyses of murine erythroleukemia (MEL) and human (U-937) cell lines, transfected with the c- and v-myc oncogene, respectively, modify the view that c-myc has to be down-regulated to allow the differentiation gene program to be executed, and furthermore suggest that only the terminal stage of differentiation requires the down-regulation of the c-myc expression (Coppola and Cole 1986, Dmitrovsky et al 1986, Prochowstik and Kukowska 1986, Larsson et al 1988).

For human B-cells similar cell line models are not available for studies of c-myc expression during the development of the unipotent B-lymphoid stem cell into mature B-cells. The relationship of c-myc to B-cell - plasma cell differentiation, however, can be examined using fresh B-type chronic lymphocytic leukemia (B-CLL) cells, as such cells can be reproducibly induced to differentiate in vitro by 12-O-tetradecanoyl-phorbol-13-acetate (TPA). We previously showed that this differentiation is associated with an increase in c-myc mRNA and protein levels (Larsson et al 1987). This appeared to contradict the findings in the HL-60 and U-937 systems but it was assumed that they might reflect a) a malignancy associated aberrant feature of CLL cells or b) an in vitro artefact, as CLL cells stimulated by TPA differentiated without proliferation, perhaps due to non-physiological in

[1]Department of Pathology, University of Uppsala, Uppsala, Sweden, and [2]Ludwig Institute for Cancer Research, University Hospital, Uppsala, Sweden

Current Topics in Microbiology and Immunology, Vol. 141

vitro conditions. Finally, c) it is also possible that the unexpected kinetics of c-*myc* expression was a correct reflection of a differentiation-associated high expression of the gene at the early stages of B-cell differentiation.

This paper describes futher analyses in B-CLL cells using an *in situ* hybridization technique to study the c-*myc* mRNA at the single cell level and to correlate the amount of the messenger to the morpho-logically definable B-cell maturation steps.

B-CLL AS A TUMOR MODEL FOR STUDIES OF B-CELL DIFFERENTIATION

Leukemic cells from about 60% of patients with chronic lymphocytic leukemia (CLL) of B-cell type, preferentially those with an active disease, are inducible to further differentiation *in vitro* by TPA (Tötterman et al 1980, Nilsson et al 1985a). Employing the morphological transition of the small B-CLL cells into lymphoblasts and plasmablasts (and a small fraction of plasma cells) and the acquired ability of immunoglobulin (Ig) secretion as hallmarks of differentiation, a pronounced heterogeneity in the inducibility to differentiation was found in the group of TPA-responsive B-CLL clones. Furthermore, the response to TPA was not constant upon repeated sampling of leukemic cells at monthly intervals but varied considerably (Tötterman et al 1982). In our recent studies we have, therefore, focused on one typical CD5+ CLL clone, obtained from the donor patient EA at two occasions, at which the inducibility to differentiation by TPA was comparable. The peripheral blood mononuclear cells, of which 94 % were monoclonal B-cells, were harvested by leukapheresis. The two cell batches were designated I-40 and I-73 and stored in liquid nitrogen until use. Most experiments have been carried out with I-73 cells (Nilsson et al 1986, Larsson et al. 1987, Carlsson et al 1988a-c) but due to cell shortage recent studies have employd also I-40 cells. One selected batch of fetal calf serum and the same type of medium have been used throughout these studies.

Induction of differentiation without proliferation by TPA in EA B-CLL cells.

The earliest recognizable event in the I-73 cells, following within an hour of TPA exposure, is an increase in the expression of adhesion molecules, making the cells adhere to each other and form clumps. This is followed by a gradual increase in size and by changes in the morphology of the small leukemic lymphocytes towards lymphoblastoid-plasmablastoid cells during 5-6 days after which the cells start dying. At the molecular level induced differentiation is reflected by an increase in RNA and protein synthesis, beginning at about 4 hours and peaking at 24-48 hours with 13- and 25-fold increased levels, respectively. (Larsson et al 1987). An increase in the ratio between secreted and membrane forms of u-chain mRNA was apparent two hours after induction (Larsson et al 1987), and IgM secretion is usually detactable at 24 hours only to increase further during the next 48-72 hours. The expression of several surface antigens (eg. CD 23, CD 25, CD 37, 4F2 and the transferrin receptor), known to be associated with activation, G_1 progression and differentiation in normal B-cells, are also ordely induced in the I-73 B-CLL cells by TPA (Carlsson et al 1988a).

The cell cycle analyses, performed mainly with flow cytometry analyses of the RNA and DNA content of acridine orange-stained cells, demonstrated that the EA B-CLL clone is homogeneous and arrested in the G_0 phase of the cell cycle (Carlsson et al 1988a). Upon TPA induction there is a synchronous transition of about 65-70% of the cells to the G_1 stage of the cell cycle where they become arrested. The small fraction (1-3%) of proliferating cells were shown to represent T-cells activated to DNA synthesis by TPA (Carlsson et al 1988a).

The TPA-induced differentiation of the EA cells appears to be strictly dependent on the presence of this small fraction of contaminating T-cells, as the leukemic B-cells do not respond to TPA by morphological and antigenic changes or by Ig secretion after their removal (Danersund et al 1985). The early steps of the TPA induced differentiation of B-CLL cells appears to involve PKC mediated signals (Nilsson et al 1985). This problably leads to activation of the G_0 cells and to the expression of receptors for the putative growth-(?) and differentiation factors provided by the contaminating T-cells simultanously activated by TPA, and by the serum.

Induction of simultaneous differentiation and DNA synthesis by TPA and a human T-cell hybridoma (MP6) derived B-cell stimulatory lymphokine (BSF-MP6) in EA B-CLL cells

The lack of proliferation in the EA cells differentiating after TPA exposure was assumed to reflect either a malignancy-associated aberrant feature of the cells, or that the tissue culture conditions failed to provide the cells with the proper growth factors. A more far-fetched possiblity was that the CLL cells in the peripheral blood might be the counterpart of a normal B-cell programmed to mature in the absence of proliferation (memory cell?).

Recent systematic studies (Carlsson et al 1988b) have, however, shown that DNA synthesis and proliferation are indeed inducible in differentiating EA-CLL cells when they are exposed to a B-cell stimulatory factor (BSF) produced by a human T-hybridoma (MP6; Rosén et al 1986, Noma et al 1987). The BSF-MP6 is distinct from any of the interleukins, interferons and other B-cell stimulatory lymphokines.

Two distinctly different induction protocols have been described, i.e. a) stimulation with TPA and BSF-MP6 (Carlsson et al 1988b) and b) preactivation of the cells with SAC followed by addition of BSF-MP6+IL-2 (Carlsson et al 1988c) The protocols offer a possibility to study differentiation/proliferation in the absence of TPA. The kinetics of TPA-free induction of differentiation is very similar to that with TPA only. However, with respect to functional maturation it appears to be superior to that employing TPA only, at least in some of the CLL clones. In many instances the morphologic maturation is more pronounced than after TPA stimulation, in that cells with typical plasma cell morphology appear and that the rate of Ig secretion is higher. The EA B-CLL model thus offers the possibility of analyzing various aspects of B-cell differentiation with or without proliferation.

EXPRESSION OF THE c-*myc* PROTOONCOGENE IN EA-CLL CELLS STIMULATED BY TPA TO DIFFERENTIATION THE ABSENCE OF DNA-SYNTHESIS

We have previously demonstrated that non-induced I-73 cells do not expess c-*myc* mRNA as examined by northern blot analysis (Larsson et al 1987). After a TPA exposure c-*myc* mRNA became detectable after 2 hours, peaked at 4 hours, then decreased to reach a plateau level of expression which then remained unchanged throughout the 72-hour experiment. Densitometric scanning of the autoradiogram showed that the increase in c-*myc* mRNA during the first 4 hours was approximately 70-fold, a level similar to that of exponentially growing HL-60 cells. The amount of mRNA at the plateau level was 20-fold that in the non-induced cells. Immunoprecipitation, using *myc*-specific anti-bodies, demonstrated that the *myc* protein was indeed produced, thus showing that in B-CLL *myc*-expression is upregulated at the initiation of differentiation inducible by TPA. It also illustrates that there is not an obligatory need for down-regulation of c-*myc* during hematopoietic differentiation as was suggested from the HL-60, U-937, and MEL cell studies (Westin et al 1982, Reitsma et al 1983, Coppola and Cole 1986, Dmitrovsky et al 1986, Prochowstik and Kukowska 1986, Larsson et al 1988).

Using an RNA-RNA in situ hybridization technique (Brahic M and Haase A 1978) we have now been able to study the differentiation-associated c-myc expression at the single cell level. The technique is powerful but is only semiquantitative. Reservations must also be made regarding the precision in the morphological classification, as the cells become extensively distorted during the preparation of the slides. Studies with the I-73 cells (Fig. 1) demonstrated induction of c-myc mRNA expression in about 70% of the cells. This correponds to the fraction of cells leaving the G_0 to enter G_1 after TPA induction (Carlsson et al 1988a). The kinetics of the c-myc expression was almost identical to that previously found in the I-73 cells and furthermore demonstrated that c-myc remained expressed for 120 hours.

As the TPA-induced cells do not enter the S-phase (Carlsson et al 1988a) the results thus agree with the previous data in that a) small, uninduced (G_0) CLL cells do not express c-myc mRNA and also show that b) activated cells undergoing G_0 to G_1 transition express high amounts 4 hours post induction and that c) a fraction of the cells (20%) keep expressing c-myc mRNA for at least 5 days. These cells resemble the large lympho- and plasmablasts which can be identified on Giemsa stained cytocentrifuge slides (Carlsson et al 1988b).

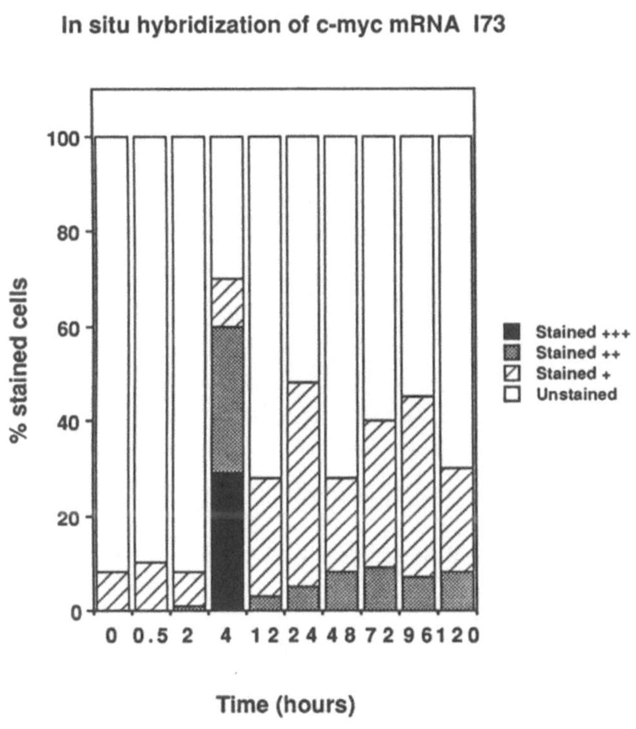

Fig. 1. The kinetics of c-myc mRNA expression in I-73 B-CLL cells cultured in the presence of TPA

EXPRESSION OF c-myc IN EA B-CLL CELLS STIMULATED BY TPA AND BSF-MP6 TO SIMULTANEOUS DIFFERENTIATION AND DNA SYNTHESIS

As depicted in Fig. 2 c-myc mRNA expression is induced also in EA B-CLL cells (batch I-40) by TPA and BSF-MP6. The basic pattern of c-myc expression is similar to that in cells stimulated with TPA only (Fig. 1), but c-myc mRNA is detectable somewhat earlier, and a slightly higher fraction of cells remain positive throughout the experiment. The in situ hybridization data, although only semiquantitative, therefore suggest that the induction of B-CLL cells with TPA and BSF-MP6 leads to a higher c-myc expression than after TPA stimulation only, at least in a fraction of the cells.

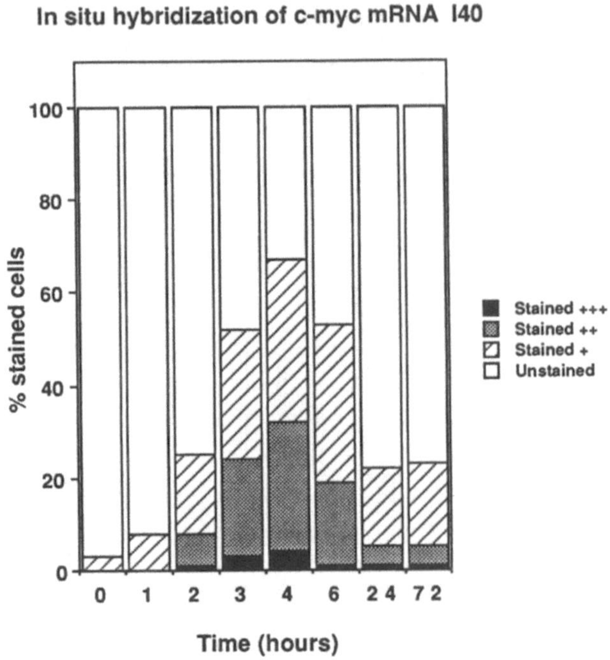

Fig.2. The kinetics of c-myc expression in I-40 B-BLL cells cultured in the presence of TPA + BSF-MP6

As the TPA plus MP6-BSF induced B-CLL cells were known to undergo a more pronounced differentiation then cells stimulated by TPA only such cultures were utilized in attempts to correlate c-myc expression with the various morphologically definable stages of B-cell differentiation. As already pointed out the morphology of cells processed for in situ RNA-RNA hybridization is difficult to assess. The results shown in Fig. 3 must therefore be regarded as preliminary. However, the data appear to demonstrate that:

a) small non-induced cells do not express c-myc mRNA and a small fraction of such cells remain negative in the culture at 72 hours. They probably represent cells that do not become activated to differentiation/proliferation,

b) medium-sized and larger lymphoblasts develop with time, most of which expressing c-myc mRNA,

c) cells with a plasmablast morphology, developing late after induction, express less c-myc mRNA, and
d) a few cells with plasma cell morphology were negative.

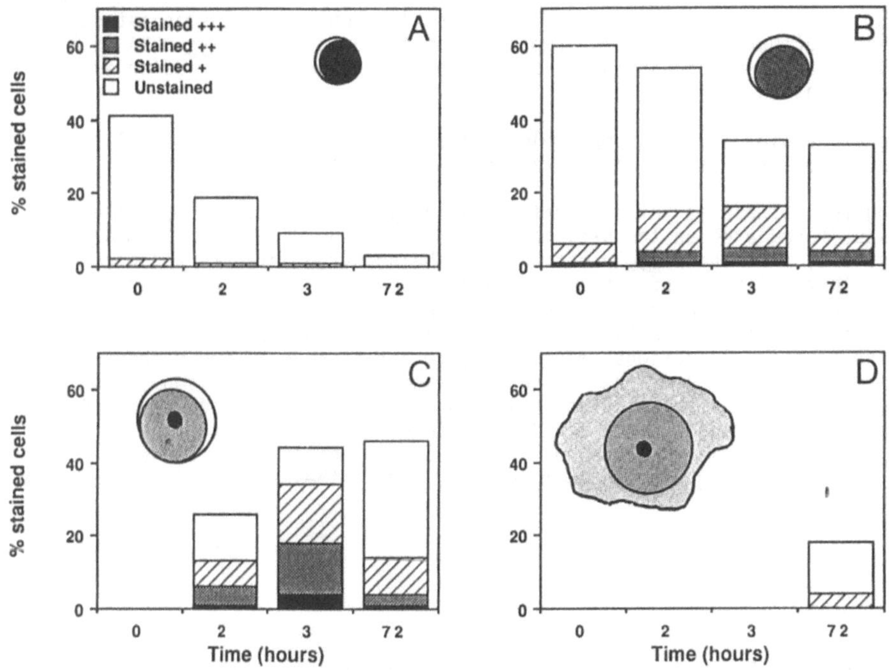

Fig.3. The kinetics of c-myc mRNA expression in I-40 B-CLL cells stimulated by TPA + BSF-MP6 related to morphology. In A-D the various morphological types are schematically depicted. A = small uninduced cell; B = activated small cell; C = medium-sized lymphoblast; D = large plasmablast

SUMMARY AND CONCLUSIONS

1. Depending on the nature of the inducing signals selected B-CLL clones can be induced either to differentiation only (e.g.TPA) or to simultaneous proliferation and differentiation (TPA plus BSF-MP6 or SAC ativation plus BSF-MP6 and IL-2). In the former situation the cells traverse from the G_0 to the G_1 phase of the cell cycle where they remain arrested throughout the differentiation process. Using the latter protocol of induction about 10% of the cells will be in S-phase at 144 hours post-induction and mitoses will occur followed by an arrest in G_1.

2. TPA-induced differentiation generates Ig secreting cells with lymphoblast-plasmablast morphology and only rare plasma cells, while stimulation with lymphokines appears to be more efficient in inducing mature plasma cells.

3. In situ hybridization studies demonstrate that 70-80% of the B-CLL cells, induced to differentiate with or without simultaneous DNA synthesis do express c-myc mRNA.

4. The c-myc protooncogene is induced rapidly (after 2-4 hours) during the G_0 to G_1 transition. The expression then declines but remains expressed in a fraction of cells (10-20 %) from 12 hours up to at least 120 hours.

5. In situ hybridization reveals that c-myc mRNA is expressed in activated cells during G_0 to G_1 transition and in cells at the lymphoblast - plasmablast stage of maturation. Furthermore, the data suggest that the amount of c-myc mRNA is highest in activated and lymphoblastoid cells, indicating that c-myc expression is related to differentiation .

6. c-myc expression is thus compatible with the initiation of the B-cell differentiation process but most likely not with its terminal phase.

Taken together, these as yet somewhat preliminary data seem to support the hypothesis, formulated on the basis of studies with the MEL and the human U-937 model systems, that the c-myc protein has to be down-regulated to allow cells to enter the terminal stage of differentiation and thereby become arrested in the quiescent G_0 phase of the cell cycle. Thus our results also support the notion that deregulation of c-myc, resulting from a translocation or some other malignancy-associated genetic alterations, may play an important role in the pathogenesis of B-cell lymphoma/leukemia by preventing the transition of the cycling cells into resting, terminally differentiated cells. However, further studies of the B-CLL and normal bone marrow model at the single cell level are required to finally prove that non-cycling plasma cells do not express c-myc. Specifically, studies relating the the presence of myc protein to morphology, Ig synthesis and cell cycle phase are required.

REFERENCES

Brahic M, Haase A T (1978) Detection of viral seqeuences of low reiteration frequency by in sity hybridization. Proc Natl Acad Sci USA 75:6125-6129.

Carlsson M, Tötterman T H, Matsson P, Nilsson K (1988a) Cell cycle progression of B-chronic lymphocytic leukemia cells induced to differentiate by TPA. Blood 71:415-421.

Carlsson M, Rosén A, Tötterman T H, Matsson P, Sundström C, Nilsson K (1988b) Phorbol ester and B cell stimulatory factor synergize to induce B-chronic lymphocytic leukemia cells to simultaneous immunoglobulin secretion and DNA synthesis. In manuscript.

Carlsson M, Tötterman T H, Rosén A, Nilsson K (1988c) Strong synergistic effect of r-IL 2 and B cell stimulatory factor on the proliferation and differentiation of human chronic lymphocytic B cells. In manuscript.

Coppola J A, Cole M D (1986) Constitutive c-myc oncogene expression blocks mouse erythroleukaemia cell differentiation but not commitment. Nature (London) 320:760-763.

Danersund A, Tötterman T H, Nilsson K, Egle-Jansson I, Kabelitz D, Sjöberg O (1985) Phorbol ester-induced differentiation of chronic B lymphocytic leukaemia cells - regulatory impact of autologous and alogeneic accessory cells. Clin Exp Immunol 59:644-652.

Dmitroswky E, Kuehl W M, Hollis G F, Kirsch I R, Bender T P, Segal S (1986) Expression of a transfected human c-myc oncogene inhibits differentiation of a mouse erythroleukaemia cell line. Nature 322: 748-750.

Einat M, Resnitzky D, Kimchi A (1985) Close link between reduction of c-myc expression by interferon and G_0/G_1 arrest. Nature (London) 313:597-600.

Greaves M F (1979) Tumour markers, phenotypes, and maturation arrest in malignancy: A cell selection hypothesis. IN: Tumour Markers (Boelsma E, Rümke P, eds). Elsevier, Amsterdam, pp 201-211.

Klein G, Klein E (1986) Conditioned tumorigenicity of activated oncogenes. Cancer Res 46:3211-3224.

Larsson L-G, Gray H E, Tötterman T, Pettersson U (1987) Drastically increased expression of myc and fos protooncogenes during in vitro differentiation of chronic lymphocytic leukemia cells. Proc Natl Acad Sci USA 84:223-227.

Larsson L-G, Ivhed I, Gidlund M, Pettersson U, Vennström B, Nilsson K (1988) Phorbol ester induced terminal differentiation is inhibited in human U-937 monoblastic cells expressing a v-myc oncogene. Proc Natl Acad Sci USA in press.

Nilsson K, Tötterman T, Danersund A, Forsbeck K, Hellman L, Pettersson U (1985a) Phorbol ester (TPA)-induced differentiation of B-type chronic lymphocytic leukemia cells. IN: Molecular Biology of Tumor Cells (Wahren B, Holm G, Hammarström S, Perlmann P, eds), Raven Press, New York pp 233-242.

Nilsson K, Ivhed I, Forsbeck K (1985b) Induced differentiation in human malignant hematopoietic cell lines. In: Gene expression during normal and malignant differentiation (Andersson L C, Gahmberg C G, Ekblom P, eds) Academic Press pp 57-72.

Nilsson K, Larsson L-G, Carlsson M, Danersund A, Forsbeck K, Hellman L, Tötterman T, Pettersson U (1986) Expression of c-myc and c-fos during phorbol ester induced differentiation of B-type chronic lymphocytic leukemia cells. IN: Mechanisms of B-cell neoplasia. Current Topics in Microbiol and Immunol (F Melchers M Potter, eds) 132: pp 280-289.

Noma T, Mizuta T, Rosén A, Hirano T, Kishimoto T, Honjo T (1987) Enhancement of the interleukin 2 receptor expression on T cells by multiple B-lymphotropic lymphokines. Immunol Lett 15:249-253.

Prochowstik E V, Kukowska J (1986) Deregulated expression of c-myc by murine erythroleukaemia cells prevents differentiation. Nature 322:848-850.

Reitsma P H, Rothberg P G, Astrin S M, Trial J, Bar-Shavit Z, Hall A, Teitelbaum S L, Kahn A J (1983) Regulation of myc gene expression in HL-60 leukaemia cells by a vitamin D metabolite. Nature (London) 306:492-494.

Rosén A, Uggla C, Szigeti R, Kallin B, Lindquist C, Zeuthen J (1986) A T-helper cell x Molt-4 human hybridoma constitutively producing B-cell stimulatory and inhibitory factors. Lymphokine Res 5:185-204.

Tötterman T, Nilsson K, Sundström C (1980) Phorbol ester-induced differentiation of chronic lymphocytic leukaemia cell. Nature 288:176-178.

Tötterman T, Gidlund M, Nilsson K, Wigzell H (1982) Natural killer (NK) cell sensitivity of phorbol ester-differentiated tumour cells correlates with disease activity in chronic B lymphocytic leukemia. Br J Haematol 52:563-571.

Westin E H, Wong-Staal F, Gelmann E P, Dalla Favera R, Papas T S, Lautenberger J A, Eva A, Reddy E P, Tronick S R, Aaronson S A, Gallo R C (1982) Expression of cellular homologues of retoviral onc genes in human hematopoietic cells. Proc Natl Acad Sci USA 79:2490 -2494.

B and T Cell Responses Induced by Interleukin-6

J. VAN SNICK[1], A. VINK[1], C. UYTTENHOVE[1], F. HOUSSIAU[1], and P. COULIE[2]

INTRODUCTION

Mouse interleukin-6 (IL6), formerly designated interleukin-HP1 or PCT-GF, was originally identified as a growth factor for mouse plasmacytomas and hybridomas (Nordan and Potter 1986, Van Snick et al. 1986 and 1987). Natural mouse IL6 has been purified from supernatants of helper T cells (Van Snick et al. 1986), macrophages (Nordan et al. 1987) and fibroblasts (Cayphas et al. 1987), but it has also been found in high titers in the serum of lipopolysaccharide-treated mice (Coulie et al. 1987) and in supernatants of a wide range of tumor cells (unpublished observations).

The functional relation of mouse IL6 with human IL6, also known as B cell stimulatory factor 2 (Hirano et al. 1986), interferon-β_2 (Zilberstein et al. 1986) or IL1-inducible 26K factor (Haegeman et al. 1986), was demonstrated last year when it appeared that human IL6 had hybridoma and plasmacytoma growth factor activity (Van Damme et al. 1987, Brakenhoff et al. 1987). More recently, comparison of cDNA sequences demonstrated a significant degree of structural homology between the two proteins, especially in their middle and C-terminal regions (Van Snick et al. 1988).

Here we describe some characteristics of the responses of normal lymphocytes to IL6.

RESULTS

B Cell Growth and Differentiation Induced by Mouse IL6.

One of the first effects ascribed to human IL6 was the induction of terminal differentiation of human B cells in the absence of proliferation (Hirano et al. 1984). It was of interest therefore to examine the response of mouse B cells to mouse IL6. These experiments were carried out before recombinant mouse IL6 became available. To minimize the risk of attributing to IL6 effects due to possible contaminants, we used natural IL6 purified from three different sources. Macrophage-derived IL6 was provided by Rick Nordan (Laboratory of Genetics, National Cancer Institute, National Institutes of Health, Bethesda, Md), fibroblast-derived IL6 was purified from supernatants provided by Jo Van Damme (Rega Institute, Catholic University of Leuven, Belgium) and T-cell-derived IL6 was prepared from a mouse helper T cell clone. As an additional control, we verified that all effects could be inhibited by a monoclonal anti-IL6 antibody. Our results can be summarized as follows.

When small splenic B cells were incubated in the presence of anti-IgM antibodies, IL6 induced a 5-fold increase in proliferation. This response culminated on day 2 and was dose-dependent, half-maximal proliferation being obtained with 1-2 ng/ml of IL6. By contrast, IL6 had little or no effect on the proliferation of unstimulated or dextran sulfate-stimulated B cells. The response of anti-IgM-stimulated B cells to IL6 was not inhibited by anti-IL4 antibodies, indicating that the effect of IL6 was due neither to contamination with IL4 nor to induction of IL4 in residual T cells possibly contaminating the B cell preparations. Contamination of our IL6 preparations with IL4 was further excluded by the fact that they had no effect on IA gene expression or on IgG1/IgG3 ratios in LPS-stimulated cultures.

[1] Ludwig Institute for Cancer Research, Brussels Branch, B1200, Brussels, Belgium
[2] Laboratory of Experimental Medicine, Université Catholique de Louvain, Brussels, Belgium.

Current Topics in Microbiology and Immunology, Vol. 141

Although the results obtained with IL6 and anti-IgM-stimulated B cells were highly reproducible, they were not very dramatic. We therefore examined possible synergies between IL6 and other cytokines. This approach revealed the existence of a unique interaction with IL1. The most striking effects were observed with dextran sulfate-stimulated B cells, whose IgM production was stimulated > 20-fold when the two cytokines were added together and ≈ 2-fold when they were added separately (Table 1).

Table 1. IgM production by dextran sulfate-stimulated B cells in response to IL1 and IL6.

Cytokines	IgM production (ng/ml)
none	44
IL1	73
IL6	102
IL1 + IL6	1140

Small C57BL/6 B cells (density between 1.070 and 1.088) were cultured in microwells (3 x 10^4 cells / well) with 10 μg/ml of dextran sulfate and IL1 (100 U/ml) or IL6 (2ng/ml) as indicated. IgM levels were measured on day 5.

Delaying the addition of one or of both cytokines for increasing periods of time resulted in a commensurate decrease in IgM production, indicating that both factors had to be present together and acted from the beginning of the culture. Synergy between IL1 and IL6 was also observed when measuring IgM production in the absence of dextran sulfate. Moreover, IL1 and IL6 induced ≈ 30-fold increases in the proliferation of anti-IgM-stimulated B cells. Several other factors including IL4, IL5, IL2, GM-CSF and TNF were tested for possible synergy with IL1 or IL6 but none yielded results comparable to those obtained with the IL1-IL6 combination.

T Cell Growth and Differentiation Induced by IL6.

Considering the broad range of activities of other interleukins, we thought it unlikely that the activity of IL6, which is made by so many different cell types, would be restricted to B lymphocytes. We therefore examined its interactions with T cells. We observed that peripheral mouse T cells proliferated extensively in response to mouse IL6 and submitogenic doses of phytohemagglutinin (PHA) (Table 2). Half-maximal responses were obtained with ≈ 0.1 ng/ml of IL6. No difference was observed between $CD4^+$ and $CD8^+$ cells and similar results were obtained when PHA was replaced by anti-T3 antibodies. In general the responses induced by IL6 were quantitatively comparable to those induced by an optimal dosage of IL2 and stronger than those elicited by IL4. Strong proliferations were also observed with mature but not with immature thymocytes.

Table 2. Comparison of the stimulatory effects of IL6 and IL2 on the proliferation of splenic T lymphocytes in response to PHA.

Factor added	Costimulus	Proliferation (cpm)
none	none	200
	PHA	900
IL2 (100 U/ml)	none	1,900
	PHA	77,000
IL6 (1 ng/ml)	none	700
	PHA	87,000

Spleen cells were incubated in microwells (5x10^4 cells/well) with or without a submitogenic dose of PHA. Thymidine incorporation was measured on day 4 after a 6 h pulse. No proliferation was observed with T cell-depleted populations.

Accessory cell-depleted T cells fail to respond to lectins (Habu and Raff 1977). These responses were completely restored by addition of IL6, suggesting that production of IL6 by macrophages plays an important role in the ability of these cells to support T cell responses. Similar results have been reported by Garman and co-workers in studies with human IL6 and mouse T cells (Garman et al. 1987). These authors also showed that the activity of human IL6 on mouse T cells was inhibited by antibodies directed against the IL2-receptor, suggesting that the action of IL6 was mediated by IL2. We have partly confirmed these results with mouse IL6, although a detailed kinetic analysis indicated that the early response to IL6 could not be inhibited by anti-IL2 receptor antibodies. The existence of an IL2-independent pathway of T cell activation by IL6 was confirmed by our studies with human IL6 and human T cells, in which we observed that anti-IL2 receptor antibodies had virtually no inhibitory effect on IL6-induced responses.

Another difference noted between the human and mouse systems was the relative importance of IL1. In the murine system, the presence of IL1 enhanced T cell responses to IL6 only with low doses of IL6. By contrast, human T cells rigorously depleted of accessory cells responded only weakly to optimal doses of IL6, unless IL1 was added to the culture.

Table 3. Proliferation of accessory cell-depleted human T cells stimulated with PHA, IL1 and IL6.

Cytokines	Proliferation (cpm)
none	108 ± 26
IL1	974 ± 476
IL6	670 ± 250
IL1 + IL6	9504 ± 2162

Accessory cell-depleted tonsillar T cells were incubated in microcultures (3×10^4 cells /well) in the presence of saturating doses of human IL1 and IL6. Thymidine incorporation was measured on day 4. Results correspond to the mean ± 1 S.E. of 5 independent experiments.

DISCUSSION

The results presented here indicate that IL6 is more than a B cell differentiation factor as was originally suggested by the work with human B lymphocytes. The results obtained in the mouse system even suggest that the primary immunological target of IL6 is the T rather than the B lymphocyte, since we consistently observed that the concentrations of IL6 required to activate B cells were ≈ 10 times higher than those required to activate T cells.

The finding that IL6, either alone or in conjunction with IL1, restores the response of accessory cell-depleted T cells to lectins suggests that IL6 is a major mediator of accessory cell function. This notion is consistent with the fact that, although both human and mouse IL6 were originally purified from T cell supernatants, the macrophage seems to be the major source of IL6 in the immune system (Aarden et al. 1987).

A final point that deserves some comment is the interaction between IL1 and IL6. Our results with mouse B and human T lymphocytes revealed the existence of a striking synergy between these two cytokines. If one recalls that IL1 is a potent inducer of IL6 in human fibroblasts (Content et al. 1985), it follows that these two molecules may form the basis of a self-amplifying signalling circuit connecting many different cell types with the immune system.

ACKNOWLEDGEMENTS

We thank Dr. R. Nordan for macrophage-derived IL6, Dr. Jo Van Damme for fibroblast supernatants, and Dr. P. T. Lomedico and Hoffmann-La Roche for recombinant human IL1α.

REFERENCES

Aarden LA, De Groot ER, Schaap OL, Lansdorp PM (1987) Production of hybridoma growth factor by monocytes. Eur. J. Immunol. 17: 1411-1416 .

Brakenhoff JPJ, de Groot ER, Evers R, Pannekoek H, Aarden LA (1987) Molecular cloning and expression of hybridoma growth factor in Escherichia coli . J. Immunol. 139: 4116-4121.

Cayphas S, Van Damme J, Vink A, Simpson RJ, Billiau A , Van Snick J (1987) Identification of an interleukin HP1-like plasmacytoma growth factor produced by L cells in response to viral infection. J.Immunol. 139 : 2965-2969.

Content JL, De Wit P, Poupart P, Opdenakker G, Van Damme J, Billiau A (1985) Induction of a 26 kDa-protein mRNA in human cells treated with an interleukin 1-related, leukocyte-derived factor. Eur. J. Biochem. 152: 253-257.

Coulie PG, Cayphas S, Vink A, Uyttenhove C, Van Snick J (1987) Interleukin-HP1-related hybridoma and plasmacytoma growth factors induced by lipopolysaccharide in vivo. Eur. J. Immunol. 17: 1217-1220.

Garman RD, Jacobs KA, Clark SC, Raulet DH (1987) B-cell-stimulatory factor 2 (ß2 interferon) functions as a second signal for interleukin 2 production by mature murine T cells. Proc. Natl. Acad. Sci. USA 84: 7629-7633.

Habu S, Raff, M (1977) Accessory cell-dependence of lectin-induced proliferation of mouse T lymphocytes. Eur. J. Immunol. 7: 451-457.

Haegeman G, Content J, Volckaert G, Derynck R, Tavernier J, Fiers W (1986) Structural analysis of the sequence coding for an inducible 26-kDA protein in human fibroblasts. Eur. J. Biochem. 159 : 625-632.

Hirano T, Teranishi T, Lin B, Onoue K (1984) Human helper T cell factor(s). IV. Demonstration of a human late-acting B cell differentiation factor acting on staphylococcus aureus Cowan I-stimulated B cells. J. Immunol. 133: 798-802.

Hirano T, Yasukawa K, Harada H, Taga T, Watanabe Y, Matsuda T, Kashiwamura SI, Nakajima K, Koyama K, Iwamatsu A, Tsunasawa S, Sakyama F, Matsui H, Takahara Y, Taniguchi T, Kishimoto T (1986) Complementary DNA for a novel human interleukin (BSF-2) that induces B lymphocytes to produce immunoglobulin. Nature 324: 73-76.

Nordan R, Potter M (1986) A macrophage-derived factor required by plasmacytomas for survival and proliferation in vitro. Science 233: 566-569.

Nordan R, Pumphrey J, Rudikoff S (1987) Purification and NH_2-terminal amino acid sequence of a plasmacytoma growth factor derived from the murine macrophage cell line, P388D1. J. Immunol. 139: 813-817.

Van Damme J, Opdenakker G, Simpson RJ, Rubira MR, Cayphas S, Vink A, Billiau A, Van Snick J (1987) Identification of the human 26-kD protein, interferon ß2 (IFN-ß2), as a B cell hybridoma/plasmacytoma growth factor induced by interleukin 1 and tumor necrosis factor. J. Exp. Med. 165: 914-919.

Van Snick J, Cayphas S, Szikora J-P, Renauld J-C, Van Roost E, Boon T, Simpson RJ (1988) cDNA cloning of murine interleukin-HP1 : homology with human interleukin 6. Eur. J. Immunol. 18: 193-197.

Van Snick J, Cayphas S, Vink A, Uyttenhove C, Coulie P, Simpson R. (1986) Purification and NH_2-terminal amino acid sequence of a new T cell-derived lymphokine with growth factor activity for B cell hybridomas. Proc. Natl. Acad. Sci. USA 83: 9679-9683.

Van Snick J, Vink A, Cayphas S, Uyttenhove C (1987) Interleukin HP1, a T cell-derived hybridoma growth factor that supports the in vitro growth of murine plasmacytomas. J. Exp. Med. 165: 641-649.

Zilberstein A, Ruggieri R, Korn JH, Revel M (1986) Structure and expression of cDNA and genes for human interferon-ß2, a distinct species inducible by growth-stimulatory cytokines. Embo J. 5: 2529-2537.

Part II:
Studies of B Cell Relevant Oncogenes

c-*myc*

Prediction of a Dimerization Surface Common to a New Class of Sequence-Specific DNA Binding Proteins

S. L. McKnight, W. H. Landschulz, and P. F. Johnson

For the past several years we have studied a heat-stable, rat liver nuclear protein that we term C/EBP. This protein is capable of specific binding to the CCAAT homologies of a number of genes that are transcribed by RNA polymerase II (Graves et al., 1986). Surprisingly, this same protein also binds to the "core homology" common to many different viral enhancers (Johnson et al., 1987). We have used standard biochemical fractionation techniques to purify C/EBP. The purified protein was used to generate a partial amino acid sequence which, in turn, facilitated retrieval of a recombinant DNA clone of the gene encoding C/EBP. These reagents have allowed us to delineate the DNA binding domain of C/EBP and determine its amino acid sequence.

Upon defining the amino acid sequence of the DNA binding domain of C/EBP we compared it with sequences in several computer data bases. This exercise revealed substantial amino acid sequence similarity between C/EBP and the products of two proto-oncogenes, myc and fos (Landschulz et al., 1988). The regions of amino acid sequence between these proteins are generally free of residues that are incompatible with a helical structure (prolines and glycines). We therefore plotted these amino acid sequences on an idealized α helix in an exercise to search for amphipathy. Figure 1 shows the result of this analysis in the case of C/EBP. The putative α helical region is highly amphipathic; one of its faces is rich in hydrophobic amino acid residues, another is rich in charged and hydrophilic residues.

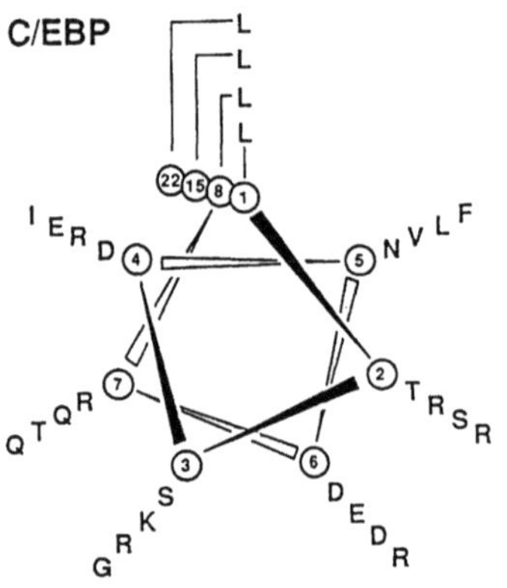

Figure 1. Helical Wheel Analysis of a 28 Amino Acid Region Within the DNA Binding Domain of C/EBP. Leucine residue located at position 1 corresponds to residue 315 of the intact C/EBP polypeptide, and is the most amino-terminal residue of the putative α helix displayed in this Figure. The most carboxyl-terminal residue is glutamine (Q) residue 342 of C/EBP (the most distal residue of helical position 7).

We were particularly surprised by the repeating array of leucine residues. Leucines occur at 7 amino acid intervals over the entire α helical region. Since we had noted amino acid sequence similarity between C/EBP and the MYC and FOS transforming proteins, we also plotted those sequences on an idealized α helix. Much to our surprise, MYC and FOS also display leucine residues at every seventh residue over at least 8 α helical turns (Landschulz et al., 1988). Indeed, this same repeating motif of leucine is also present in the product of the jun proto-oncogene, as well as in GCN4, a yeast transcriptional regulatory protein.

Department of Embryology, Carnegie Institution of Washington, 115 West University Parkway, Baltimore, Maryland 21210

Current Topics in Microbiology and Immunology, Vol. 141

What role do these leucine repeats play in the function of these diverse nuclear proteins? If the regions containing these leucine repeats indeed exist in an α helical conformation, then the leucines would project as an ordered scaffold. Computer modeling studies carried out with the help of Dr. Hamilton Smith indicated that the repeating leucine residues on a single α helix were too far apart to provide continuous hydrophobic stabilization. That is, the van der Waals radii of adjacent leucine side chains did not come in close contact. This observation forced us to search for a hydrophobic surface that might match the leucine scaffold, since the surface would clearly be incompatible with an aqueous environment. The matching surface could occur within the same polypeptide, or on a separate polypeptide. Since earlier experiments had indicated that C/EBP could carry out its role in specific recognition of DNA after having been trimmed down to as little as 14 kilodaltons (including the leucine repeat), and since this 14 kilodalton region contained no other regions of substantial hydrophobicity, we reasoned that the matching hydrophobic interface must be donated by a separate polypeptide. Finally, knowing that homogeneously purified C/EBP is capable of specific recognition of DNA, we reasoned that the matching surface for the leucine repeat would not be conferred by a different polypeptide. On these grounds we have predicted that the matching hydrophobic surface for the leucine repeat would be the leucine repeat of a second C/EBP polypeptide. That is, two leucine repeats would constitute a dimerization interface. We call this hypothetical structure the *leucine zipper*.

At the time that this prediction was made, we were unaware of the multimeric form of C/EBP. Recent experiments have shown, satisfyingly, that C/EBP exists in solution as a stable dimer (W.H. Landschulz, P.B. Sigler and S.L. McKnight, unpublished observations). Moreover, the region of the protein necessary for dimerization occurs within the minimal DNA binding domain (which houses the leucine repeat motif). These experiments follow earlier work on GCN4, which had provided very clear evidence of a dimeric conformation for that sequence-specific DNA binding protein (Hope and Struhl, 1987).

Sequence-specific DNA binding by C/EBP is not established by the leucine repeat motif alone. Recent experiments have indicated that a region immediately adjacent to the leucine repeat must remain intact in order for C/EBP to bind DNA. This additional region contains three pockets that exhibit a net positive charge of at least +3 (Landschulz et al., 1988). We call this the *basic region,* and predict that it will constitute the surface of the protein that actually contacts DNA. Figure 2 shows a schematic diagram of the C/EBP polypeptide.

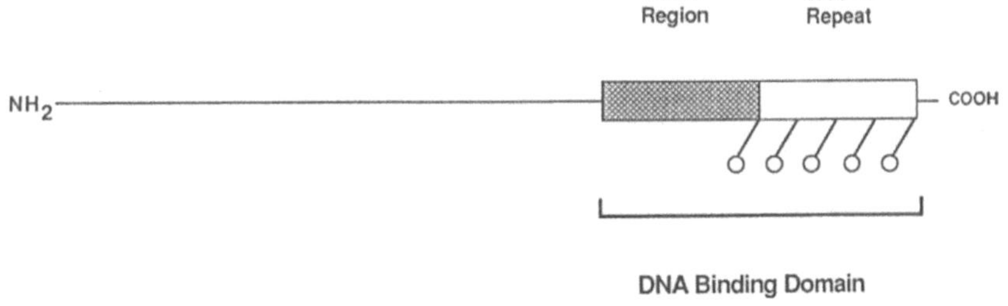

Figure 2. Schematic Diagram of C/EBP Showing the Relative Locations of the Leucine Repeat Motif and the Basic Region. Hatched region denotes 30 amino acid segment of C/EBP that is rich in positively charged residues. Projections terminating with circles denote repeating leucine residues.

A final speculation that we offer concerns the possibility that the leucine repeat domains of different polypeptides might facilitate the formation of heterotypic dimers. If we are correct in predicting that the *basic region* actually represents the contact surface for interaction with DNA, then the mixing of two different *basic regions* via the *leucine zipper* might establish a DNA binding specificity different from either of the parental homodimeric molecules. It is obvious that such a scenario would expand the repertoire of DNA binding specificities encoded by a finite number of regulatory genes.

REFERENCES

Graves BJ, Johnson PJ, McKnight SL (1986) Homologous recognition of a promoter domain common to the MSV LTR and the HSV tk gene. Cell 44: 565-576

Hope IA, Struhl K (1987) GCN4, a eukaryotic transcriptional activator protein, binds as a dimer to target DNA. EMBO Journal 6:2781-2784

Johnson PF, Landschulz WH, Graves BJ, McKnight SL (1987) Identification of a rat liver nuclear protein that binds to the enhancer core element of three animal viruses. Genes and Development 1:133-146

Landschulz WH, Johnson PF, McKnight SL (1988) The leucine zipper: A hypothetical structure common to a new class of DNA binding proteins. Science, in press

Landschulz WH, Johnson PF, Adashi E, Graves BJ, McKnight SL (1988) Isolation of a recombinant copy of the gene encoding C/EBP. Genes and Development, in press

Immunological Probes in the Analysis of *myc* Protein Expression

G.I. EVAN[1], J.P. MOORE[2], J.M. IBSON[3], C.M. WATERS[1], D.C. HANCOCK[1], and T.D. LITTLEWOOD[1]

Imperial Cancer Research Fund Laboratories, Dominion House, Bartholomew Close, London EC1A 7BE, England.

THE MYC PROTO-ONCOGENE FAMILY

Three well-defined genes have been identified within the human genome, c-myc, N-myc and L-myc (Alt et al., 1986). c-myc is the proto-oncogene homologue of the retroviral v-myc genes found in avian myelocytic leukemia viruses and feline leukaemia viruses. It is expressed in many tissues during embryogenesis (Ohlsson et al., 1986; Zimmerman et al., 1986; Ruppert et al., 1986) and is a mitogen response gene in many, perhaps all, adult cell types (Kelly et al., 1983; Rabbitts et al., 1985). c-myc has also been found amplified in transformed cell lines derived from several lineages (Little et al., 1983; Alitalo et al., 1987; Dalla Favera et al., 1982). In contrast to c-myc, both N-myc and L-myc show a much more restricted pattern of expression (Alt et al., 1986). Both are expressed during embryogenesis in certain tissues. N-myc is also expressed in certain lineage cells in the adult, and is found amplified in more serious childhood neuroblastomas. It is not known which, if any, cells in the adult express L-myc, but the gene is found amplified in some small cell lung carcinomas. c-myc , N-myc and L-myc genes are related by sequence homology which is particularly marked in certain regions (Figure 1) - the so-called "myc boxes" (Schwab, 1985; Alt et al., 1986). The three genes also share a similar structural organisation within the genome, and all are active in transformation assay when co-transfected with the ras oncogene (Alt et al., 1986).

Despite a fairly detailed characterisation of the proteins encoded by the c-myc and N-myc genes (Persson and Leder, 1984; Evan and Hancock, 1985; Hann et al., 1988; Slamon et al, 1986; Evan et al., 1986) surprisingly little is known about the functions of myc proteins in normal and transformed cells. Both the c-myc and N-myc gene products are short-lived phosphoproteins which localise to the cell nucleus, and both bind DNA, although the binding shows no sequence specificity. So far, the L-myc gene product has not been identified. The fact that both the c-myc and N-myc proteins are so biochemically similar, and that they also posses similar transforming activities (Small et al., 1987), argues that they share similar functional roles. We believe that a comparative study of the biochemical characteristics and patterns of expression of the different myc proteins may provide valuable information as to the functions of myc proteins in normal and transformed cells. We have sought to identify and characterise the myc proteins using site-directed antibodies raised against synthetic peptide sequences found within myc proteins. We are attempting to use these antibodies for several purposes: to quantitate accurately myc protein levels in

---

[1] Imperial Cancer Research Fund Laboratories, Dominion House, Bartholomew Close, London EC1A 7BE, England.
[2] Dept. Veterinary Pathology, University of Glasgow, Scotland.
[3] Dept. Microbiology and Immunology, University of California, San Francisco, California 94143, USA.

Current Topics in Microbiology and Immunology, Vol. 141

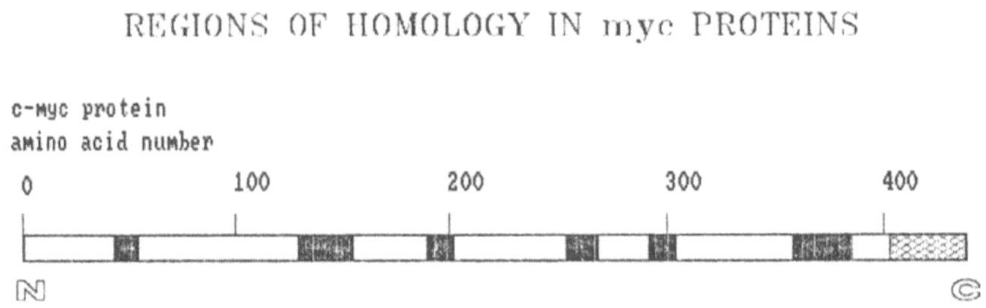

Figure 1. Regions of homology shared amongst _myc_ proteins. Regions of homology between _myc_ proteins are depicted on a linear map of the human c-_myc_ protein

cells; to identify and compare each of the different members of the _myc_ protein family, including the L-_myc_ protein; and to examine the functions and interactions of the various _myc_ proteins.

A QUANTITATIVE ASSAY FOR c-_myc_ and N-_myc_ PROTEINS

The functions of _myc_ proteins in normal and transformed cells remain obscure, although several studies suggest a role for the c-_myc_ protein in the proliferation of normal cells (e.g. Kelly _et al_., 1983). A significant part of the problem with determining the biochemical role of _myc_ proteins in normal and transformed cells is the lack of accurate quantitation of _myc_ protein levels within cells. Preliminary studies, using semi-quantitative immunoblotting analysis, suggested that many tumour cell lines expressed substantially elevated levels of c-_myc_ protein when compared with non-transformed fibroblasts proliferating at a comparable rate (Evan _et al_., 1985). However, such assays are inadequate for accurate quantitation and insufficiently sensitive to study the very low levels of c-_myc_ protein expression in non-transformed fibroblasts or lymphocytes, arguably the best types of cell in which to investigate the normal role of _myc_ proteins. Because of this lack of sensitivity in existing assays, we designed a quantitative assay for _myc_ proteins based upon a novel enzyme linked immunosorbence assay (ELISA) (Moore _et al_., 1987) whose components are outlined in figure 2. In this assay, _myc_ proteins are captured from a cell or tissue lysate by an antibody adsorbed onto a microtitre well. The capture antibody is specific for a peptide sequence conserved in all known _myc_ proteins (i.e. c-, N-, L- and v-_myc_) from all tested species (i.e. vertebrates). We call this sequence con-_myc_-1 (see below). The capture antibody ("pan-_myc_ antibody") recognises all _myc_ antigens irrespective of whether they are in native or denatured form. Captured _myc_ protein is then detected using a monoclonal antibody specific for another peptide epitope distant from the pan-_myc_ sequence. The detection antibody (specific for a particular class of protein) is conjugated to alkaline phosphatase which catalyses the dephosphorylation of NADP to NAD. The amount of NAD generated is proportional to the amount of _myc_ protein captured. This "dephosphorylation signal" is then amplified using an extremely sensitive coupled enzymatic system (AMPAK, IQ(Bio) Ltd., Cambridge, UK) in which the NAD product of the phosphatase reaction is used to drive a pair of cycling redox reactions, ultimately reducing a colourless precursor to a red formazan dye. The assay is calibrated against known amounts of a c-myc protein standard expressed in E.coli assuming a molecular weight of 50 kDa. For the purposes of the assay, all myc proteins are first denatured, reduced and alkylated. This not only renders all myc proteins immunologically identical with respect to the antibodies used, but also ensures complete solubilisation of myc proteins from cells or tissues. In our hands, the myc ELISA is at least 100 fold more sensitive than immunoblotting and is extremely rapid and reproducible.

Measurement of myc Protein Levels in Cells by ELISA

We have used the myc ELISA to determine the number of c-myc protein molecules in different cells, and to compare the levels of c-myc protein in transformed and non-transformed cells (Table 1). From the results shown in Table 1, it is clear that levels of myc proteins vary enormously between different cells. Cells with amplified myc genes (e.g. Colo 320, N417, HL60) not surprisingly express elevated levels of myc protein. However, many transformed cells without myc gene amplification also express high levels (e.g. HeLa, RAMOS, etc). Thus, amplification is only one mechanism by which cells acquire high levels of myc protein. In general, almost all transformed cells we have investigated express substantially elevated myc protein levels as compared to the levels found in non-transformed fibroblasts or lymphocytes. The most prominent exceptions are cells expressing elevated levels of another member of the myc protein family. In such cells, c-myc expression is invariably very low (e.g. in Kelly and IMR-32 neuroblastoma cells which have amplified N-myc genes or in COR-L88 small cell lung carcinoma cells which have an amplified L-myc gene).

In contrast to transformed cells, levels of c-myc protein in non-transformed fibroblasts or lymphocytes are always extremely low. In such cells, c-myc expression is rapidly induced by mitogenic stimuli, but even at peak levels, there are only a few thousand c-myc protein molecules per cell. Despite this low level, both fibroblasts and lymphocytes proliferate quite rapidly, with doubling times not too dissimilar from that of HeLa or Colo 320 cells. Clearly, then, cells do not require high levels of c-myc protein to proliferate rapidly. Rather, high c-myc protein levels appear to correlate more with some aspects of transformation than with proliferative rate.

These measurements thus pose an obvious question. Namely, what is the function, if any, of the elevated levels of myc protein in transformed cells?

It might be argued, that high myc protein levels in transformed cells are simply the result of earlier lesions within the proliferative machinery of the cell. The excess myc protein molecules need have no actual function, and are simply degraded. Evidence against this argument is that specific mechanisms which cause deregulation of myc genes, and hence elevated levels of myc protein, are common in many tumours; examples of this being chromosomal translocation, gene amplification and retroviral transduction. Clearly, then, elevated myc expression, over and above that found in non-transformed cells, must confer some form of growth advantage on the affected cell. So little is known of the function of myc proteins in normal cells that it is difficult to predict what the biochemical effect of an excess of myc protein may

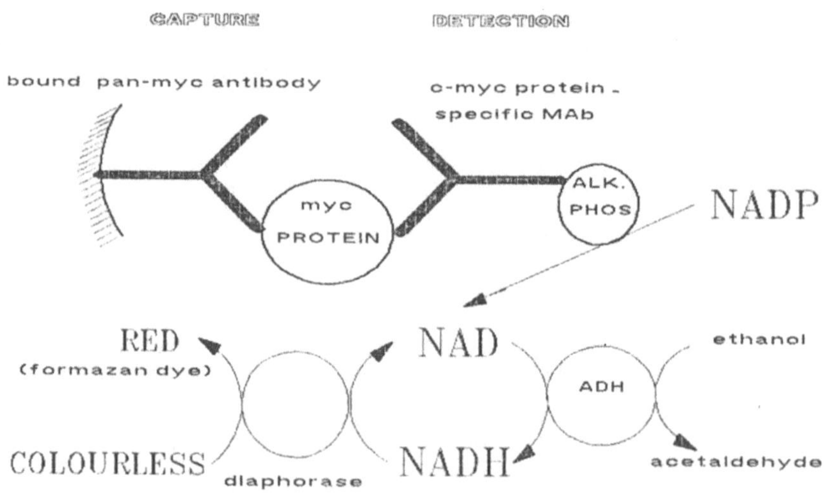

Figure 2. Schematic diagram of the components of the <u>myc</u> protein ELISA

Table 1. Levels of c-*myc* protein in cells

Cell Line	Cell Type	c-*myc* protein (molecules/cell)
HUMAN		
MRC-5 (quiescent)	lung fibroblast	400
MRC-5 (serum stimulated)	lung fibroblast	6300
MRC-5 (log phase)	lung fibroblast	3000
Kelly	neuroblastoma	140
IMR32	neuroblastoma	280
COR-L88	small cell lung ca.	310
HeLa S3	cervical carcinoma	16000
HeLa (adherent)	cervical carcinoma	97000
Molt 4	T cell leukemia	34000
DAUDI	Burkitt's lymphoma	62000
RAMOS	Burkitt's lymphoma	84000
RAJI	Burkitt's lymphoma	104000
HL60	promyelocytic leuk.	88000
COLO 320 HSR	colonic apudoma	124000
N417	small cell lung ca.	91000
MURINE		
thymocyte (quiescent)		120
thymocyte (stimulated)		1860
Swiss 3T3 (quiescent)	fibroblast	150
Swiss 3T3 (stimulated)	fibroblast	1980
F9	teratocarcinoma	10800
SP2/0	hybridoma	65800

be. For example, it is unknown whether all myc protein within the nucleus is actively engaged in myc protein functions all of the time, or whether there exists an intranuclear pool of unassociated myc protein. Conceivably, only a fraction of myc protein within the cell nucleus may be associated with specific targets. If this is the case, then biochemical analysis of the properties and associations of the bulk of myc protein in tumour cells expressing high levels of myc

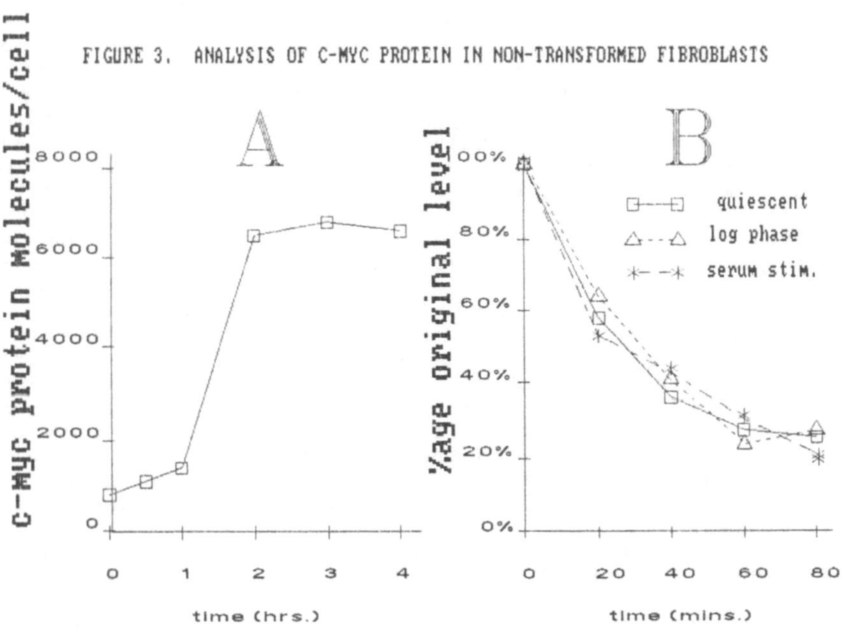

Figure 3. Analysis of c-myc protein in non-transformed human fibroblasts using the myc ELISA

Panel A: Induction of c-myc protein expression in mitogenically induced fibroblasts. MRC-5 human embryonic lung fibroblasts (passage 10-15) were arrested in medium containing 0.5% serum for two days. Cells were then stimulated by the addition of 10% serum to the growth medium. At the time points shown, equal aliquots of cells were removed, lysed and assayed for c-myc protein levels against a bacterially expressed c-myc protein standard.

Panel B: Measurement of c-myc protein half-life in non-transformed fibroblasts. MRC-5 cells were obtained from quiescent, mitogenically treated (2 hours after stimulation) and log phase cultures. Cells were treated with 50 mcg/ml cycloheximide, and equal aliqouts of the cells then measured for c-myc protein levels at 20 minute intervals.

protein may be misleading because the proportion of myc protein within the "active" fraction will be much smaller than in an non-transformed cell. For this reason, we have attempted to use the extreme sensitivity of the myc ELISA to investigate the properties of c-myc protein in normal cells which express low levels of c-myc.

Analysis of c-myc Protein in Non-Transformed Fibroblasts

Reliable analysis of c-myc protein in non-transformed fibroblasts has hitherto been virtually impossible because of the extremely low levels of the protein in these cells. With the sensitivity of the myc ELISA, however, such studies are straightforward. We have been able to use the ELISA to quantitate the average number of c-myc protein molecules in human fibroblasts during mitogenic stimulation (Figure 3A). Under the conditions of the experiment, about 98% of the cells become positive for c-myc protein expression with little apparent heterogeneity in staining, as judged by immunocytochemical analysis (Figure 4b). Thus, the average number of c-myc protein molecules per cell is a good indication of the actual number of c-myc protein molecules in each cell. Mitogenic stimulation of subconfluent quiescent human fibroblasts causes an approximately 10 fold increase in c-myc protein level, from 500-800 molecules per cell in quiescent cells to a peak of 6,000-8,000 molecules some 120 minutes after stimulation. Subsequently, c-myc protein level declines to about half the peak level, and is then maintained as long as the cells remain in log phase. Interestingly, the pattern of c-myc protein expression is different in confluent quiescent fibroblasts, which do not enter S phase or divide when serum stimulated. Although confluent cells exhibit a similar initial rapid induction of c-myc protein to subconfluent cells, c-myc protein levels then rapidly decline to the baseline level seen in quiescent cells. Thus, maintenance of c-myc protein levels at about 3-4,000 molecules per cell appears to be a requirement for passage through the cell cycle.

All existing studies indicate that the c-myc protein has an extremely short biological half life, a property thought to be an essential component of the protein's function. However, all of these studies have been conducted in tumour cells expressing elevated levels of c-myc. The observed short half-life of the c-myc protein might, therefore, be a consequence of its overproduction. Because of this we were interested to determine the biological half life of the c-myc protein in non-transformed fibroblasts which express very low levels of c-myc protein. To do this, we incubated cultures of quiescent, mitogenically stimulated and log phase MRC-5 human fibroblasts with cycloheximide to prevent protein synthesis and then monitored the rate of disappearance of the c-myc protein from the cells using the myc ELISA. The results, shown in Figure 3B, show the half life of the c-myc protein to be about 20-30 minutes in quiescent, log phase and mitogenically stimulated non-transformed fibroblasts. Thus, the reported short biological half life for the c-myc protein is indeed a general property of the protein, and not just an artefact of overexpression in tumour cells. We conclude that rapid turnover of the c-myc protein is a characteristic of the protein in both normal and tumour cells which must reflect some aspect of its function.

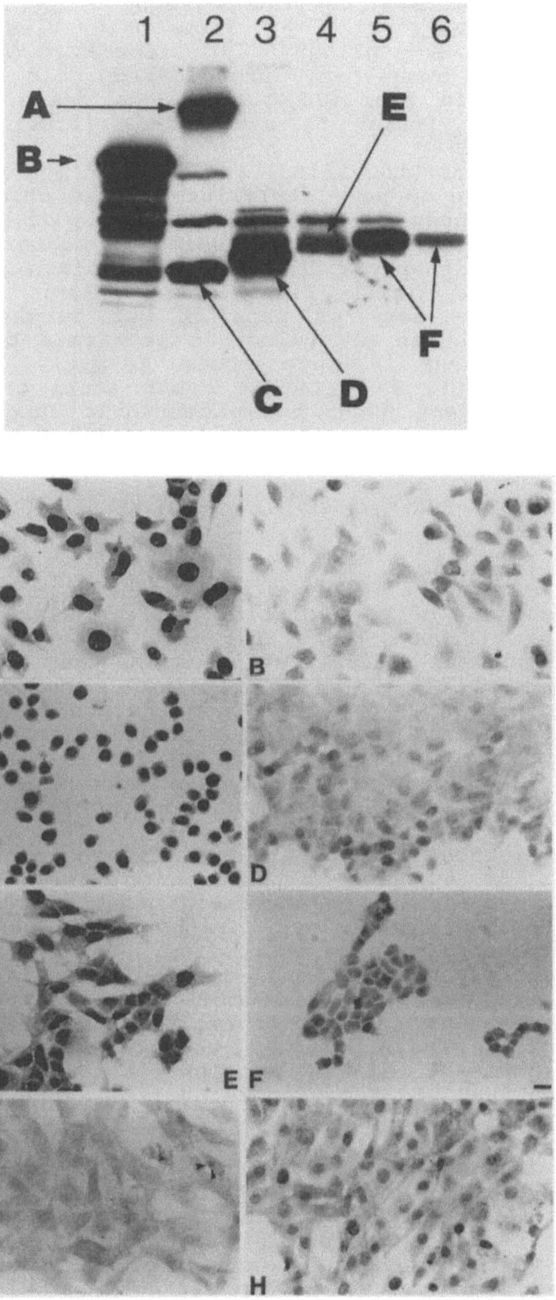

Figure 4. Recognition of _myc_ proteins by polyclonal pan-_myc_ antibody

Panel 1. All cells were labelled for one hour with ^{35}S-methionine, lysed in RIPA buffer and immunoprecipitated with antibody as described (Evan _et al._, 1985). Immunoprecipitated _myc_ proteins were fractionated on an SDS polyacrylamide gel, and labelled myc proteins visualised by fluorography. *myc* proteins are as follows: A - $p200^{gag-pol-v-myc}$ from OK10 virus infected quail fibroblasts (VIQF); B - $p110^{gag-v-myc}$ from MC29 VIQF; C - $p58^{v-myc}$ from OK10 VIQF; D - human c-*myc* protein from Colo 320 HSR cells; E - murine c-*myc* protein from SP2/0 hybridoma cells; F - human N-*myc* protein from Kelly (track 5) and IMR-32 (track 6) neuroblastoma cells.

Panel 2. Immunocytochemical analysis of *myc* proteins in cells. All cells were grown on poly -L-lysine-treated chamber slides, fixed in 4% paraformaldehyde in PBS and stained with pan-*myc* antibody. Bound antibodies were visualised using a commercial biotinylated second antibody-avidin-peroxidase kit (Vectastain). Panels are: A - MC29 virus-infected quail fibroblasts (VIQF) (expressing avian $p110^{gag-v-myc}$); B - same as panel A but blocked with con-*myc*-1 peptide; C - HL60 human promyelocytic leukaemia cells (c-*myc*); D - HeLa xenograft grown in nude mouse (c-*myc*); E - Kelly human neuroblastoma (N-*myc*); F - COR-L88 human small cell lung carcinoma (L-*myc*); G - quiescent MRC-5 human embryonic lung fibroblasts; H - MRC-5 cells, as in panel G, but 120 minutes after serum stimulation.

ANTIBODIES SPECIFIC FOR *myc* PROTEIN HOMOLOGY REGIONS - PAN-*myc* ANTIBODIES.

Several regions of the c-*myc* protein are especially well conserved throughout c-*myc* evolution and are also present in N- and L-*myc* proteins (Figure 1). The high degree of evolutionary conservation suggests that these regions may comprise important functional domains of the *myc* proteins. Accordingly, antibodies specific for conserved *myc* protein domains are likely to be valuable reagents both for identifying new *myc* proteins in various species and also as potential reagents to block *myc* protein functions. We have been attempting to generate antibodies specific for conserved *myc* protein domains by immunisation with appropriate synthetic peptide immunogens.

Residues 44-55 of the human c-*myc* protein comprise the first *myc* homology box. The sequence of this region, A-P-S-E-D-I-W-K-K-F-E-L, is one of the most highly conserved continous peptide sequences found in *myc* proteins. We synthesised this peptide, "con-*myc*-1", and used it as an immunogen to prepare antibodies. Polyclonal antibodies were prepared in rabbits and monoclonal antibodies in BALB/c x B6 F1 mice. We than tested these antibodies for their abilities to recognise various *myc* proteins in immunoprecipitation, immunoblotting and immunocytochemical assays. All of the polyclonal rabbit antibodies ("pan-*myc* antibodies") and one monoclonal antibody (PM-3E7) were able to recognise c-, v-, and N-*myc* proteins from every vertebrate species tested in each of these assays (Figure 4) (Evan *et al*., Mol. Cell. Biol., manuscript in press).

We next tested the pan-*myc* antibodies for their ability to recognise the previously unidentified human L-*myc* protein. Recent sequencing of the human L-*myc* gene (Legouy *et al*., 1987) shows that the first *myc* homology region is perfectly conserved in the L-*myc* protein. We therefore tested whether our antibodies recognised a candidate L-*myc* protein in L-*myc* expressing cells. As an L-*myc* expressing cell we chose the small cell lung carcinoma cell line COR-L88 which contains an amplified L-*myc* gene and expresses high amounts of L-*myc* mRNA (Ibson *et al*., 1987). In COR-L88 cells, the pan-*myc* antibody recognises a complex of three nuclear polypeptides with molecular weights between 60 and 66 kDa, as estimated on SDS gels. These putative L-myc products are not recognised by established c-myc or N-myc protein-specific antibodies, and competition with the appropriate peptide immunogen completely abolishes binding of these proteins by the pan-myc antibodies (Figure 5). Identical 60-66 kDa triplets of proteins are observed in other L-myc expressing cell line tested (data not shown), a pattern clearly different from that observed in c-myc and N-myc expressing cells. Further biochemical characterisation shows that these L-myc

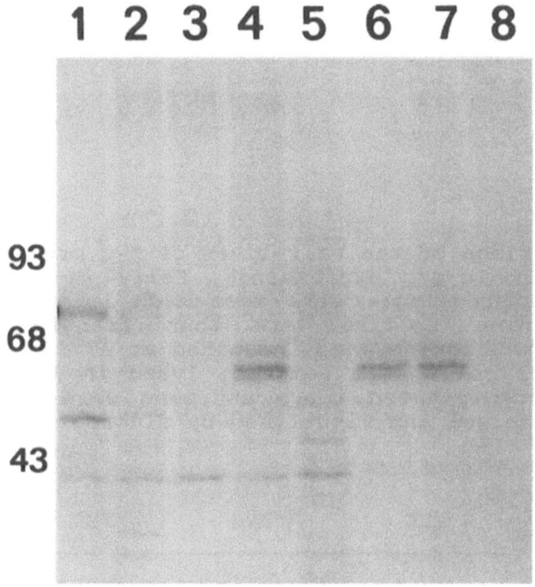

Figure 5. Recognition of the human L-myc protein by myc antibodies.
COR-L88 small cell lung carcinoma cells (Ibson et al., 1987) were labelled with 35S-methionine for 1 hour as described (Evan et al., 1985). Half of the labelled cells were lysed in RIPA buffer, the lysate divided into five equal fractions, and each fraction immunoprecipitated with antibody as follows: track 1 - PAB-421 anti-p53 MAb (Harlow et al., 1981); track 2 - CT14-G4 anti-$p62^{c\text{-}myc}$ MAb (Evan et al., 1985); track 3 - NM-1A8 anti $p66^{N\text{-}myc}$ MAb; track 4 - pan-myc polyclonal antibody (see above); track 8 - pan-myc polyclonal antibody plus 20 mcg/ml con-myc-1 peptide. The other half of the cells lysed in a lysis buffer consisting of 20mM TrisHCl/20 mM NaCl, 0.5% NP40/0.5% Na deoxycholate/1mM MgCl2 pH 7.4, and cytoplasmic (track- 5) and nuclear fractions prepared (Evan and Hancock, 1985). The nuclei were extracted with the same buffer containing 0.4M NaCl to give a soluble "chromatin" fraction (track- 6) and a residual "nuclear matrix" fraction (track- 7). Fractions were then immunoprecipitated with polyclonal pan-myc antibody

polypeptides are all phosphorylated, localise to the cell nucleus, and bind DNA (Evan et al., Mol. Cell. Biol., in press). The L-myc proteins, like those encoded by the c-myc and N-myc genes, have a short biological half life, although at 90 minutes the L-myc protein is significantly more stable than its c-myc and N-myc cousins (Figure 6). Comparative biochemical properties of the three human myc proteins are summarised in Table 2.

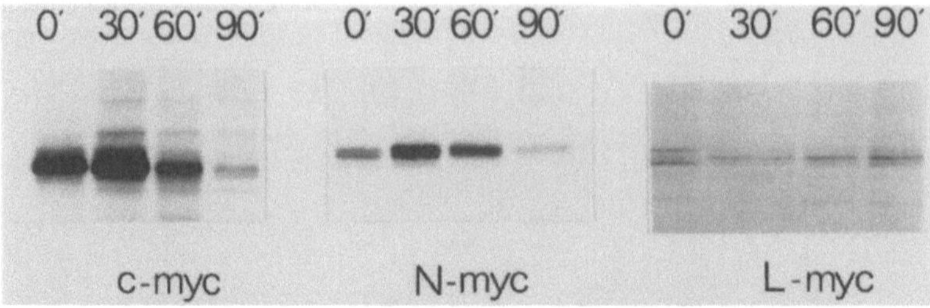

Figure 6. Comparison of the half-lives of myc proteins.

COLO 320 HSR (c-myc expressing), Kelly (N-myc expressing) and COR-L88 (L-myc expressing) cells were each labelled for 10 minutes with ^{35}S-methionine. Cells were then transferred into medium containing cold methionine, and incubated at 37°C. At various times, equal aliquots of cells were removed, lysed in RIPA buffer, and myc proteins immunoprecipitated. myc antigens were fractionated on an SDS polyacrylamide gel and visualised by fluorography

	c-myc	N-myc	L-myc
MOLECULAR WEIGHT	62/64 kDa	66/68 kDa	60/62/64 kDa
LOCATION	nucleus	nucleus	nucleus
HALF LIFE (mins)	20	25	90
PHOSPHORYLATED?	yes	yes	yes
BINDS DNA?	yes	yes	yes

Table 2. Comparison of biochemical properties of human myc proteins

The Immunological Analysis of a Conserved myc Protein Domain

The con-myc-1 peptide epitope recognised by the myc antibodies described above is highly evolutionarily conserved. There is thus a strong possibility that the con-myc-1 sequence comprises a functional domain involved in some conserved myc protein function - presumably a site of interaction between the myc protein and some target molecule in the cell nucleus. A number of recent studies have suggested a functional association between the c-myc protein and DNA replication (Iguchi-Ariga et al., 1987; Classon et al., 1987). Other data suggest roles for c-myc protein in regulation of gene expression

(Kaddurah-Daouk *et al.*, 1987) and, perhaps in RNA processing (Sullivan *et al.*, 1986).

In collaboration with Drs. Dean Jackson and Peter Cooke (Dunn School of Pathology, University of Oxford, England) we examined the ability of our *myc* protein-specific antibodies, including pan-*myc* antibodies, to interfere with DNA replication in isolated but fully functional HeLa cell nuclei (Jackson and Cooke, 1986). However, none of the antibodies used had any effect on DNA synthesis under conditions where anti-DNA polymerase antibodies caused substantial inhibition. Nor did any c-*myc* protein antibodies affect global RNA synthesis in the same nuclei. The same negative results were also obtained in N-*myc* and L-*myc* expressing cells using appropriate antibodies. We are thus unable to confirm any direct role for *myc* proteins in these nuclear functions using these assays.

If the con-*myc*-1 epitope is a functional *myc* protein domain, it is most likely that it lies on the surface of the native *myc* protein molecule. In order to determine whether or not this was so, we extracted c-*myc* and N-*myc* proteins from cell nuclei under a variety of conditions, so as to obtain *myc* proteins under native and denatured conditions. We then tested our pan-*myc* antibodies' abilities to recognise the extracted *myc* proteins (Figure 7). All of the antibodies used, including the pan-*myc* monoclonal antibody PM-3E7, were able to recognise c-*myc* and N-*myc* proteins whether the antigens were native or denatured. Binding of the antibodies to the putative L-*myc* protein from COR-L88 cells and to the $p110^{gag-v-myc}$ protein from MC29 virus-infected quail fibroblasts was similarly independent of the state of the antigen (not shown). The fact that the pan-*myc* monoclonal antibody PM-3E7 binds *myc* proteins independently of their conformation is of particular interest. Unlike the pan-*myc* polyclonal antibodies, which presumably comprise a heterogeneous population of combining sites capable of recognising various conformations of the con-*myc*-1 peptide, the pan-*myc* MAb PM-3E7 possesses a single binding site conformation. Thus, the binding of MAb PM-3E7 to *myc* proteins, independent of their comformational state, strongly suggests that the con-*myc*-1 epitope is indeed exposed on the surface of the native *myc* protein molecule and that it possesses an intrinsic structure which is largely independent of the conformational status of the rest of the oncoprotein.

The results from these experiments are consistent with the idea that the highly conserved con-*myc*-1 epitope could comprise a functional domain common to all *myc* proteins. We are currently preparing polyclonal and monoclonal antibodies against the other highly conserved regions of the *myc* protein family. If conserved peptide epitopes in *myc* proteins represent important functional domains, then antibodies specific for such regions should make good idiotypic determinants against which to raise antibodies whose binding sites mimic those *myc* protein domains. We have already prepared anti-idiotypic antibodies specific for the pan-*myc* monoclonal antibody PM-3E7 and using them to probe for potential *myc* protein targets. We believe that anti-idiotypic reagents of this type will prove useful in the identification of molecules with which *myc* proteins interact within the cell nucleus.

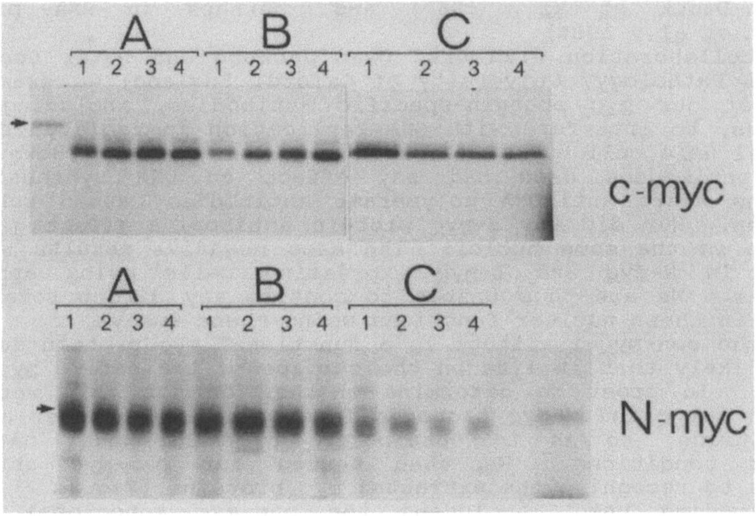

Figure 7. Stability and accessibility of the con-myc-1 epitope in c-myc and N-myc proteins

COLO 320 HSR (c-myc) and Kelly (N-myc) cells were labelled with ^{35}S-methionine. Cells were then washed and divided into four equal fractions. Cells in fraction 1 were lysed in 200 mM NaCl containing 20 mM TrisHCl pH 7.4 and 0.5% NP40. Nuclei were removed by centrifugation and the supernatant used for immunoprecipitation. Cells in fraction 2 were lysed in RIPA buffer, and the clarified lysates used for immunoprecipitations. Cells from fraction 3 were lysed in Tris buffered saline pH 7.4 containing 2% SDS. Lysates were sheared through a fine gauge needle until no longer viscous, boiled for 5 minutes, clarified by centrifugation, diluted 20X in RIPA buffer lacking SDS and used for immunoprecipitations. Cells in fraction 4 were lysed in Tris buffered saline containing 2% SDS and 10% 2-mercaptoethanol. Lysates were then treated exactly as for fraction 3 except that 5 mM ammonium persulphate was also added to the final solution.

Fractions were precipitated with A: polyclonal rabbit pan-myc antibody; B: monoclonal pan-myc antibody PM-3E7; C: either CT14-G4 MAb (recognising C terminus of c-myc protein) for COLO 320 lysates or NM-1A8 MAb (recognising N-myc protein) for Kelly cells.

REFERENCES

Alitalo K, Koskinen P, Makela TP, Saksela K, Sistonen L and Winqvist R (1987) oncogenes: activation and amplification. Biochem. Biophys. Acta 907: 1-32

Alt FW, DePinho R, Zimmerman K, Legouy E, Hatton K, Ferrier P, Tesfaye A, etc. (1987) The human myc gene family. Cold Spring Harb. Symp. Quant. Biol. 51: 931-941

Classon M, Henriksson M, Sumegi J, Klein G and Hammaskjold M-L (1987) Elevated c-myc expression facilitates the replication of SV40 DNA in human lymphoma cells. Nature (London) 330: 272-274
Evan GI and Hancock DC (1985) Studies on the interaction of the human c-myc protein with cell nuclei: p62$^{c\text{-}myc}$ as a member of a discrete subset of nuclear proteins. Cell 43: 253-261
Evan GI, Hancock DC, Littlewood TD and Gee NS (1986) Characterisation of human myc proteins. Curr. Top. Micro. Immunol. 132: 362-374
Evan GI, Lewis GK, Ramsay G and Bishop JM (1985) Isolation of monoclonal antibodies specific for the human c-myc protooncogene product. Mol. Cell. Biol. 5: 3610-3616
Hann SR, King MW, Bentley DL, Anderson CW and Eisenman RN (1988) A non-AUG translational initiation in c-myc exon 1 generates an N-terminally distinct protein whose synthesis is disrupted in Burkitt's lymphoma. Cell 52: 185-195
Ibson JM, Waters JJ, Twentyman PR, Bleehen NM and Rabbitts PH (1987) Oncogene amplification and chromosomal abnormalities in small cell lung cancer. J. Cell. Biochem. 33: 267-288
Iguchi-Ariga SMM, Itani T, Kiji Y and Ariga H (1987) Possible function of the c-myc product: promotion of cellular DNA replication. EMBO J. 6: 2365-2372
Kauddurah-Daouk R, Greene JM, Baldwin AS, Kingston RE (1987) Activation and repression of mammalian gene expression by the c-myc protein. Gen. Dev. 1: 347-357
Kelly K, Cochran BH, Stiles CD and Leder P (1983) Cell-specific regulation of the c-myc gene by lymphocyte mitogens and platelet-derived growth factor. Cell 35: 603-610
Legouy E, DePinho R, Zimmerman K, Collum R, Yancopoulos G, Mitsock L, Kriz R and Alt FW (1987) Structure and expression of the murine L-myc gene. EMBO J. 6: 3359-3366
Little CD, Nau MM, Carney DN, Gazdar AF and Minna JD (1983) Amplification and expression of the c-myc oncogene in human lung cancer cell lines. Nature (London) 306: 194-196
Moore JP, Hancock DC, Littlewood TD and Evan GI (1987) A sensitive and quantitative enzyme-linked immunosorbence assay for the c-myc and N-myc oncoproteins. Oncogene Res. 2: 65-80
Ohlsson RI and Pfeifer-Ohlsson S (1986) Myc expression in vivo during human embryogenesis. Curr. Top. Micro. Immunol. 132: 272-279
Persson H and Leder P (1984) Nuclear localization and DNA-binding properties of a protein expressed by human c-myc oncogene. Science 225: 718-721
Rabbitts PH, Watson JV, Lamond A, Forster A, Stinson MA, Evan G, Fischer W, etc. (1985) Metabolism of c-myc gene products: c-myc mRNA and protein expression in the cell cycle. EMBO J. 4: 2009-2015
Ruppert C, Goldowitz D and Wille W (1986) Proto-oncogene c-myc expression in cerebellar neurons at different developmental stages. EMBO J. 5: 1897-1901
Schwab M (1985) Amplification of N-myc in human neuroblastomas. Trends in Genet. 1: 271-275
Slamon DJ, Boone TD, Seeger RC, Keith DE, Chazin V, Lee HC and Souza LM (1986) Identification and characterisation of the protein encoded by the human N-myc oncogene. Science 232: 768-772
Small MB, Hay N, Schwab M and Bishop JM (1987) Neoplastic transformation by the human gene N-myc. Mol. Cell. Biol. 7: 1638-1645
Sullivan NF, Watt RA, Delannoy MR, Green CL and Spector DL (1986) Colocalisation of the oncogene protein and small nuclear ribonucleoprotein particles. Cold Spring Harb. Symp. Quant. Biol. 51: 943-947

The Replication Activity of SV40 DNA Correlates with the Level of c-*myc* Expression in Human Tumor Cell Lines

M. Henriksson, M. Classon, M.-L. Hammarskjöld, G. Klein, and J. Sumegi

INTRODUCTION

As a rule c-myc appears to be expressed in all proliferating cells and is turned off in quiescent (G0) cells. After induction, myc mRNA and protein appear within hours in both fibroblasts and lymphoid cells, prior to the onset of cellular DNA synthesis (Campisi et al., 1984, Kelly et al., 1983). The correlation between c-myc expression and growth state suggests that c-myc is involved in growth control.

Transgenic mice bearing c-myc coupled to the immunglobulin heavy-chain enhancer show an abnormal expansion of pre-B cell compartment (Alexander et al., 1987). Perturbation of c-myc expression occurs in many B-cell malignancies in chicken, mice, rat and humans. In Burkitt`s lymphomas with the t(8:14), mouse plasmacytomas t(12:15) and rat immunocytomas with t(6:7) translocation (Sumegi et al., 1983) the Ig/c-myc juxtaposition causes dysregulated myc expression which is implicated in the development of the neoplastic state.

RESULTS AND DISCUSSION

In order to demonstrate a possible relationship between the dysregulated high level of c-myc expression and enhanced replication of cellular DNA we designed the following experiment; i.) construction of plasmid DNA molecules with eukaryotic origin of replication sequences, ii.) transfection of such plasmid DNA molecules into normal cells and neoplastic cells with high c-myc expression using DEAE-dextran and iii.) measuring the amount of replicated DNA in the transfected cells. One of the vectors, the pSVEpR4 (Fig. 1), bears the SV40 origin of replication, enhancer and T antigen coding sequences cloned into pR4 (Hammarskjöld et al., 1986).

Department of Tumor Biology, Karolinska Institute, Box 60 400, S-10401 Stockholm, Sweden

Current Topics in Microbiology and Immunology, Vol. 141

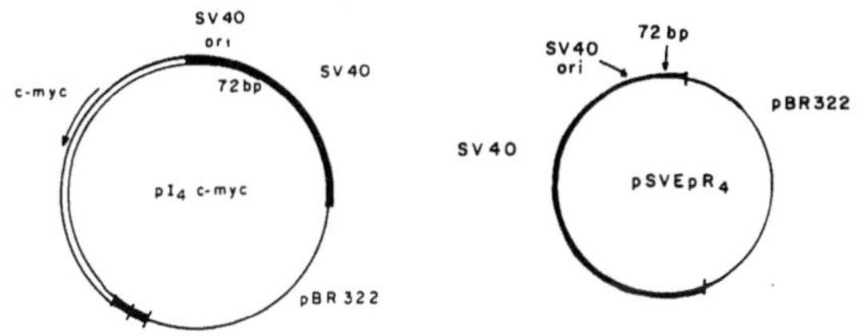

Figure 1. Schematic representation of the pSVEpR4 and pI4-P6 vector molecules. Heavy lines represent SV40 sequences, open line shows the c-myc second and third exons and single line displays the pR4 plasmid DNA sequences

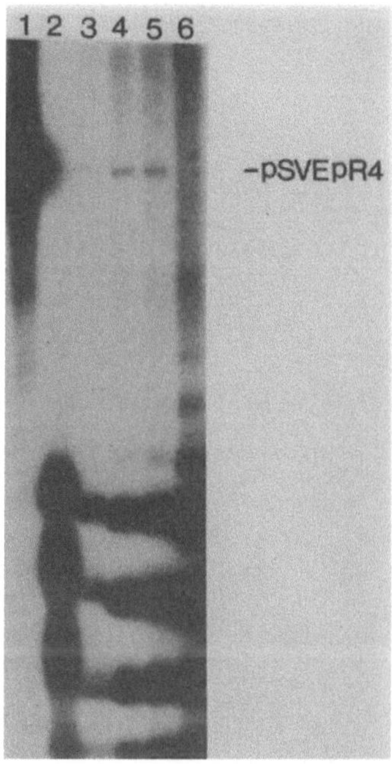

Figure 2. Southern blot analysis of pSVEpR4 in Hirt-lysates obtained from Ramos cells transfected with 5 ug of pSVEpR4 (lanes 1, 2) and from HEK cells (lanes 3, 4, 5, 6). The HEK cells were transfected with 1 ug (lane 3), 2 ug (lane 4) 5 ug (lane 5) and 10 ug (lane 6) of pSVEpR4 DNA. In lane 2 the DNA was cleaved with MboI in addition to DpnI/BamHI

This vector replicated well in monkey cells but its replication was barely detectable in human embryo fibroblasts (data not shown) and human embryo kidney cells (HEK) (Fig.2) as judged by the restriction analysis of Hirt lysate DNA with the isoschizomers DpnI and MboI. DpnI can only cleave the DNA at the GATC recognition sequence if the adenine is methylated on both DNA strands by the bacterial dam methylase whereas MboI cleaves only unmethylated DNA. The 5.2 kb long pSVEpR4 plasmid contains one BamHI site. Therefore the replicated plasmid molecules appear as a 5.2 kb linear fragment after DpnI/BamHI double digestion. The DNA transfected into primary human kidney cells and human fibroblasts was sensitive to DpnI due to the lack of effective replication in these eukaryotic cells. This is in line with previous observations of SV40 replication in HEK cells (Lebkowski et al., 1985).

A significant difference was seen when certain human neoplastic cell lines were tested for their ability to promote replication of the SV40 based vector . Fig. 3 shows that the pSVEpR4 DNA molecules replicated well in two Epstein-Barr virus (EBV) carrying and two EBV negative Burkitt`s lymphoma lines (Raji, Daudi, Ramos and Loukes).

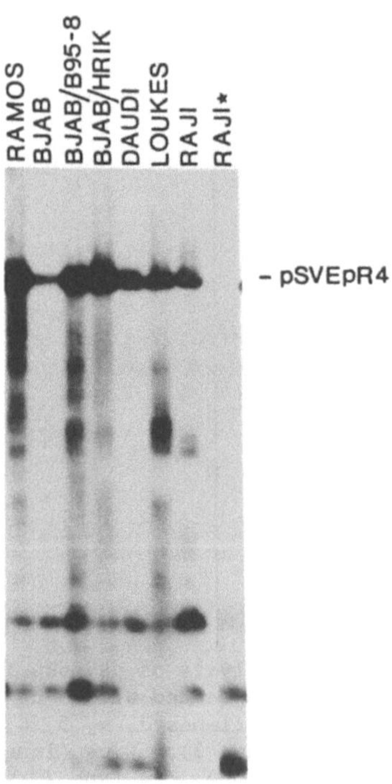

Figure 3. Replication of pSVEpR4 in human lymphoma lines. The cells were transfected with 5 ug of pSVEpR4 DNA. The DNA in the last lane was cleaved with MboI in addition to DpnI/BamHI cleavage

Only a small amount of replicated pSVEpR4 DNA was found in the human lymphoma cell line BJAB. The four Burkitt`s lymphoma lines contain high levels of c-myc transcripts due to the (8:14) translocation associated Ig/c-myc juxtaposition. BJAB is EBV negative, lacks the Ig/c-myc translocation and has very low level of c-myc expression. Two EBV converted BJAB sublines BJAB HRIK and BJAB B95-8 express c-myc at a level comparable to the BL`s lines. They supported the replication of the pSVEpR4 20-50 fold better than the parental BJAB line.
Adherent HeLa cells also supported replication of the SV40 DNA (data not shown). It is noteworthy that HeLa contains similar amounts of c-myc protein as HL60 that carries approximately 25-50 copies of c-myc (G. Evan , personal communication).

These findings suggested a possible relationship between a high c-myc expression and the ability of pSVEpR4 to replicate effectively in HeLa and human lymphoma cells.

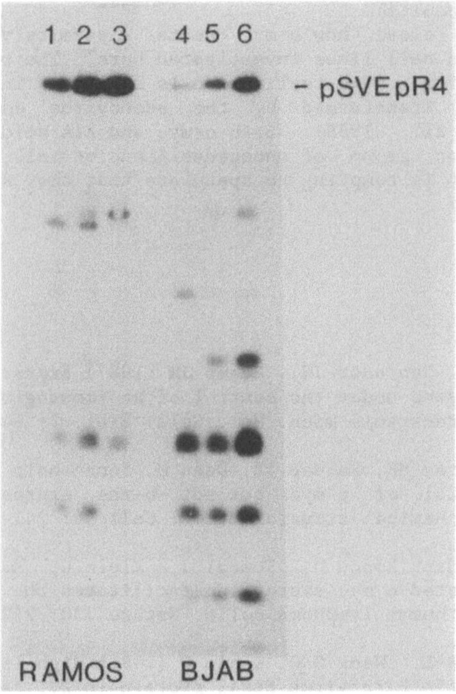

Figure 4. Level of pSVEpR4 replication in function of increasing amounts of pI4-P6 transfected into Ramos (lanes 1, 2, 3) and BJAB (lanes 4, 5, 6) cells. The total amount of DNA cotransfected with pSVEpR4 vector was kept constant by addition of empty pI4 vector to give the following molar ratios of pI4/pI4-P6; 2,5 pmol/0,0 pmol (lanes 1 and 4), 1.25 pmol/1.25 pmol (lanes 2 and 5) and 0.6 pmol/1.9 pmol (lanes 3 and 6)

To confirm the dependence of the pSVEpR4 replication upon the the high level of c-myc we measured the replication of the vector cotransfected with a c-myc expressing vector in Ramos and Bjab lymphoma cells. The vector pI4-P6 contains the second and third coding exons of c-myc under the control of the early SV40 promoter and enhancer (Fig. 1). The vector governs the synthesis of c-myc protein (Classon et al., 1987). The result of the cotransfection is shown in Fig.4. The replication of pSVEpR4 was proportional to the amount of c-myc expressing vector transfected into both human lymphoma lines suggesting a direct relation between the level of c-myc expression and the replication of pSVEpR4.

An increase of the T antigen level in BJAB cells was not sufficient to enhance the replication of pSVEpR4 DNA (Classon et al., 1987) indicating that the myc-dependent enhancement of replication was not due to an increased SV40 T antigen level.

The high c-myc level did not abrogate the requirement for T antigen since the pI4-P6 vector, that does not express T antigen, failed to replicate in Ramos cells (Henriksson unpublished results).

It is not clear how c-myc creates a permissive environment in the human cell lines investigated here. The restriction of human cells for SV40 replication is abolished in human embryo kidney cells transformed by the adenovirus early genes E1 (Lebkowski et al., 1985). Both c-myc and E1A belong to the same complementation group of oncogenes (Land et al., 1983, Ruley 1983) and it is tempting to speculate that they act through a common pathway.

REFERENCES

Alexander WS, Schrader JW, Adams JM (1987) Expression of the c-myc oncogene under the control of an immunoglobulin enhancer in Eu-myc transgenic mice. Mol. Cell. Biol. 7: 1436-1444.

Campisi J, Gray HE, Pardee AB, Dean M, Sonenshein GE (1984) Cell cycle control of c-myc but not c-ras expression is lost following chemical transformation. Cell 36: 241-247

Classon M, Henriksson M, Sumegi J, Klein G, Hammarskjöld M-L (1987) Elevated c-myc expression facilitates the replication of SV40 DNA in human lymphoma cells. Nature 330: 272-274.

Hammarskjöld M-L, Wang S-C, Klein G (1986) High level expression of the Epstein Barr virus EBNA1 protein in CV1 cells and human lymphoid cells using a SV40 late replacement vector. Gene 43:41- 50.

Kelly K, Cochran BH, Stiles CD, Leder P (1983) Cell specific regulation of the c-myc gene by lymphocyte mitogens and platelet derived growth factor. Cell 35: 603-610.

Land H, Parada LF, Weinberg RA (1983) Tumorigenic conversion of primary embryo fibroblasts requires at least two cooperating oncogenes. Nature 304: 596-602.

Lebkowski JS, Clancy S, Calos MP (1985) Simian Virus 40 replication in adenovirus-transformed human cells antagonizes gene expression. Nature 317: 169-171.

Ruley HE (1983) Adenovirus early region 1A enables viral and cellular transforming genes to transform primary cells in culture. Nature 304: 602-606.

Sumegi J, Spira J, Bazin H, Szpirer J, Levan G, Klein G (1983) Rat c-myc oncogene is located on chromosome 7 and rearranged in immunocytomas with t(6:7) chromosomal translocation. Nature 306: 497-499.

Elevated Levels of the c-*myc* Protein in Bloom's Syndrome and Induction of c-*myc* by DNA Strand Breakage

N. F. Sullivan and A. E. Willis

Abstract

We demonstrate elevated levels of the c-*myc* protein in lymphoblastoid cell lines derived from Bloom's syndrome patients and, furthermore, show that treatment of normal lymphocytes with DNA damaging agents results in induction of c-*myc* protein. We present a model whereby increased DNA strand breakage in Bloom's syndrome, as a consequence of the reduced activity of DNA ligase I, results in the constitutive induction of c-*myc*. Increased c-*myc* protein may thus contribute towards the predisposition to cancer observed in Bloom's syndrome.

Introduction

A number of rare inherited syndromes are associated with apparent defects in DNA repair and a greatly increased frequency of cancer. One of these, Bloom's syndrome (BS), is of particular interest since patients are predisposed to a variety of malignancies which generally occur in patients less than 20 years of age (German et al., 1984). In addition, BS is unique in exhibiting a particularly high frequency of sister chromatid exchange (SCE) (Chaganti et al., 1974) and a deficiency in DNA ligase I (Willis and Lindahl, 1987; Willis et al., 1987). The DNA ligase I deficiency in BS would be expected to promote homologous recombination events and hence also SCE via retarded joining of DNA strand interruptions (Willis and Lindahl, 1987). Along a similar theme, if either normal fibroblastoid or lymphoblastoid cell lines are treated with alkylating agents to induce single strand breaks in the DNA via excision repair, the frequency of SCE can be shown to increase (Krepinski et al., 1979). It was thus of interest to note recent reports which have indicated that transfection of established rat fibroblasts either the c-*myc* oncogene or certain other immortalizing viral oncogenes increases the frequency of SCE (Cerni et al., 1987). The experiments presented investigate the levels of c-*myc* protein in BS and a possible association between strand breakage and c-*myc* expression.

Results

Expression of c-*myc* in Bloom's syndrome

Several lines of lymphoblastoid origin (Table 1) were examined for

Imperial Cancer Research Fund, Clare Hall Laboratories, Blanche Lane, South Mimms, Herts., EN6 3LD, U.K.

Current Topics in Microbiology and Immunology, Vol. 141

Table 1

Line¥	Characteristic	Ligase genotype+	SCE/cell*	Source/ Reference
1004	Bloom's syndrome; Mennonite	lig-/-	80	R. Weksberg
W67-4	Bloom's syndrome; Ashkenazi	lig-/-	60-80	E.Henderson
D86-1-2	Bloom's syndrome; Ashkenazi	lig-/-	10-15	E.Henderson
AA87-4	Bloom's syndrome; Ashkenazi	lig+/-	3-8	E.Henderson
GM1953	Normal	lig+/+	3-5	HGMCR
AG3829	Werner's syndrome	lig+/+	Normal*	HGMCR
GM1526	Ataxia Telangiectasia	lig+/+	Normal	HGMCR
GM1712	Cockayne's syndrome	lig+/+	Normal	HGMCR
Raji	Burkitts lymphoma	lig+/+	3	HGMCR

¥ All lines have been immortalized by EBV
+ Homozygote for DNA ligase I deficiency (lig-/-), heterozygote lig+/-, normal lig+/+
* Normal SCE represents 3-5 SCE per cell (10).
HGMCR Human Genetic Mutant Cell Repository, Camden NJ 08103, U.S.A.

expression of the c-*myc* protein (Fig. 1) using the recently described enzyme linked immunosorbent assay for c-*myc* (Moore et al., 1988).

Figure 1 Expression of c-*myc* in Bloom's syndrome

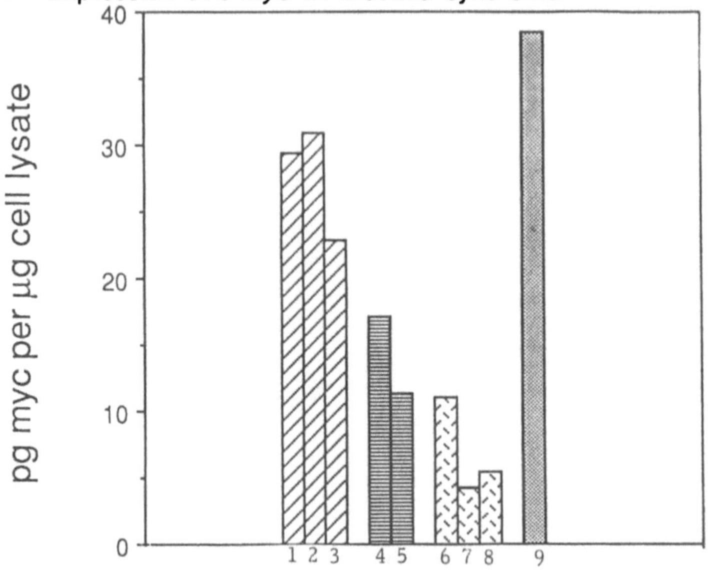

Fig. 1. 20mls (at $1x10^6$ cells/ml) of Bloom's syndrome and other lymphoblastoid cell lines, grown in RPMI (1640) containing 15% foetal calf bovine serum, were harvested by centrifugation, washed once in TBS (20mm Tris-HCl pH7.5, 150mm NaCl) and stored at -70°C. Elisas were performed according to the procedure described (Moore et al., 1988) with the following modifications: **1)** for some assays we used alkaline phosphatase conjugated goat anti mouse second antibody instead of a directly conjugated 3C7 (anti c-*myc*) antibody. No significant differences were observed. **2)** The results are expressed in terms of the amount of c-*myc* protein present in the total cell lysate. Thus, to approximately $1\text{-}2x10^7$ cells, 40µl of lysis buffer containing protease inhibitors (Moore et al., 1988) were added and the mixture briefly sonicated. The sample was boiled for 5 minutes and iodoacetamide was added to 100mM, followed by incubation on ice for 30 minutes. Samples were centrifuged for 15 minutes at 15,000g and the supernatant collected. Protein concentrations in the supernatant were typically 2mg/ml. For Elisa, samples were diluted with lysis buffer such that each well received 7µg of protein. As calculated from a standard curve using recombinant c-*myc* protein (Watt et al., 1985) 10pg of c-*myc*/µg lysate corresponds to approximately 10,000 molecules per cell. The results presented are the average of at least 4 independent experiments, and the average error was within 10% of the stated value in all cases.

Lane: 1, W67-4; 2, 1004; 3, D86-1-2; 4, W87-4; 5, GM1953; 6, AG3829; 7, GM1526; 8, GM1712, 9, Raji.

Three lines derived from BS patients (W67-4, 1004 and D86-1-2) and homozygous for a type I-1 DNA ligase I mutation (Willis et al., 1987) all express elevated levels of the *c-myc* protein (32, 34 and 25pg/µg respectively) when compared to either a DNA ligase I heterozygote (AA87-4, 17pg/µg). Lymphoblastoid cell lines derived from other inherited diseases (Werner's syndrome, 11pg/µg; Ataxia telangiectasia, 6.5pg/µg, and Cockayne's syndrome 6pg/µg) show reduced amounts of the *c-myc* protein. The increased expression of *c-myc* observed in lines derived from BS patients in similar to that seen in a representative Burkitts lymphoma line (Raji, 41pg/µg). Of the lines used, only Raji was capable of forming tumours in immunodeficient mice, which suggests that besides elevated *c-myc* expression, other factors must be required in order to elicit the fully transformed phenotype in EBV immortalized BS lymphoblasts. The expression of another immortalizing oncogene, p53, was examined using an Elisa system similar to that used for detection of *c-myc*. However, there was no clear relationship between the levels of p53 and the inherited diseases tested here (data not shown).

Correlation of *c-myc* expression with SCE and the activity of DNA ligase I

The BS lines which are homozygous for a DNA ligase I mutation (lig-/-) and show the highest level of SCE (W67-4, 1004) are also those lines which show the highest amounts of the *c-myc* protein (32 and 34 pg/µg respectively). We have examined one BS line which is homozygous for a DNA ligase I mutation yet spontaneously exhibits reduced SCE (D86-1-2). This line expresses a lower amount of the *c-myc* protein (25pg/µg). However, all BS homozygotes show both elevated SCE and *c-myc* protein when compared to a line heterozygous for a DNA ligase I mutation (AA87-4) or a control lymphoblastoid line (GM1953). Other inherited diseases (represented by the cell lines: AG3829, GM1526 and GM1712; Table 1) show normal frequencies of SCE (Evans, 1982) and quantities of *c-myc* protein similar to those observed in the control line GM1953.

In addition to the correlation with SCE, *c-myc* expression is approximately inversely related to the activity of the DNA ligase I (Fig. 2). The deficiency in DNA ligase I may result in an increase in the steady state level of endogenous DNA strand breakage, and therefore, further experiments were designed to investigate the potential relationship between DNA strand breakage and *c-myc* expression.

Induction of *c-myc* by DNA damaging agents

Normal lymphoblastoid cells (GM1953) were treated with both methylating (dimethylsulphate, DMS) and ethylating (ethylmethanesulphonate, EMS) agents in order to induce DNA strand breakage, the result of repair of alkylation damage. The final stages of excision repair involve generation of a single strand gap in the DNA with subsequent filling and resealing of the gaps by the action of DNA polymerase I and DNA ligase I (Lindahl, 1982).

Figure 2. DNA ligase activity in Bloom's syndrome

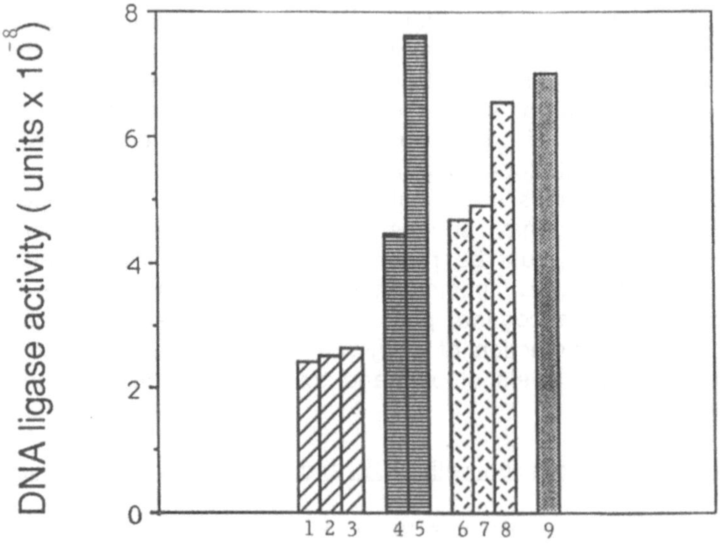

Fig. 2. The activity of DNA ligase I was measured as described (Arrand et al., 1986). Lanes 1-9 are as Figure 1

The c-*myc* protein is induced approximately six fold by 0.1μM DMS (to 60pg/μg; 60,000 molecules/cell) (Fig. 3a). Similarly, induction with EMS is also dose dependent with a five fold increase in the amount of c-*myc* protein (Fig. 4a). Using the optimum dose of DMS a time course reveals maximal c-*myc* expression at 50 minutes followed by a return to near normal levels after approximately 2 hours (Fig. 3b). Treatment with EMS (0.1mM) gave maximal c-*myc* induction at 6 hours returning to a basal level by 72 hours (Fig. 4b). With both DMS and EMS, induction of c-*myc* protein exactly parallels the kinetics of excision repair. Similar results have also been obtained with mitomycin C and irradiation (not shown). Thus expression of the c-*myc* protein can be stimulated by a variety of agents which induce strand breaks in DNA.

Discussion

The data presented here shows an increased constitutive expression of c-*myc* in Bloom's syndrome. In addition, c-*myc* can be induced in normal lymphoblasts by the action of DNA damaging agents. The DNA ligase I deficiency in BS may result in an increase in the steady state level of DNA strand breaks thereby providing potential substrates for SCE and inducing

Figure 3. Induction of c-*myc* by DMS

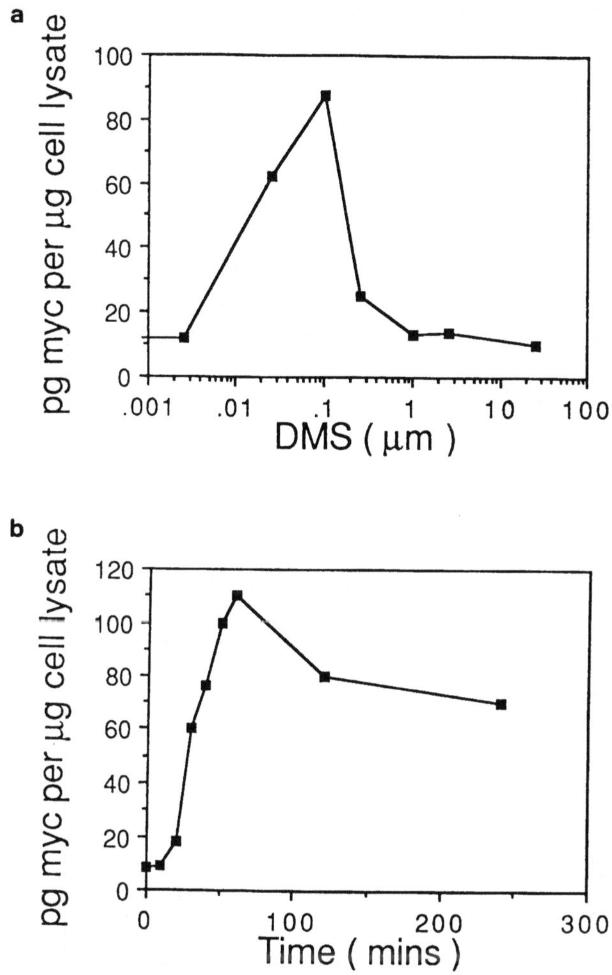

Figs. 3a, b. 20mls of GM1953 at 1x10^6 cells/ml were grown as described in Fig.1. Cells were (a) treated with dimethylsulphate (DMS) at various concentrations and harvested at 1hr and (b) treated with an optimum concentration of DMS (0.1µM) and harvested at predetermined time intervals

c-*myc*. However, in a normal cell endogenous strand breakage will be low, a substrate for SCE will become limiting and the signal for c-*myc* induction will be reduced. The increase in c-*myc* levels concurrent with DNA repair processes may indicate a role for c-*myc* in DNA repair. Experiments are in progress to investigate this possibility.

As previously mentioned, BS patients are predisposed to various forms of cancer. Our finding of increased c-*myc* protein in non tumorigenic BS lymphoblasts leads us to propose that the c-*myc* protein may provide a step in the progress of BS lymphoblasts towards malignancy

Figure 4 Induction of c-*myc* by EMS

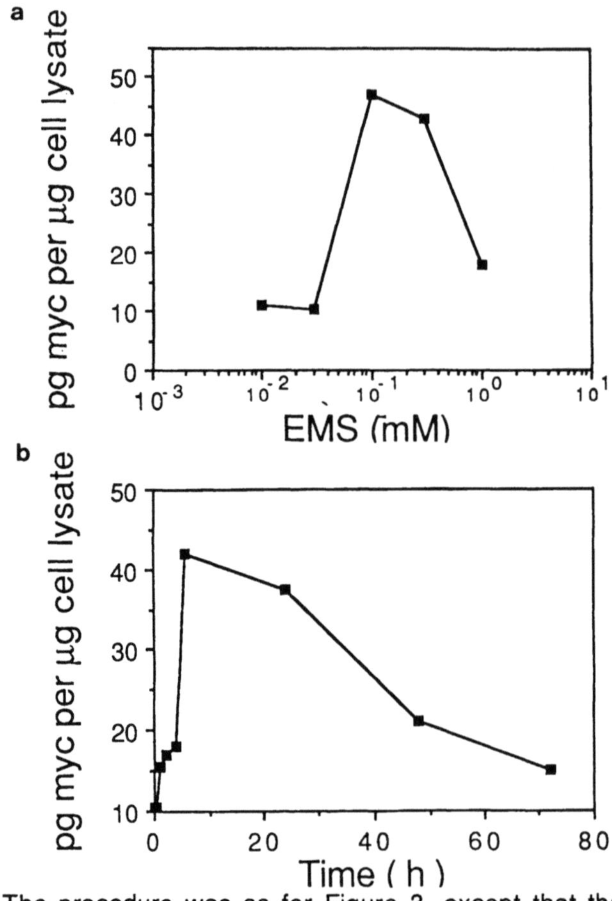

Figs. 4a, b. The procedure was as for Figure 3, except that the EMS dose dependence (a) was harvested at 6hrs and (b) the concentration of EMS used in the time course was 0.1mM

Acknowledgements

We wish to thank Sally Tomlinson and the staff of Cell Production at Clare Hall for tissue culture, Julian Gannon who kindly provided anti p53 antibodies and the Secretarial staff of Clare Hall, Derval Byrne, Anna Ives and Jyotsna Dalal for preparation of this manuscript. We would also like to thank Tomas Lindahl, Peter Karran and Richard Wood for helpful discussions.

References

Arrand JE, Willis AE, Goldsmith I and Lindahl T (1986) J. Biol. Chem. **261**, 9097-9082.

Cerni C, Mougneau E and Curzin F (1987) Exp. Cell Res., **168**, 439-446.

Chaganti RSK, Schonberg S and German J (1974) Proc. Natl. Acad. Sci. U.S.A., **71**, 4505-4512.

Evans HJ (1982) Chap. 7 in "Sister Chromatid Exchange" 184-214, ed. Sheldon Wolff, Wiley and Sons.

German J, Bloom D and Passarge E(1984) Clin. Genet., **25**, 166-174.

Henderson E and Ribeckey R(1980) Chem. Biol. Interactions, **33**, 63-81.

Krepinski AB, Heddle JA and German J (1979) Hum. Genet., **50**, 151-156.

Lindahl T (1982) Ann. Rev. Biochem., **51,** 61-87.

Moore JP, Hancock DC, Littlewood TD and Evan GI (1988) Oncogene Res., **2**, 65-80.

Watt RA, Shatzman A and Rosenberg R (1985) Mol. Cell. Biol., **5,** 448-456.

Weksberg R, Smith C, Anson-Cartwright L and Maloney K. Am. J. Hum. Genet., in press.

Willis AE and Lindahl T (1987) Nature, **325**, 355-357.

Willis AE, Weksberg R, Tomlinson S. and Lindahl T (1987) Proc. Natl. Acad. Sci. U.S.A., **84**, 8016-8020.

Deregulation of the c-*myc* and N-*myc* Genes in Transformed Cells

M. DEAN[1], J. CLEVELAND[1], H.-Y. KIM[2], J. CAMPISI[2], R. A. LEVINE[2], J.N. IHLE[1], and U. RAPP[1]

INTRODUCTION

Expression of the c-myc proto-oncogene is regulated by mitogens in a number of cell types, including fibroblasts and lymphoid cells. In growth arrested fibroblasts, serum, epidermal growth factor, or platelet derived growth factor are all capable of inducing c-myc mRNA(Campisi et al., 1984; Kelly et al.,1983). The mechanism of this induction appears to require either protein kinase C- or protein kinase A-dependent events(Ran et al., 1986). We have begun to examine these events in greater detail by studying fibroblasts showing de-regulated c-myc expression and myeloid cells containing exogenously introduced c-myc constructs.

A31 cells are clonally-derived Balb/c-3T3 murine fibroblasts having well described growth characteristics(Campisi et al., 1984). Benzo(a)pyrene and dimethylbenz(a)anthracene transformed derivatives of A31 cells (BPA31 and DA31) have been isolated(Holley et al., 1976; Oshiro and DiPaolo, 1974). Although BPA31 and DA31 cells can be growth arrested by serum starvation, c-myc expression is not regulated as in A31 cells(Campisi et al.,1984). FDC-P1 cells are a mouse myeloid cell line that is dependent on interleukin 3 (IL-3) for both growth and survival in culture(Dexter et al., 1980).

RESULTS

c-myc is Selectively De-regulated in Chemically Transformed Fibroblasts

To further investigate the mechanisms leading to de-regulated expression of the c-myc gene previosly observed in chemically transformed Balb/c-3T3 cells (BPA31, DA31) we examined c-myc mRNA levels in relation to that of c-fos and actin. We found that whereas the expression of c-myc is constituitive in the two chemically transformed derivatives, the levels of c-fos and actin are growth regulated in A31, BPA31 and DA31 cell lines (data not shown). In BPA31 cells, serum induces a 3-fold increase in the already high levels c-myc mRNA after 2 hours of serum stimulation; in DA31 cells, c-myc mRNA was constituitively expressed at a high level and was not modulated by serum. We conclude that the lesions responsible for deregulation of c-myc expression do not detectably affect the expression of c-fos or actin, two genes normally induced in parallel with myc.

The Structure of the c-myc Gene in Transformed Fibroblasts

We have previously shown by Southern analysis that the structure of the c-myc locus is not grossly altered in BPA31 and DA31 cells nor is the gene amplified in these cells(Campisi et al.,1984). However, human Burkitt's lymphoma cell lines showing unregulated c-myc expression and mutations in c-myc exon 1 have been recently described(Cesarman et al., 1987). To more closely examine the structure of exon 1 in the transformed A31 derivatives, we compared the restriction maps of this region in the three cell lines. No alterrations were observed in over 15

[1]NCI-Frederick Cancer Research Facility, Frederick, MD. [2]Boston University School of Medicine, Boston, MA

Current Topics in Microbiology and Immunology, Vol. 141

sites tested in and around exon 1 (data not shown). In addition, we performed S1 nuclease digestion of RNA from these cell lines with an exon 1 probe and found that the transformed lines show normal transcription from the two c-myc promoters (data not shown) Thus the aberrant regulation of c-myc is not associated with any alteration in the structure of the gene, within the limits of these analyses. Although we can not rule out the presence of a small mutation in the c-myc locus that results in unregulated expression in BPA31 or DA31 cells, the data are consistent with a mutation in a trans-activating factor or pathway in these cell lines.

Unregulated c-myc Expression is Independent of Protein Kinase C Activity

We have shown in quiescent A31 cells that c-fos and c-myc mRNA can be induced by either of two intracellular growth factor dependent pathways. One of these pathways is dependent on protein kinase C (PKC) and is utilized by PDGF or phorbol ester tumor promoters; the other pathway is dependent on cAMP, and presumably protein kinase A, and is utilized by EGF (Ran et al. 1987, McCaffrey et al., 1987). Serum, being a mixture of growth factors, activates both pathways.

BPA31 cells proliferate in platelet poor plasma (data not shown), which is essentially serum that is deficient in platelet factors, including PDGF. In addition, BPA31 cells have about three fold higher PKC activity than either A31 or DA31 cells (Table 1A). It was therefore possible that, at least for BPA31 cells, elevated PKC activity, or increased phosphorylation of the PKC substrate, could be responsible for the unregulated c-myc expression shown by these cells. To test this idea, we treated all three cell lines with high concentrations of phorbol dibutyrate (PDBu), which when chronically administered, depletes the cells of PKC activity (Table 1B). Despite a greater than 20 fold reduction in PKC activity, as a result of PDBu treatment, c-myc mRNA levels remained elevated in serum-deprived BPA31 and DA31 cells (Fig. 1). In addition, all three cell lines proliferated at nearly the control rate in medium containing 10% serum and PDBu (data not shown). We conclude that cell proliferation and the unregulated c-myc expression shown by the transformed cells are independent of protein kinase C.

Activation of c-myc and N-myc in a Myeloid Cell Line

We recently reported the construction of a retrovirus vector expressing the mouse c-myc gene, and the infection of IL-3 dependent FDC-P1 cells with this virus (Dean et al.,1987). Constitutive c-myc expression led to the appearance of IL-3 independent cells and the activation of N-myc expression in several of the independent cell lines. One cell line, FDmyc-28R contains activated c-myc and N-myc genes, suggesting that a single mutation may be able to activate more than one myc family gene.

To examine the structure of the c-myc gene in FDmyc cells we have analyzed exon 1 sequences using both S1 and RNase A protection techniques. Figure 2 shows that RNase A protected the full length of an exon 1 probe in both FDC-P1 and FDmyc cells, demonstrating that no mutations, detectable by this method, are present in these cells.

Table 1. Protein kinase C activity in A31, BPA31 and DA31 cells

A. RELATIVE PROTEIN KINASE C ACTIVITY

Cell Line	PKC Activity*	% A31 Cells
A31	6.174	100
BPA31	19.890	322
DA31	6.810	110

*cpm/10^6 cells/h

B. DOWN-REGULATION OF PROTEIN KINASE C ACTIVITY

Cell Line	PDBu Treatment*	% Activity
A31	None	100
	200 ng/ml	9
	400 ng/ml	4
BPA31	None	100
	400 ng/ml	16
	1200 ng/ml	8
	4800 ng/ml	2
DA31	None	100
	200 ng/ml	26
	400 ng/ml	13
	1200 ng/ml	4

* Phorbol dibutyrate (PDBu) was added at the indicated concentrations in 0.2% serum for 72 h

Cells were made quiescent by serum deprivation, treated with PDBu where indicated and lysed in 20 mM Tris HCl (pH 7.5). 2 mM EDTA, 0.5 mM EGTA, 10 mM DTT, 1% Triton-X 100, 10 µg/ml aprotinin and 1 mM PMSF. The lysates were clarified by centrifugation, incubated with DE-52 resin and PKC eluted from the washed resin with 100 mM NaCl. The eluate was assayed for calcium (1mM)- and phospholipid (160 µg/ml phosphatidylserine, 200 ng/ml PDBu)-dependent histone kinase activity using 400 µg/ml histone, 2µCi gamma-^{32}P-ATP (5-25 Ci/mmol) and 50 µM ATP for 10 min at 30 C. Phosphorylated histone was separated from ATP on PI-cellulose paper and detected by counting Cerenkov radiation using a beta counter.

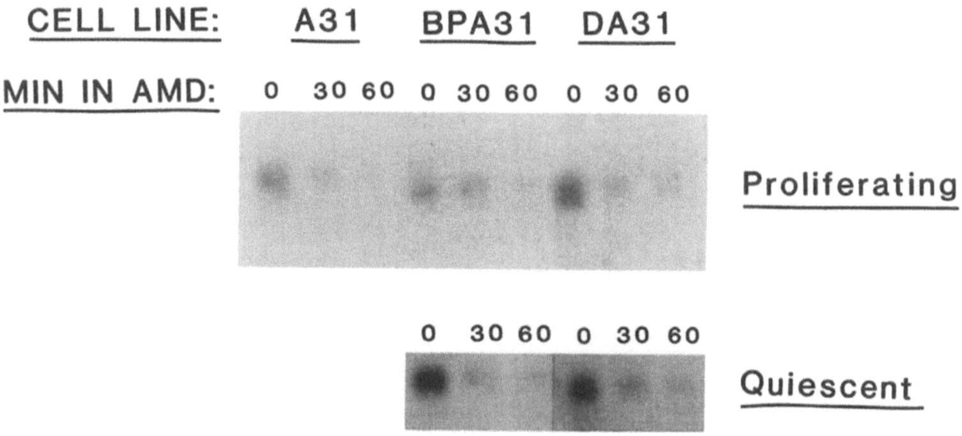

Fig. 1 c-myc mRNA levels remain elevated after transformed cells are made deficient in protein kinase C activity. Cells were made quiescent by serum deprivation (control) and then given the indicated concentrations of phorbol dibutyrate (PDBu) in 0.2% serum for 72 hrs. Total cellular RNA was isolated from control and PDBu-treated cultures and the prevalence of c-myc mRNA was determined by Northern blot analysis

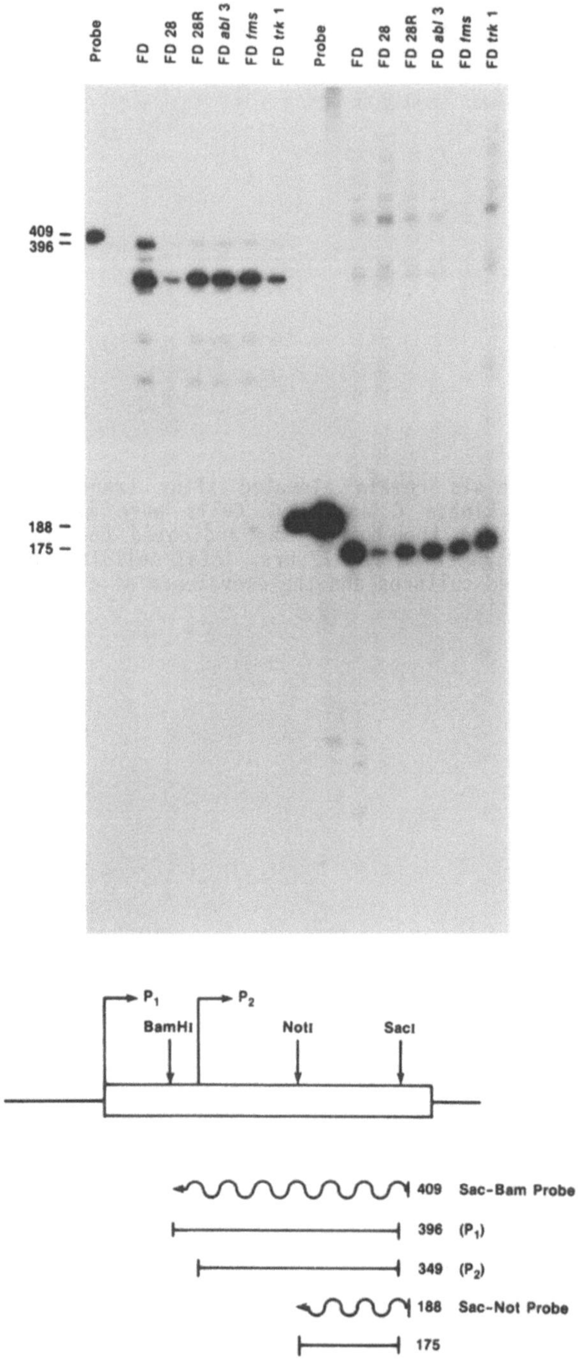

Fig. 2 RNase A protection of c-myc mRNA. RNA probes from exon 1 were hybridized to mRNA from FDmyc cells, digested with RNase A and run on a denaturing gel

DISCUSSION

The fact that the c-myc gene is alterred in many different types of tumor cells argues that the gene plays an important role in the control of cell growth. In addition, the regulation of this gene is complex, and tightly coupled to the growth state in most cell types (Kelly et al., 1983; Campisi et al., 1984). We present here two systems in which the regulation of myc genes are altered in the absence of detectable structural alterations in the myc genes themselves. Although we have not assayed the full length of the myc genes for mutations, two pieces of evidence support the idea that myc is trans-activated. the FDmyc-28R line arose after only several months in culture, yet shows de-regulation of both N-myc and c-myc genes. It seems unlikely that critical mutations in both genes could have occurred in this time. In BPA31 and DA31 cells, we have recently found that the combined inhibition of both PKC and A kinase results in the shut-off of c-myc expression (data not shown), suggesting that the gene can be regulated in these cells under certain conditions and is not irreversibly activated.

We suggest that the aberrant regulation of myc genes observed in these two cell systems is due to the activation of cellular trans-activating factors. Furthermore, this represents a new class of cells showing myc activation and suggests that the percentaqge of tumors with activated nyc genes may be higher than previously thought.

ACKNOWLEDGEMENTS

"This project has been funded at least in part with Federal funds from the Department of Health and Human Services under contract number N01-CO-74102 with Program Resources, Inc., and by NIH grant NO.R01-CA45695 to J.C. The content of this publication does not necessarily reflect the views or policies of the Department of Health and Human Services, nor does mention of trade names, commercial products, or organizations imply endorsement by the U.S. Government."

REFERENCES

Campisi J, Gray EG, Pardee AB, Dean M, Sonenshein GE (1984) Cell-cycle control of c-myc but not c-ras expression is lost following chemical transformation. Cell 36: 241-247

Cesarman E, Dalla-Favera R, Bentley D, Groudine M (1987) Mutations in the first exon are associated with altered transcription of c-myc in burkitt lymphoma. Science 238: 1272-1275

Dean M, Cleveland JL, Ulf RR, Ihle, JN (1987) Role of myc in the Abrogation of IL3 dependence of myeloid FDC-P1 cells. Oncogene Research 1: 279-296

Dexter TM (1980) Growth of factor-dependent hemopoietic precursor cell lines. J Exp Med 152: 1036-1047

Holley RW, Baldwin JH, Kiernan JA, Messmer TO (1976) Control of growth of benzo(a)pyrene-transformed 3T3 cells. Proc Nat Acad Sci 73: 3229-3232

Kelly K, Cochran BH, Stiles CD, Leder P (1983) Cell-specific regulation of the c-myc gene by lymphocyte mitogens and platelet-derived growth factor. Cell 35: 603-610

McCaffrey P, Ran W, Campisi J, Rosner, MR (1987) Two independent growth factor-generated signals regulate c-fos and c-myc mRNA levels in Swiss 3T3 cells. J Biol Chem 262: 1442-1445

Oshiro Y, DiPaolo JA (1974) Changes in the uptake of 2-deoxy-D-glucose in Balb/3T3 cells chemically transformed in culture. J Cell Physiol 83: 193-202
Ran W, Dean M, Levine RA, Henkle C, Campisi J (1986) Induction of c-fos and c-myc mRNA by epidermal growth factor or calcium ionophore is cAMP dependent. Proc Natl Acad Sci 83: 8216-8220

Stimulation of the Antigen Receptor Complex Leads to Transcriptional Activation of the c-*myc* Gene in Normal Human T Cells

T. LINDSTEN[1], C.H. JUNE[2], and C.B. THOMPSON[1]

INTRODUCTION

The expression of the cellular proto-oncogene c-myc has been implicated in growth regulation of both normal and neoplastic cells (for review see Eisenman and Thompson 1986). Quiescent lymphocytes have been shown to express low levels of both c-myc mRNA and protein (Smeland et al. 1987). Stimulation of lymphocytes with mitogens results in the rapid accumulation of c-myc mRNA to levels 10-20 times above background within 2 hours following stimulation (Kelly et al. 1983). Thereafter, c-myc mRNA and protein continue to be expressed throughout the proliferative response of mitogen-activated cells (Hann et al. 1985, Thompson et al. 1985, Rabbitts et al. 1985). The ability of mitogen-activated cells to enter S phase of the cell cycle can be blocked by anti-sense myc oligonucleotides (Heikkila et al. 1987). Constitutive expression of c-myc in murine 3T3 cells has been shown to lead to partial growth factor independence of cell proliferation (Armelin et al. 1984). Finally, agents such as γ-interferon which arrest cell proliferation in vitro are associated with downregulation of c-myc gene expression (Einat et al. 1985). Taken together, these results suggest that the c-myc proto-oncogene may encode a regulatory nuclear protein required for the proliferation of lymphoid cells. In view of the central role c-myc appears to play in cell proliferation, it is not surprising that deregulation of c-myc expression can also contribute to neoplastic transformation of cells. Previous studies have suggested that c-myc expression can be regulated at multiple molecular levels. The expression of c-myc following stimulation of quiescent cells has been reported to involve both transcriptional and posttranscriptional mechanisms (Blanchard et al. 1985, Smeland et al. 1987). The abrogation of c-myc expression during differentiation of the human myeloid cell line HL-60 has been shown to be the result of an increase in a block to transcriptional elongation within the first intron of the gene (Bentley and Groudine 1986a). Alterations of the c-myc chromatin structure by chromosomal translocation and of mRNA structure by deletion of both 5' and 3' untranslated regions have also been reported to result in deregulated expression of the gene (Saito et al. 1983, Dyson and Rabbitts 1985, Jones and Cole 1987).

RESULTS

Expression of c-myc Following Activation of the T Cell Receptor/CD3 Complex

In order to understand how receptor-ligand interactions at the cell surface contribute to the regulation of c-myc gene expression, we have investigated the role of stimulation of the T cell receptor (TCR)/CD3

[1]Howard Hughes Medical Institute and Departments of Medicine and Microbiology/Immunology, University of Michigan, 1150 W. Medical Center Dr., 4510 MSRB I, Ann Arbor, MI 48109
[2] Naval Medical Research Institute, Bethesda, MD 20814

Current Topics in Microbiology and Immunology, Vol. 141

surface receptor complex in regulating the expression of c-myc in normal, resting peripheral blood T cells. Previous studies have shown that activation of both murine (Buckler et al. 1988) and human (Smeland et al. 1987) B lymphocytes by crosslinking of surface immunoglobulin results in a rapid induction of c-myc mRNA expression. Since the TCR/CD3 complex is thought to be the T cell equivalent of the immunoglobulin receptor, we have sought to generalize a role for the antigen receptor in the induction of the c-myc expression during the normal lymphoid proliferative response. We have also initiated studies to investigate the molecular mechanisms by which stimulation of the antigen receptor on normal peripheral blood human T cells can lead to the induction of c-myc mRNA expression. In order to perform these studies, it was necessary to obtain a population of highly purified resting T cells devoid of accessory cell populations that might produce lymphokines involved in secondary activation events following TCR/CD3 stimulation. We have previously reported the purification of human peripheral blood T lymphocytes by negative selection using magnetic immunoabsorption with monoclonal antibodies directed against monocytes, natural killer cells, suppressor cells, and B lymphocytes (June et al. 1987). The remaining cells are over 98% T cells as judged by surface staining with CD3 and CD2 monoclonal antibodies. This population of T cells does not proliferate in vitro in response to stimulation with phytohemagglutinin (PHA), a mitogen known to require accessory cells for the induction of proliferation (Table 1).

Table 1. Initiation of DNA synthesis in purified, resting T cells by CD3sp monoclonal antibodies or PMA + Ionomycin costimulation

Stimulus[a]	Thymidine Incorporation[b]
Medium	0.1 ± 0.1
PHA	0.3 ± 0.1
CD3sp Ab	25.0 ± 3.1
PMA	0.8 ± 0.4
Ionomycin	0.1 ± 0.1
PMA+Ionomycin	19.7 ± 1.7

[a] $CD2^+$,$CD3^+$ T cells were cultured at 1 x 10^5/well with either medium alone or medium plus PHA at 3 µg/ml, PMA at 1 ng/ml, ionomycin at 0.1 µg/ml, or CD3sp Ab at 50 ng/well as indicated above
[b] Thymidine incorporation (cpm x 10^{-3}, mean ± S.E.M.) was determined on day 3 of culture

This population also fails to proliferate in response to phorbol myristate acetate (PMA), a potent activator of protein kinase C, or the calcium ionophore, ionomycin. However, solid phase CD3 (CD3sp) monoclonal antibodies or the combination of PMA and ionomycin can efficiently induce the proliferation of this population of cells. When RNA was isolated at different time points following CD3sp stimulation of this resting T cell population and c-myc expression analyzed by Northern blot analysis, we found that c-myc mRNA levels were induced within 1 hour following stimulation and were maintained throughout the proliferative response (Fig. 1).

Both Phorbol Esters and Calcium Ionophores can Induce c-myc mRNA Expression in Resting T Cells

Previous studies have shown that antibodies to the TCR/CD3 receptor

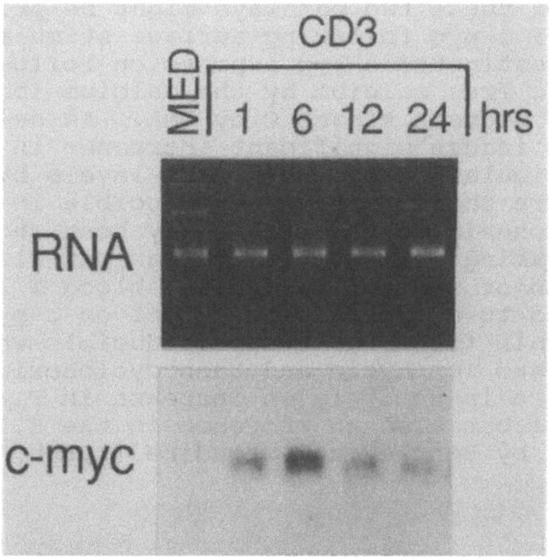

Figure 1. c-*myc* mRNA levels following stimulation of resting peripheral blood T cells with CD3sp antibodies, which crosslink the TCR/CD3 antigen receptor complex. MED = medium control, upper panel - ethidium bromide staining of the total RNA, lower panel - autoradiogram following hybridization with a c-*myc*-specific probe

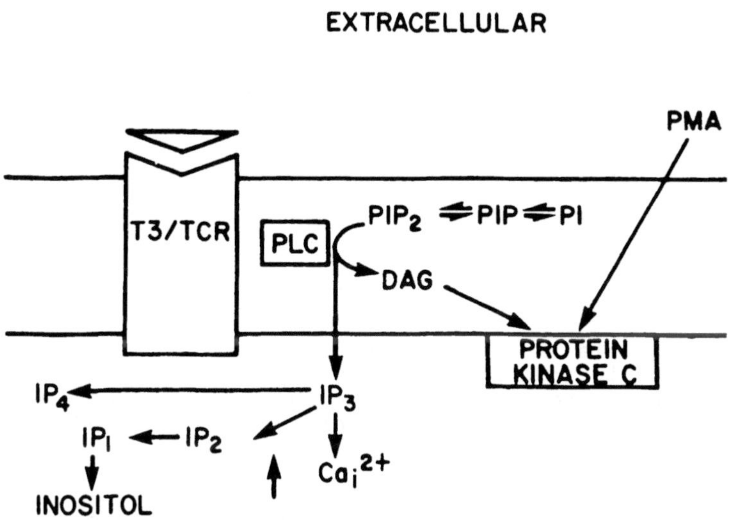

Figure 2. Activation cascade for induction of protein kinase C and increased intracellular calcium following stimulation of the T3(CD3)/T cell receptor complex

complex initiate their intracellular effects by activating phospholipases to release phosphoinositol (PI) and diacylglycerol (DAG) from the membrane. These intracellular secondary messages in turn lead to a rapid rise in cytoplasmic free calcium and the activation and membrane translocation of protein kinase C (Truneh et al. 1985, Ledbetter et al. 1987, Fig. 2).

To determine which of these two pathways might be primarily responsible for the activation of c-myc following surface stimulation of the TCR/CD3 complex, we have investigated c-myc expression following stimulation of increased cytoplasmic free calcium by the calcium ionophore, ionomycin, and/or activation of protein kinase C by PMA. As seen in Fig. 3, both agents were found to induce significant increases in the steady-state c-myc mRNA levels. Stimulation of c-myc mRNA levels by both PMA and calcium ionophore were shown to be superinducible in the presence of cycloheximide. Cycloheximide has previously been shown to induce c-myc mRNA levels by increasing mRNA stability (Dani et al. 1984). In addition, stimulation of resting peripheral blood T lymphocytes by both PMA and ionomycin led to a synergistic effect on c-myc mRNA levels. This response was again found to be superinducible when cells were pretreated with PMA and ionomycin and then cycloheximide added. Under these conditions there is not only an increase in P1/P2 promoter initiated transcripts but also an increase in the 3.1 kb P0 initiated transcripts reported by Bentley and Groudine (1986b, Fig. 3).

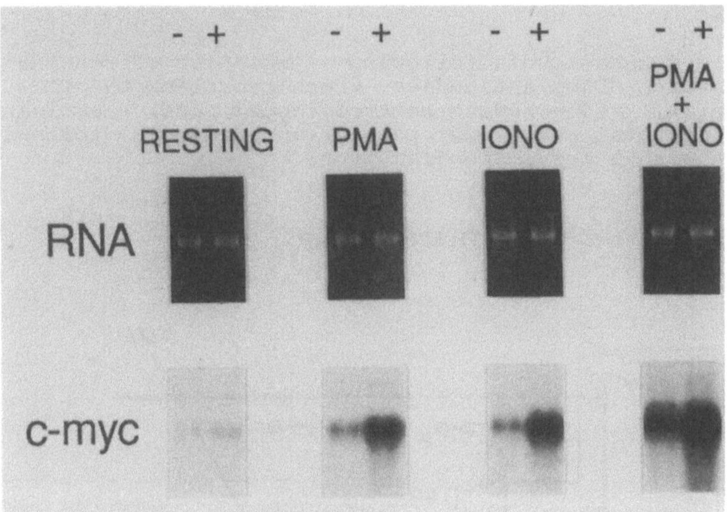

Figure 3. c-myc mRNA levels in resting, PMA induced, ionomycin (IONO) induced and PMA + Ionomycin induced T cells. The effect of cycloheximide (+) on pretreated cells is also shown. Upper panels - ethidium bromide staining of total RNA. Lower panels - autoradiogram following hybridization with a c-myc specific probe

Ionomycin and PMA Induce c-myc Expression by Independent Transcriptional Mechanisms

We next sought to investigate the individual roles PMA and ionomycin played in inducing c-myc gene expression in quiescent peripheral blood T cells. Experiments examining the half-life of the c-myc mRNA induced by ionomycin or PMA revealed the c-myc half life to be approximately 15 minutes in both instances (data not shown). In parallel experiments, it was shown that the c-myc mRNA levels induced by cycloheximide could be entirely accounted for by an increase in the mRNA half life to over 1 hour following treatment of resting cells with cycloheximide. These results suggest that both PMA and ionomycin are having their primary effects on the transcription rate and/or processing of the primary c-myc transcripts. To investigate the role of alterations in transcriptional regulation in the induction of c-myc gene expression by PMA and/or ionomycin, we have begun to analyze the regulation of c-myc transcription as measured by run-on transcription assays (data not shown). In these preliminary experiments, nuclei were isolated from resting peripheral T lymphocytes following stimulation with media alone, PMA for 12 hours, ionomycin for 2 hours, or PMA alone for the first 12 hours of culture followed by the addition of ionomycin for an additional 2 hours. Nuclei were then incubated with [^{32}P]-radiolabelled UTP under conditions that allowed for elongation of initiated transcripts by polymerases already bound to DNA. Radiolabelled nascent RNA transcripts were then isolated from the reaction mix and hybridized with filters containing single stranded M13 probes for exon 1 and exon 2 in both the sense and anti-sense orientations. As reported previously by others (Bentley and Groudine 1986, Nepveu and Marcu 1986, Eick and Bornkamm 1986), resting peripheral blood T cells demonstrated a significant transcriptional rate of the sense strand of exon 1 demonstrating that even in the resting state, the c-myc major promoters, P1 and P2, are being utilized by polymerases. However, significantly less transcription was observed using a similarly sized exon 2 template. Comparison of the band intensities between exon 1 sense strand and the exon 2 sense suggest that there is 4-5 fold block to transcriptional elongation between exon 1 and exon 2 in resting cells. Studies previously reported by others have suggested that this block to elongation occurs at the exon 1/intron 1 boundary (Cesarman et al. 1987) and we have performed no further studies to locate this region. When resting peripheral blood T cells are incubated for 12 hours with PMA, the ratio of exon 1 to exon 2 transcription was significantly lowered (<2:1), suggesting that there has been an alleviation of the block to transcriptional elongation observed between exon 1 and exon 2 in resting peripheral blood T cells. In contrast to these results, when resting peripheral blood T cells are stimulated for 2 hours with the calcium ionophore, ionomycin, a 3-5-fold increase in the rate of transcription within both exon 1 and exon 2 was observed. The increase in exon 2 transcription can entirely account for the increase in c-myc mRNA that occurs following ionomycin treatment of the cells. Despite the increase in overall transcription within the locus following ionomycin treatment alone, there is still a significant block to transcription elongation between exon 1 and exon 2. We next examined the transcription of cells stimulated with both PMA and ionomycin. The transcription rate of exon 1 in cells pretreated with PMA and then stimulated with ionomycin appears to be no different than transcription induced by treatment with ionomycin alone. However, in contrast to ionomycin treatment, the transcription rate within exon 2 appears to occur at a rate equivalent to that of exon 1 when cells are pretreated with PMA and then stimulated with ionomycin, suggesting that the block to transcriptional elongation has been alleviated.

CONCLUSIONS

Surface activation events mediated by the TCR/CD3 receptor can initiate the expression of c-myc mRNA that appears to be mimicked in both the level and kinetics of c-myc induction by the combination of protein kinase C activation by PMA and the release of intracellular calcium by ionomycin. Both PMA and ionomycin appear to have their primary effects at the level of transcription and/or mRNA processing. Treatment of resting peripheral T cells by calcium ionophore alone leads to an increased rate of promoter utilization at the major promoters of the c-myc proto-oncogene. This induction in transcription rate does not appear to have any effect on the level of the block to transcriptional elongation that is present within the first intron of the gene.

Treatment of resting peripheral blood T lymphocytes with PMA alone appears to decrease the block to transcriptional elongation present within the first intron of the gene. When resting peripheral blood T lymphocytes are stimulated with a combination of PMA and ionomycin, one finds that the primary transcriptional effects of these two agents synergize, leading to increases in promoter utilization as a result of ionomycin treatment of the cells and a decrease in the block to transcriptional elongation within the first intron, presumably as a result of PMA pretreatment of the cells. The independence of the PMA and ionomycin effects on the transcriptional regulation of the gene can be demonstrated by the fact that the effect of PMA can only be seen when cells are pretreated with the agent and can be completely abolished by pretreatment with PMA in the presence of cycloheximide. This suggests that the PMA effect requires the induction of new protein synthesis. In contrast, the ionomycin effect can be observed in cultures in which cycloheximide is present from the onset of stimulation, demonstrating that the effect of ionomycin on increasing promoter utilization in the 5′ end of the gene is independent of new protein synthesis. Finally, both the PMA- and ionomycin-mediated effects on transcription can be independently augmented by stabilization of mRNA in the cytoplasm by addition of cycloheximide to PMA, ionomycin, or PMA+ionomycin pretreated cultures.

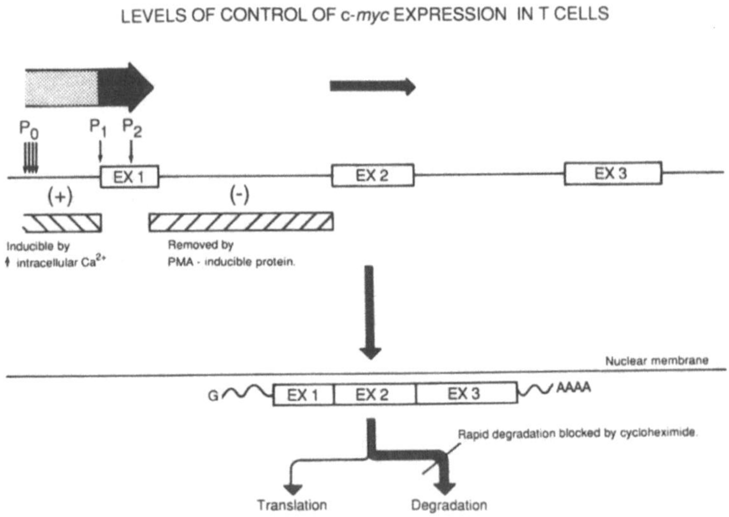

Figure 4. Proposed model for regulation of c-myc gene expression in normal cells. Points of potential transcriptional control by increased intracellular calcium and PMA-inducible proteins are identified

In summary, our results suggest that a major effect of antigen receptor stimulation of resting peripheral blood T cells is the induction of c-myc mRNA expression. The two intracellular pathways induced by TCR/CD3 receptor stimulation as mimicked by PMA and ionomycin both appear to play a significant role in the induction of c-myc gene expression by the antigen receptor complex. The presence of multiple levels of control of c-myc gene expression during cellular activation is consistent with the hypothesis that c-myc plays a central role in regulating the proliferative response (Fig. 4). In fibroblasts, constitutive expression of c-myc has been shown to abrogate the growth requirement for the serum competence factor PDGF (Armelin et al. 1984). Similarly, the deregulation of c-myc gene expression seen in a variety of lymphoid neoplasms may partially account for the ability of these cells to proliferate in an antigen-independent manner (Dyson and Rabbitts 1985, Erikson et al. 1986, Cesarman et al. 1987).

ACKNOWLEDGEMENTS

We would like to thank our colleagues, Mark Groudine, David Bentley, James Battey, Geoffrey Krystal, Tim Bender, Wayne McCormack, and Jeffrey Leiden for numerous thoughtful discussions and Jeanelle Clark for assistance in the preparation of this manuscript. This work was supported by funds from both the Howard Hughes Medical Institute and the Naval Medical Research and Development Command. Tullia Lindsten is a Leukemia Society of America Special Fellow.

REFERENCES

Armelin HA, Armelin MCS, Kelly K, Stewart T, Leder P, Cochran BH, Stiles CD (1984) Functional role of c-myc in mitogenic response to platelet-derived growth factor. Nature 310:655-660

Bentley DL, Groudine M (1986) A block to elongation is largely responsible for decreased transcription of c-myc in differentiated HL60 cells. Nature (London) 321:702-706

Bentley DL, Groudine M (1986) Novel promoter upstream of the human c-myc gene and regulation of c-myc expression in B-cell lymphomas. Mol. Cell. Biol. 6:3481-3489

Blanchard JM, Piechaczyk M, Dani C, Chambard JC, Franchi A, Pouyssegur J, Jeanteur P (1985) c-myc gene is transcribed at high rate in G0-arrested fibroblasts and is posttranscriptionally regulated in response to growth factors. Nature (London) 317:443-445

Buckler AJ, Rothstein TL, Sonenshein GE (1988) Two-step stimulation of B lymphocytes to enter DNA synthesis: Synergy between anti-immunoglobulin antibody and cytochalasin on expression of c-myc and a G_1-specific gene. Mol. Cell. Biol. 8:1371-1375

Cesarman E, Dalla-Favera R, Bentley D, Groudine M (1987) Mutations in the first exon are associated with altered transcription of c-myc in Burkitt lymphoma. Science 238:1272-1274

Dani C, Blanchard JM, Piechaczyk M, El Sabouty S, Marty L, Jeanteur P (1984) Extreme instability of myc mRNA in normal and transformed human cells. Proc. Natl. Acad. Sci. USA 81:7046-7050

Dyson PJ, Rabbitts TH (1985) Chromatin structure around the c-myc in Burkitt lymphomas with upstream and downstream translocation points. Proc. Natl. Acad. Sci. USA 82:1984-1988

Eick D, Bornkamm GW (1986) Transcriptional arrest within the first exon is a fast control mechanism in c-myc gene expression. Nucleic Acids Res. 14:8331-8346

Eick D, Piechaczyk M, Henglein B, Blanchard JM, Traub B, Kofler E, Wiest S, Lenoir GM, Bornkamm GW (1985) Aberrant c-myc RNAs of Burkitt's lymphoma cells have longer half-lives. EMBO J. 4:3717-3725

Einat M, Resnitzky D, Kimchi A (1985) Close link between reduction of c-myc expression by interferon and G_0/G_1 arrest. Nature 313:597-600

Eisenman RN, Thompson CB (1986) Oncogenes with potential nuclear function: myc, myb, and fos. Cancer Surveys 5:309-327.

Erikson J, Finger L, Sun L, Ar-Rushdi A, Nishikura K, Minowada J, Finan J, Emanuel BS, Nowell PC, Croce CM (1986) Deregulation of c-myc by translocation of the α-locus of the T-cell receptor in T-cell leukemias. Science 232:884-886

Hann SR, Thompson CB, Eisenman RN (1985) c-myc oncogene protein synthesis is independent of the cell cycle in human and avian cells. Nature 314:366-369

Heikkila R, Schwab G, Wickstrom E, Loke SL, Pluznik DH, Watt R, Neckers LM (1987) A c-myc antisense oligodeoxynucleotide inhibits entry into S phase but not progress form G_0 to G_1. Nature 328:445-449

Jones TR, Cole MD (1987) Rapid cytoplasmic turnover of c-myc mRNA: Requirement of the 3′ untranslated sequences. Mol. Cell. Biol. 7:4513-4521

June CH, Ledbetter JA, Gillespie MM, Lindsten T, Thompson CB (1987) T-cell proliferation involving the CD28 pathway is associated with cyclosporine-resistant interleukin 2 gene expression. Mol. Cell. Biol. 7:4472-4481

Kelly K, Cochran BH, Stiles CD, Leder P (1983) Cell-specific regulation of the c-myc gene by lymphocyte mitogens and platelet-derived growth factor. Cell 35:603-610

Ledbetter JA, Gentry LE, June CH, Rabinovitch PS, Purchio AF 1987 Stimulation of T cells through the CD3/T-cell receptor complex: Role of cytoplasmic calcium, protein kinase C translocation, and phosphorylation of $pp60^{c-src}$ in the activation pathway. Mol. Cell. Biol. 7:650

Nepveu A, Marcu KB (1986) Intragenic pausing and anti-sense transcription within the murine c-myc locus. EMBO J. 5:2859-2865

Rabbitts PH, Watson JV, Lamond A, Forster A, Stinson MA, Evan G, Fischer W, Atherton E, Sheppard R, Rabbitts TH (1985) Metabolism of c-myc gene products: c-myc mRNA and protein expression in the cell cycle. EMBO J. 4:2009-2015

Saito H, Hayday AC, Wiman K, Hayward WS, Tonegawa S (1983) Activation of the c-myc gene by translocation: A model for translational control. Proc. Natl. Acad. Sci. USA 80:7476-7480

Smeland E, Godal T, Ruud E, Beiske K, Funderud S, Clark EA, Pfeifer-Ohlsson S, Ohlsson R (1985) The specific induction of myc proto-oncogene expression in normal human B cells is not a sufficient event for acquisition of competence to proliferate. Proc. Natl. Acad. Sci. USA 82:6255-6259.

Smeland EB, Beiske K, Ek B, Watt R, Pfeifer-Ohlsson S, Blomhoff HK, Godal T, Ohlsson R (1987) Regulation of c-myc transcription and protein expression during activation of normal human B cells. Exp. Cell Res. 172:101-109

Thompson CB, Challoner PB, Neiman PE, Groudine M 1985 Levels of c-myc oncogene mRNA are invariant throughout the cell cycle. Nature 314:363-366

Truneh A, Albert F, Golstein P, Schmitt-Verhulst A (1985) Early steps of lymphocyte activation bypassed by synergy between calcium ionophores and phorbol ester. Nature 313:318-320

A Repressor of c-*myc* Transcription Is Found Specifically in Plasmacytomas

E. Kakkis, K. Riggs, and K. Calame

INTRODUCTION

Since the discovery that translocation of *c-myc* genes results in altered *c-myc* gene expression (for review see Cory 1986), a great deal of effort has been directed toward understanding how regulation of *c-myc* occurs under both normal and altered conditions. Work by many groups has shown that translocated *c-myc* is activated in plasmacytomas, but little insight has been gained into the mechanism of this activation (Cory 1986, Kakkis, Mercola and Calame 1988). Another interesting finding is that *c-myc* genes in their normal chromosomal context are not transcribed at detectable levels in plasmacytomas (Fahrlander et al 1985, Kakkis, Mercola and Calame 1988). Although many interpretations of this result are possible, we considered it likely that the *c-myc* gene could be normally repressed at the terminal stage of B-cell differentiation and that translocation could release *c-myc* from repression and activate its transcription. The repression of *c-myc* in plasmacytomas may represent the cell's attempt to turn off cell division. Studies on a number of *in vitro* differentiation systems suggest that *c-myc* expression correlates with proliferation and *c-myc* repression reflects the cessation of cell division (e.g. Bentley and Groudine 1986, Lachman and Skoultchi 1984, Resnitsky et al 1986). The mechanism for this repression in plasmacytomas is thus of considerable importance.

Although increased levels of myc protein may autoregulate expression in plasmacytomas, it seems more plausible to propose that an independent mechanism downregulates *c-myc* in the terminal stage of B-cell differentiation, repressing *c-myc* in a permanent fashion and thus blocking further cell division. The lability of *c-myc* mRNA and protein makes *c-myc* an unlikely candidate for autoregulation in the terminal stage of differentiation, since the presence of appropiate levels of protein would be required continuously. An autoregulatory role may exist in earlier stages of development when non-dividing stages are encountered (e.g. resting B cells). Considering the importance that repression of *c-myc* in plasmacytomas might play in normal B cell development, we sought to

Department of Biological Chemistry , UCLA School of Medicine
and the Molecular Biology Institute, UCLA, Los Angeles, CA 90024

Current Topics in Microbiology and Immunology, Vol. 141

assess the presence of proteins which might downregulate *c-myc* in plasmacytomas.

RESULTS AND DISCUSSION

Myc-PRF and myc-CF1 bind the c-myc promoter: Based on our hypothesis that repression of *c-myc* in plasmacytomas represented end-stage specific regulation, we searched for DNA-binding proteins that bound conserved regions of the *c-myc* promoter, using a mobility shift assay (Kakkis and Calame 1987). We found that within a 1 kb region of human/murine conserved promoter sequences, there was a single binding site for a protein which was specific to plasmacytomas and not found in cell lines representing earlier stages of B cell development . The prominent and specific DNA-binding activity in plasmacytoma nuclear extracts was unique; there were no other prominent binding activites which showed plasmacytoma specificity. Methylation interference, footprinting and sequence analysis demonstrated that a sequence "AGAAAGGGAAAGGA" was the sequence recognized by the plasmacytoma-specific factor. Gel shift experiments with deletion mutants and competition experiments with a synthetic double-stranded oligonucleotide confirmed that this was the binding site (not shown). The factor was originally named myc-PCF, but because of results described below, we have changed its name to myc-PRF for "plasmacytoma repressor factor".

Footprinting analysis on another factor common to all B cell stages showed that this factor protects a conserved sequence "AGAAAATGGT" which is located approximately 10 bp 3' of the myc-PRF binding site (Kakkis and Calame 1987). We have named this common factor myc-CF1 (common factor #1). These two factors bind to the same DNA probe molecule and so generate a larger gel-shift complex denoted the A complex (Kakkis and Calame 1987).

Myc-PRF is abundant only in plasmacytomas; myc-CF1 is found in nearly all cell types: We tested a number of cell types for the presence of these two DNA binding factors using the gel mobility shift assay. Our conclusion from tests of 4 plasmacytomas (M603, P3X63-Ag8 [P3X], J558L, S107), 4 early B cell lines (18-81, 70Z/3, WEHI279, 38B9), 2 fibroblast lines (3T3, L cells), 2 T cell lines (EL4, AOIT), brain, liver and kidney are that myc-CF1 is found in all cells except possibly for brain and that myc-PRF is found in abundance exclusively in plasmacytomas (not shown). The only exception is a Ly1+ B cell line 1414, which contains highly amplified *c-myc* genes and a trace amount of myc-PRF. The presence or absence of myc-PRF was confirmed by oligonucleotide competition experiments (not shown). These data suggest the interesting possibility that myc-PRF is restricted to terminally differentiated B-Cells.

Additional experiments with human cell lines including Daudi, Manca, BJAB, HL60 (with and without DMSO treatment for up to 6 days), and COLO320 showed no evidence of myc-PRF activity. We also tested these extracts with a human myc promoter DNA probe containing the homologous myc-PRF site and found that none of these lines, including the two Burkitt lymphomas, contain myc-PRF. We also found that myc-CF1 from a human or mouse source binds either human or mouse DNA interchangeably, whereas mouse myc-PRF from M603 was unable to bind the human PRF sequence (AGAAAGGGAGAGGG). The differences between the two core sequences or perhaps between closely surrounding sequences is sufficient to prevent binding. This raises interesting questions regarding the divergence and coevolution of DNA-binding factors and binding sites and the differences in regulation which may exist between mouse and human *c-myc* genes.

Myc-PRF represses transcription of myc/CAT fusion genes: We wished to establish the functional role of myc-PRF in *c-myc* regulation in plasmacytomas. If myc-PRF does repress transcription of *c-myc*, deletion of the myc-PRF binding site should activate *c-myc* transcription in plasmacytomas. We constructed a specific deletion of the myc-PRF binding site using oligonucleotide-directed mutagenesis. The deletion of 14 bp and addition of one "T" lead to the formation of an Acc I restriction site which we could use to follow and confirm the presence of the deletion. Five different pairs of wildtype and mutant constructs were made which fused different portions of the *c-myc* promotor to the CAT gene. When transiently transfected into the plasmacytoma P3X Ag63.8, only *c-myc* promoters containing $^{-}424$ to $^{+}568$ or $^{-}1150$ to $^{+}568$ (Sma-Bgl and Bgl-Bgl) gave significant activity and the myc-PRF deletion had no effect in any of the constructs. Although at first surprising, we considered the possibility that the apparent low abundance of myc-PRF in nuclear extracts (only ~5 fold excess of cold competitor DNA competes for the factor) might result in titration of the putative negative factor under transient transfection conditions in which thousands of DNA copies were present in a cell. To address this potential technical problem, we made stable transfectants.

Stable transfectants were made using the Bgl-Bgl promoter construct either in its wildtype form or with the myc-PRF binding site deleted. Constructs were cotransfected with pSV2NEO into P3X and L cells followed by selection with G418. The Bgl-Bgl construct was chosen because it had previously been shown to be transcribed correctly (Remmers et al 1986) and because it gave good CAT activity under transient conditions. Five pools of P3X transfectants for wildtype and mutant constructs were made from three separate transfection experiments. Four separate pools of L cell transfectants were made from two transfection experiments. All pools had similar copy number of the transfected gene, estimated to be between 50-150 copies per cell (not shown). We found that extracts from the P3X lines containing the mutant promoter had 31 fold higher CAT activity than those from the P3X lines with the wildtype construct (Table 1). The L cell pools showed no significant difference between wildtype and mutant constructs. These data confirm that myc-PRF represses the c-myc promoter in plasmacytomas but not in L cells, a role consistent with its distribution in plasmacytoma cells which have repressed the *c-myc* gene at the transcriptional level.

TABLE 1: STABLE TRANSFECTANT DATA

P3X plasmacytoma				L cell fibroblast			
PRF mutant		Wildtype		PRF mutant		wildtype	
cell line	activity	cell line	activity	cell line	activity	cell line	activity
1. 2A	9.65	1A	0.19	2AL	68.6	1AL	51.5
2. 2B	5.65	1B	0.56	2BL	58.2	1BL	37.5
3. 4A	5.0	3A	1.1	4AL	78.0	3AL	127
4. 4B	10.9	3B	0.1	4BL	134	3BL	153
5. 6B	28.5	5C	0.013				
AVE.=	11.9		0.39	AVE.=	84.7		92.3
Rel. activity	mut./wt = **31**			Rel. activity	mut/wt= **0.92**		

*activity is expresssed as picomoles chloramphenicol acetylated / hour / $5x10^6$ cells

Myc-PRF and myc-CF1 may interact on the c-myc promoter: We have also begun to investigate the role that myc-CF1 might play in the regulation process. We have used partially purified myc-PRF protein in mixing experiments with EL4 and 18-81 extracts which contain myc-CF1 but no myc-PRF. Addition of increasing amounts of myc-PRF (Fig.1A) causes an increase in complex A and a decrease in the myc-CF1 complex. These data suggest that the two proteins may bind synergistically to *c-myc* promoter DNA. We have further investigated the possible interaction between these two proteins by measuring dissociation rates for each protein binding alone or in combination using the gel mobility shift assay and P3X nuclear extract. The experiments show that the complex due to myc-CF1 binding alone has a half-life that is less than 30 seconds in the presence of a vast excess of unlabeled DNA competitor (Fig. 1B). The complexes specific to myc-PRF alone have a half-life that is greater than 10 minutes (Fig. 1B). The A complex, representing binding by both myc-PRF and myc-CF1, also has a halflife of 10 minutes of more (Fig. 1B). These data suggest that myc-CF1 is more tightly bound in the presence of myc-PRF. Although the role of myc-CF1 has not been established, one possibility is that it is a positive factor for c-myc transcription and that interaction with myc-PRF prevents its stimulatory action.

The ultimate understanding of the importance of myc-PRF and myc-CF1 in *c-myc* regulation within B cells will depend on future detailed assessment of the mechanism of c-myc repression and the relationship of this repressor to other regulatory regions present in the *c-myc* locus. The large, highly conserved *c-myc* promoter suggests that the transcriptional regulation of *c-myc* expression will be complex. We hope that an understanding of myc-PRF and myc-CF1 will explain a portion of the regulatory influences controling transcription of the c-myc gene.

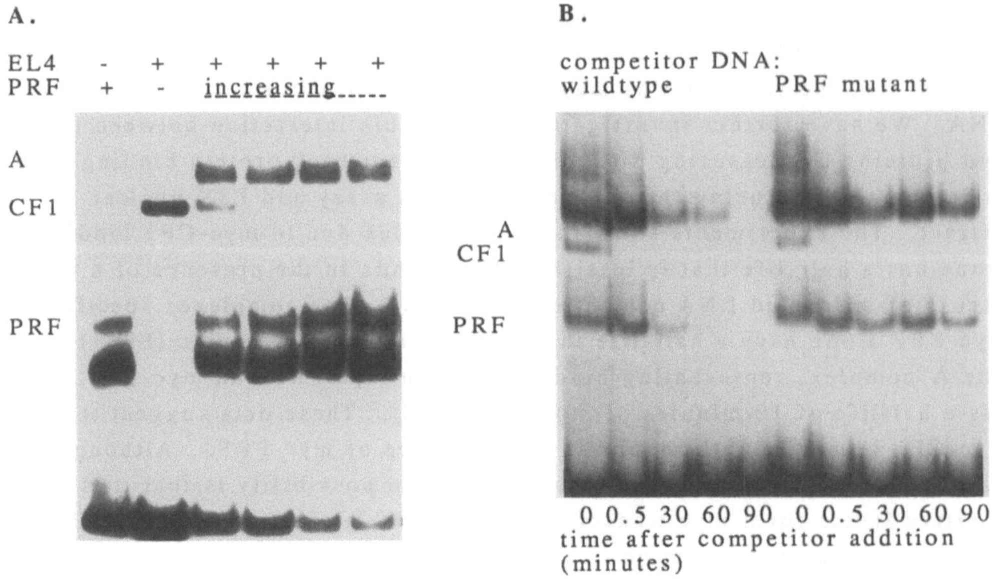

Fig. 1. A complex containing myc-PRF and myc-CF1 bound to the same DNA molecule. (A) A mixing experiment in which increasing amounts of partially purified myc-PRF were added to fixed amounts of EL4 extract containing myc-CF1. (B) An off-rate experiment in which the proteins in P3X crude extract were allowed to bind to equilibrium with labeled probe, followed by addition of a 50 fold molar excess of cold DNA as competitot at appropriate times prior to loading of the gel. In the first half of the panel, the competitor used was wildtype, and contains both the myc-PRF and the myc-CF1 binding sites. In the second half of the panel, the competitor used had the binding site for myc-PRF mutated, and contains only the binding site for myc-CF1

REFERENCES

Bentley, D. and Groudine, M. (1986) A block to elongation is largely responsible for decreased transcription of *c-myc* in differentiated HL60 cells. Nature(London) 321:702-706.

Cory, S. (1986) Activation of Cellular Oncogenes in hemopoietic cells by chromosome translocation. Adv. Cancer Res. 47: 189-234.

Fahrlander, P., Piechaczyk, M., and Marcu, K. (1985) Chromatin structure of the murine *c-myc* locus: implications for the regulation of normal and chromosomally translocated genes. EMBO J. 4: 3195-3202.

Kakkis, E., Mercola, M., and Calame, K.(1988) Strong transcriptional activation of translocated *c-myc* genes occurs without a strong nearby enhancer or promoter. Nucl. Acids Res. 16:77-96.

Kakkis, E.,and Calame, K. (1987) A plasmacytoma-specific factor binds the *c-myc* promoter region. Proc. Natl. Acad. Sci. USA 84: 7031-7035.

Lachman, H. and Skoultchi, A. (1984) Expression of *c-myc* changes during differentiation of mouse erythroleukemia cells. Nature (London) 310:592-594.

Regulation of c-*myc* Gene Transcription in B Lymphocytes: Mechanisms of Negative and Positive Control

A. J. BUCKLER[1], D. J. KESSLER[2], M. P. DUYAO[2], T. L. ROTHSTEIN[1,3], and G. E. SONENSHEIN[2]

SUMMARY

Transcription of the c-*myc* oncogene is stringently regulated. To begin to elucidate the mechanisms of transcriptional control, we have studied growth arrest of the B cell lymphoma WEHI 231 and mitogenic stimulation of quiescent splenic B lymphocytes. In these systems, transcription of c-*myc* is subject to negative and positive regulation, respectively. We have used DNA gel-retardation assays employing nuclear extracts prepared from exponentially growing or growth arrested WEHI 231 cells to characterize protein-DNA interactions within segments of the putative dehancer region, which has been implicated in the transcriptional down-modulation of c-*myc*. We find that decreased transcription is associated with specific changes in a relatively complex pattern of binding of proteins to sequences within this regulatory region. Furthermore, evidence suggests that some of this binding may involve protein-protein interactions. For activation of quiescent murine splenic B lymphocytes, two signals are required. As noted previously, these can be delivered in a competence/progression type manner by goat anti-mouse immunoglobulin antibody (GaMIg) and cytochalasin D (CD). Here we demonstrate that the synergy of action of these two signals on c-*myc* RNA levels can be accounted for by changes in the rate of transcription. We find that GaMIg "primes" the gene in such a manner that its responsiveness to CD is enhanced. These systems now allow us to dissect the signals involved in transcriptional control of this proto-oncogene.

INTRODUCTION

The c-*myc* proto-oncogene has been strongly implicated in control of normal cell growth and differentiation. Furthermore, many lines of evidence indicate that the c-*myc* oncogene can play a direct role in the neoplastic transformation of lymphocytes of the B cell lineage (Shen-Ong et al., 1982). Alterations in c-*myc* gene structure associated with B cell transformation would severely limit the cells capacity to stringently regulate this gene. In quiescent untransformed cells the level of c-*myc* mRNA is low and it increases upon mitogenic activation; on the other hand, growth arrest and differentiation of a number of cell types is preceded by decreased expression of this mRNA (Kelly et al., 1983, Campisi et al., 1984). These alterations in mRNA levels can often be related to changes in the rate of c-*myc* gene transcription (Greenberg and Ziff, 1984), suggesting that both positive and negative genetic regulatory elements are involved in the control of c-*myc* gene expression. We have been employing two model systems-- growth arrest of the murine

Departments of Microbiology[1], Biochemistry[2], and Medicine[3]
Boston University Medical School, 80 East Concord Street, Boston, Massachusetts 02118

Current Topics in Microbiology and Immunology, Vol. 141

B cell lymphoma WEHI 231 cell line and activation of normal murine splenic B lymphocytes-- to study the down- and up-modulation of c-*myc* gene transcription, respectively.

Growth Arrest of WEHI 231 Cells

The WEHI 231 line was derived from an early B cell lymphoma; these cells express surface IgM, but do not secrete immunoglobulin protein. Growth of WEHI 231 cells can be inhibited, within 24 hours, by incubation with an anti-mouse Ig antisera, e.g., goat anti-mouse immunoglobulin (GaMIg) (Boyd and Schrader, 1981). Previously we demonstrated that this growth arrest was accompanied by dramatic changes in c-*myc* gene expression (McCormack et al., 1984); an initial increase in the level of c-*myc* mRNA is followed by a dramatic decrease. Following 24 hours of treatment, the RNA level is significantly below control values seen in exponentially growing cells. This selective decrease in c-*myc* RNA was shown to be related to a drop in the rate of transcription of the gene (Levine, et al., 1986). Using various fragments of the c-*myc* gene in nuclear run-off assays, we demonstrated that by 24 hours both a decrease in the rate of initiation of transcription, as well as of elongation was evident (data not shown). To evaluate the protein-DNA interactions mediating these changes in the rate of transcription of the gene we have employed DNA gel retardation assays (Strauss and Varshavsky, 1984). By comparing binding seen with nuclear extracts from exponentially growing cells to those prepared 24 hours after GaMIg treatment, we have a measure of the potential physiological significance of any observed protein-DNA interaction. We have concentrated on the putative negative regulatory element, termed the dehancer region, originally identified by Marcu and coworkers (Remmers et al., 1986). CAT transfection assays indicate that this region has a negative influence on transcription.

Protein-Dehancer DNA Interactions are Complex:

For study by gel shift analysis, this region, located between -1141 to -426 bp upstream of the first promoter, was divided into three domains (fragments A, B, and C) as shown in Fig. 1. For each fragment, binding to exracts prepared from nuclei of exponentially growing cells yielded multiple bands indicative of multiple protein-DNA interactions. Fragment A displayed the most intense binding; and fragment C the least. In comparing lysates prepared from these control cultures with those from cells treated with GaMIg for 24 hours, clear differences in the fragment A and B banding patterns are evident (Fig. 1). With fragment A, two changes are apparent; there is less shift to the position of the upper bands, and the relative intensity of binding of the lowest band to the one above it becomes greater with lysates of GaMIg treated as compared to control cells. Similarly, a shift in band intensity is observed with fragment B. In contrast, no reproducible changes are seen with fragment C. (The apparent decrease in intensity of one of the bands with the lysates from GaMIg treated cells was not observed routinely.) Binding of proteins in control extracts to the sequences between fragment C and the first promoter (P1) was also detected; while an overall decrease in intensity was seen with extracts from GaMIg treated cells, no selective changes in the banding patterns were observed (data not shown). Thus extracts prepared from cells with decreased c-*myc* transcriptional activity

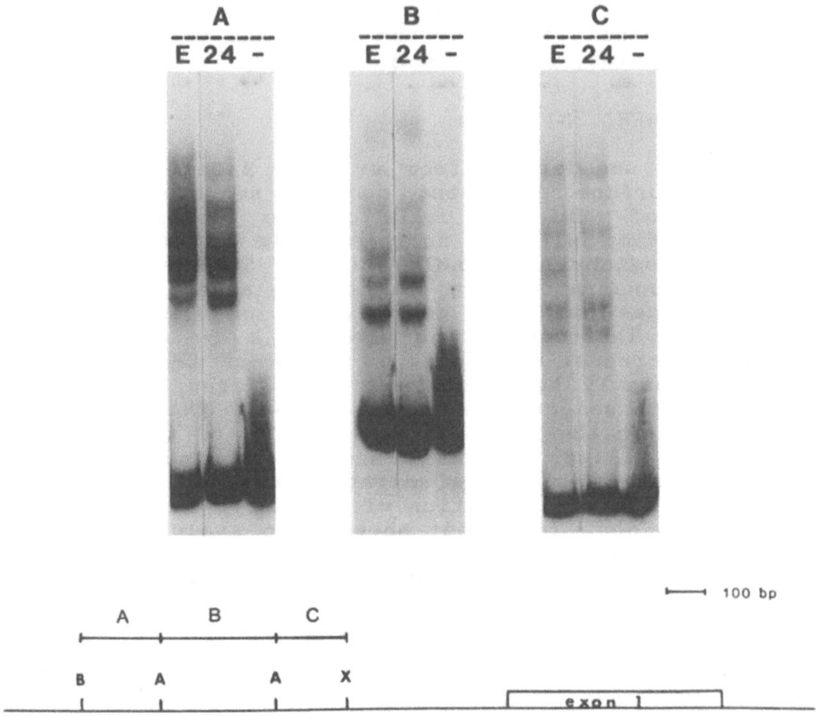

Fig. 1. Comparison of the interactions of "dehancer" region DNA with proteins within nuclear extracts of exponentially growing (E) and GaMIg treated (24) cells. The three subfragments (A,B,C), shown in the map, were incubated with the indicated extract and subjected to gel retardation analysis. (-, labelled DNA alone).
[B, Bgl II; A, Acc I; X, Xma I]

displayed selective changes in interaction with sequences within the promoter element.

Mapping Protein-Fragment A Interactions:

Additional effort was concentrated on elucidating the nature of the protein-DNA interactions within Fragment A since Marcu and coworkers have mapped one of two negative domains to this region within the "dehancer" (personal communications). Fragment A was labeled at its 5' end, and various subfragments (A1 through A5) were prepared by restriction digestion as indicated in Fig. 2a. These were isolated and tested for binding with lysates from exponentially growing cells. A minor band is lost with Hae III digestion (A1 vs A). Upon digestion with BstU I (A2) and then PflM I (A3), the intensity of upper bands decreases dramatically, while the two lower bands remain unaffected. Further digestion of fragment A3 with Hpa II results in the disappearance of these two bands, i.e. essentially no binding is seen with Fragment A4. This result suggests that the protein binding which yields the two lower bands is occurring within the PflM I to Hpa II fragment; this conclusion has been confirmed in studies with the purified

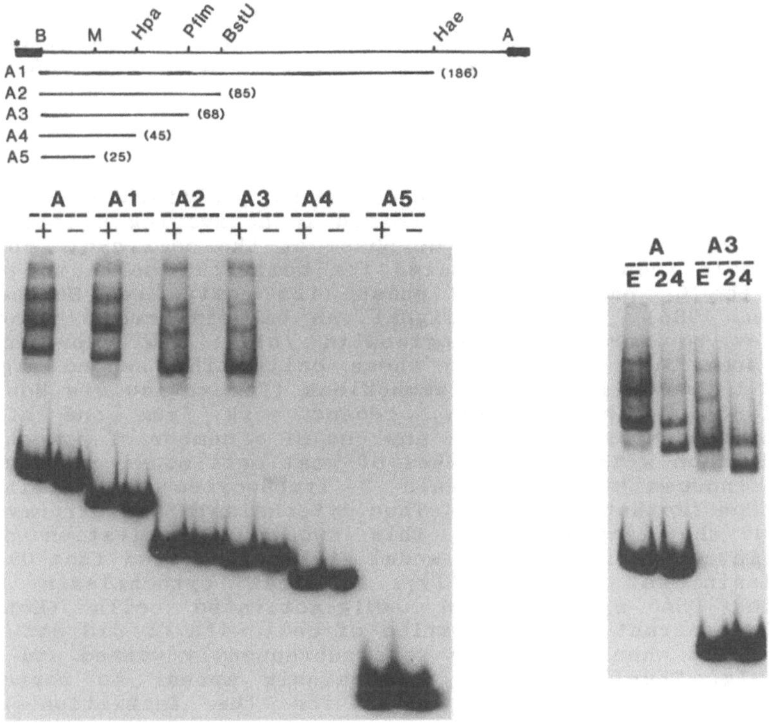

Fig. 2. Mapping of sites of interaction of nuclear proteins within Fragment A of the "dehancer" region.

a. (left) Domains of Fragment A, as illustrated in top panel, were subjected to gel retardation analysis in the presence (+) or absence (-) of extracts from cells in exponential growth.

b. (right) Comparison of nuclear factor binding to fragments A and A3 (performed as described in Fig. 1)

fragment (data not shown). The sequences within this domain that mediate binding have been further investigated in a DNAase protection assay. Using lysates from control cells, the AAGTCCGGTTTTCCCCAACC sequence surrounding the Hpa II site was protected. Having mapped the binding of the two lower bands to within fragment A3 with control lysates, we compared the binding seen with lysates from GaMIg-treated cells (Fig. 2b). As expected a dramatic difference in the intensity of binding of the two lower bands is evident, confirming the localization of the altered protein-DNA interaction.

In order to more completely define the boundaries of the protein-DNA interactions within Fragment A, a series of digestions were performed with DNA labeled at the 3' end of the fragment. Following digestion with various enzymes, the fragment was similarly analyzed for binding with lysates derived from control cells. Digestion with PflM I resulted in loss of not only the lower bands as expected, but also of most of the upper ones as well (data not shown). The remaining bands appeared unaffected by subsequent Bst UI digestion; only 1 band was seen with the 3' end Hae III fragment. This result suggests that the proteins interacting with sequences between the PflM I and Hpa II sites may be interacting directly with proteins; however, alternate hypotheses including overlapping DNA binding sites can also explain these results.

Activation of Splenic B Lymphocytes

Stimulation of resting B lymphocytes to proliferate involves at least two steps. The signals involved in activating B cells from a quiescent state to enter the G_1 phase of the cell cycle appear to be separable from those required for activated cells to progress through the G_1 phase into S phase (for review see Melchers and Andersson, 1986). The first signal can be delivered by binding of antibodies to surface immunoglobulin (sIg); sIg represents the antigen-specific receptor for these cells. The second signal is believed to be delivered by lymphokines (for review see Howard and Paul, 1983). Interestingly, recent work from one of these laboratories demonstrated that any one of a number of cytochalasins in combination with modest doses of goat anti-mouse immunoglobulin (GaMIg) induces murine splenic B lymphocytes to initiate DNA replication (Rothstein, 1985). Thus cytochalasin can circumvent the need for the lymphokine in this two step activation process. Further investigation of this model system indicated that GaMIg and cytochalasin act sequentially, in that cytochalasin D (CD) stimulates DNA synthesis in GaMIg-activated cells (Rothstein, 1986). In contrast, pre-incubation of cell with CD did not lead to DNA synthesis when such cells were subsequently washed and treated with GaMIg. Thus GaMIg and cytochalasin appear to represent a competence-progression type model for the initiation of DNA synthesis in B cells.

Synergy of GaMIg and CD on c-*myc* RNA Expression:

The effects of GaMIg and CD on the expression of c-*myc* during B cell stimulation were measured by Northern blot analysis as shown in Fig. 3a. Spleen cells from 6- to 12-week old A/J male mice were purified, and incubated in RPMI 1640 culture medium alone, or in medium supplemented with either GaMIg (5 µg/ml), CD (1 µg/ml) or GaMIg plus CD, as described previously (Rothstein, 1985). Incubation with either GaMIg or CD caused an increase in c-*myc* RNA; however, it was only of a transitory nature. By 7 and 4 hours respectively, the levels returned to those seen in cells incubated in medium alone. In contrast, when spleen cells were incubated with the combination of GaMIg plus CD, there was a very large increase in c-*myc* RNA level; with time the level of c-*myc* RNA declined from the peak value but it remained elevated above control values throughout the time course. Cells asynchronously enter the S phase between 24 and 48 hours after such co-stimulation (T.L.R., unpublished observation). Control experiments demonstrated that it is the B cells within the lymphocyte population which are responding to these signals (Buckler et al., 1988). Thus in addition to the increased level of mRNA which may be necessary to lead to a competent state, our results suggest that further expression of c-*myc* in the G_1 phase of the cell cycle may be required for progression of activated B cells to DNA synthesis.

It has also been shown that B cells which have been preincubated with GaMIg remain responsive to CD several hours after removal of GaMIg (Rothstein, 1986). We next examined the effects of this

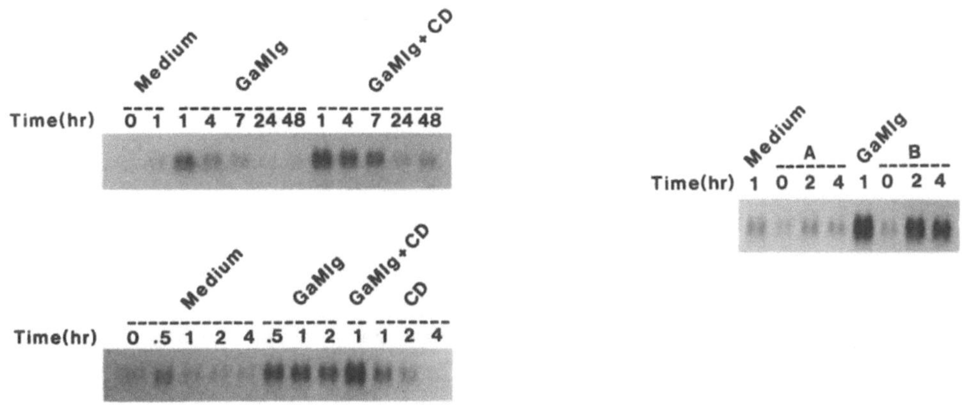

Fig. 3. Time course of c-*myc* mRNA expression in murine splenic lymphocytes.
a. (left) Effects of treatment with GaMIg, CD, or GaMIg plus CD for the indicated times.
b. (right) Effects of CD in cells preincubated with GaMIg. Lanes A, preincubation in medium; Lanes B, preincubation in medium supplemented with GaMIg

treatment on c-*myc* RNA expression. Spleen cells which had been preincubated for 1 hour in medium supplemented with GaMIg were washed and incubated in medium alone for 3 hours. As expected by 3 hours, the increased c-*myc* RNA level had returned to the control value (Fig. 3b; GaMIg vs 0 hour, lanes B). Following subsequent addition of CD to the cultures, the c-*myc* RNA level increased very dramatically. In cultures which have been mock-treated, i.e. preincubation with medium alone (Fig. 3b, lanes A), CD had only a minor influence on the c-*myc* RNA level. Thus GaMIg alters c-*myc* gene expression, rendering it susceptible to subsequent action by CD; furthermore, there are synergistic effects of competence- and progression-like factors on c-*myc* gene expression.

GaMIg Alters c-*myc* Gene Response to CD:

To investigate the nature of the changes mediated by GaMIg which render expression of the c-*myc* gene "primed" for the signals delivered by CD, the influence of these factors on gene transcription was measured using nuclear runoff analysis (Fig. 4a). Incubation of spleen cells with either GaMIg or CD resulted in a transient increase in c-*myc* gene transcription; the initial increase. with GaMIg was greater and the effect was of longer duration than that observed with CD, as seen at 4.5 h. These results are consistent with the RNA values above (Fig. 3a). Treatment with GaMIg plus CD results in a much larger increase in the rate of transcription than the increases observed upon treatment with the individual agents alone; furthermore, increased transcription was maintained at a higher level over the time course, as seen at 4.5 h (Fig. 4a).

To determine whether the changes induced by GaMIg were stable, cells pretreated with GaMIg were washed, incubated in medium for 3h, and then subsequently treated with CD. As seen in Fig. 4b, the

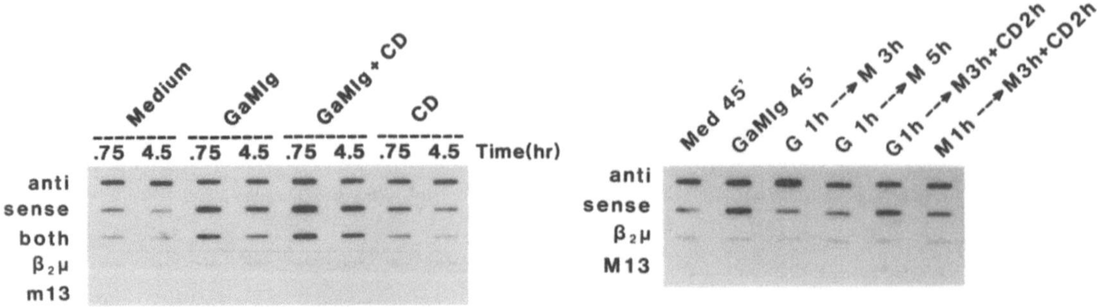

Fig. 4. Nuclear run-off analysis of c-*myc* gene transcription in murine splenic lymphocytes.
a. (left) Effects of treatment with GaMIg, CD, or GaMIg plus CD.
b. (right) Effects of CD treatment in cells pretreated with GaMIg and then incubated in medium.
Cloned DNAs include: anti and sense, M13 clones of the Bam HI-Hind III restriction fragment of mouse c-*myc* genomic DNA containing exons 2 and 3, which detect transcription of noncoding and coding strands, respectively (Kindy et al., 1987); both, 5.5 kb Bam HI genomic fragment of S107 mouse myeloma containing exons 2 and 3 in pBR322 (Kirsch et al., 1981); β2-μ, mouse β_2-microglobulin cDNA clone (Parnes et al., 1981); M13, vector DNA

increase in transcription induced by GaMIg decreased to control levels following incubation in medium for 3h. CD treatment caused a reinduction of transcription in GaMIg pretreated cells. In contrast, CD had only minimal effects on cells that had been mock-treated with medium for 1 h prior to washing and incubation in medium for 3h. Thus GaMIg, which functions as a competence factor, stably alters the c-*myc* gene in such a manner that its response to the signals mediated by CD, the progression-like factor, are enhanced. Furthermore there is a sequential activation of c-*myc* gene transcription that is associated with the two-step stimulation of B cells leading to S phase entry. "Priming" of c-*myc* associated with competence induction is likely to be mediated by nuclear factors which interact with regulatory regions of the c-*myc* gene; these will be analyzed using the approach described above.

CONCLUSIONS

DNA gel shift analysis indicates a complex interaction of nuclear proteins with putative regulatory elements located upstream of the c-*myc* mRNA transcription unit. Decreased transcriptional activity of this gene in WEHI 231 cells is accompanied by changes in protein binding within these DNA sequences, and possibly in protein-protein interactions as well. Transcriptional activation of c-*myc* in murine splenic B lymphocytes can occur in two stages; signals from competence-like factors alter the transcriptional response to progression-like factors. Thus this system allows us to dissect the steps leading to complete activation of the c-*myc* gene.

ACKNOWLEDGEMENTS

This work was supported by Public Health Service grants CA36355 (G.E.S.), AI23454 (T.L.R.), and CA40424 (T.L.R.) from the National Institutes of Health. M.P.D. is supported by National Institutes of Health Training Grant T32.

REFERENCES

Boyd, A and Schrader, JW (1981) The regulation of growth and differentiation of a murine B-cell lymphoma: II. The inhibition of WEHI 231 by anti-immunoglobulin antibodies. J Immunol 126:2466- 2469

Buckler, AJ, Rothstein, TL, and Sonenshein, GE (1988) Two-step stimulation of B lymphocytes to enter DNA synthesis: Synergy between anti-immunoglobulin antibody and cytochalasin on expression of c-*myc* and a G_1-specific gene. Mol Cell Biol 8:1371-1375

Campisi, J, Gray, HE, Pardee, AB, Dean, M and Sonenshein, GE (1984) Cell-cycle control of c-*myc* but not c-*ras* expression is lost following chemical transformation. Cell 36:241-247

Greenberg, ME, and Ziff, EB (1984) Stimulation of 3T3 cells induces transcription of the c-*fos* proto-oncogene. Nature (London) 311:433-438

Howard, M, and Paul, WE (1983) Regulation of B-cell growth and differentiation by soluble factors. Ann Rev Immunol 1:307-333

Kelly, K, Cochran, BH, Stiles, CD, and Leder, P (1983) Cell-specific regulation of the c-*myc* gene by lymphocyte mitogens and platelet-derived growth factor. Cell 35:603-610

Kindy, MS, McCormack, JE, Buckler, AJ, Levine, RA, and Sonenshein, GE (1986) Independent regulation of transcription of the two strands of the c-*myc* gene. Mol Cell Biol 7:2857-2862

Kirsch, IR, Ravetch, JV, Kwan, S-P, Max, E, Ney, R, and Leder, P (1981) Multiple immunoglobulin switch region homologies outside the heavy chain constant region locus. Nature (London) 293:585-587

Levine, RA, McCormack, JE, Buckler, A, and Sonenshein, GE (1986) Transcriptional and posttranscriptional control of c-*myc* gene expression in WEHI 231 cells. Mol Cell Biol 6:4112-4116

McCormack, JE, Pepe, VH, Kent, RB, Dean, M, Marshak-Rothstein, A, and Sonenshein, GE (1984) Specific regulation of c-*myc* oncogene expression in a murine B-cell lymphoma. Proc Natl Acad Sci USA 81:5546-5550

Melchers, F, and Andersson, J (1986) Factors controlling the B-cell cycle. Ann Rev Immunol 4:13-36

Parnes, JR, Velan, B, Felsenfeld, A, Ramanathan, L., Ferrini, U, Appella, E, and Seidman, JG (1981) Mouse β_2-microglobulin cDNA clones: a screening procedure for cDNA clones corresponding to rare mRNAs. Proc Natl Acad Sci USA 78:2253-2257

Remmers, EF, Yang, JQ, and Marcu, KB (1986) A negative transcriptional control element located upstream of the murine c-*myc* gene. EMBO J 5:899-904

Rothstein, TL (1985) Anti-immunoglobulin in combination with cytochalasin stimulates proliferation of murine B lymphocytes. J Immunol 135:106-110

Rothstein, TL (1986) Stimulation of B cells by sequential addition of anti-immunoglobulin antibody and cytochalasin. J Immunol 136:813-816

Shen-Ong, G, Keath, E, Piccoli, S, and Cole, M (1982) Novel *myc* oncogene RNA from abortive immunoglobulin-gene recombination in mouse plasmacytomas Cell 31:443-452

Strauss, F and Varshavsky, A (1984) A protein binds to a satellite DNA repeat at three specific sites that would be brought into mutual proximity by DNA folding in the nucleosome. Cell 37:889-901

Multifactorial Regulation of the Human c-*myc* Oncogene

M. Zajac-Kaye[1], M. Avigan[2], M. Takimoto[2], S. Pittaluga[2], J. Quinn[2], E. Gelmann[1], and D. Levens[2]

INTRODUCTION

The deregulation of the human c-myc oncogene is associated with a variety of molecular defects, including translocations, gene amplification, and point mutations (Cole 1986). The relative contributions of these genetic alterations to the pathogenesis and progression of malignancy remain to be determined. Identification of genetic defects at the DNA sequence level helps to define the cis-elements which physiologically regulate c-myc expression. Conversely, elucidation of the components that normally control c-myc may provide insight into the pathogenetic mechanisms disrupting c-myc regulation. We have used a sensitive exonuclease protection assay to study the interaction of sequence-specific DNA binding proteins with potential regulatory sequences from the normal c-myc gene and from the abnormal c-myc allele derived from Burkitt's lymphoma cells.

A LONG-RANGE ASSAY FOR DNA BINDING PROTEINS

Many assays for sequence-specific binding proteins employ small DNA probes. The choice of the probe is often predicated upon definition of regulatory elements by transfection experiments or an assumption of the location of such elements. However, interpretation of transfection experiments could be difficult if cis-regulatory elements are scattered over a large region or if individual elements are synergistic, or antagonize or reduntantly reinforce one another. An alternate approach we have employed assumes that many cis-regulatory sequences are the loci at which particular proteins bind. Correlation of the biochemical interactions between cis- and trans- elements with physiological alterations and with sequence analysis of abnormal genes may allow specific, defined plasmid constructions to verify the in vivo functions of these elements. Figure 1 outlines the exonuclease assay, in which 3' end-labeled DNA fragments up to 2 kb are bound to a solid support via a lac-repressor-beta-galactosidase fusion protein (Levens and Howley 1985; Quinn et al. 1987). After incubation with the protein extract, the labeled DNA is hydrolyzed with a 5'-3' exonuclease. Digestion is halted either by specifically bound factors or by the fusion protein. Displaying the products by gel electrophoresis reveals the loci of protein-DNA interactions. With this analysis numerous binding sites for potential regulatory proteins have been mapped upstream of, and within, the c-myc oncogene.

---

[1]Medicine Branch, Clinical Oncology Program, [2]Laboratory of Pathology, National Cancer Institute, Bethesda, MD 20892

Current Topics in Microbiology and Immunology, Vol. 141

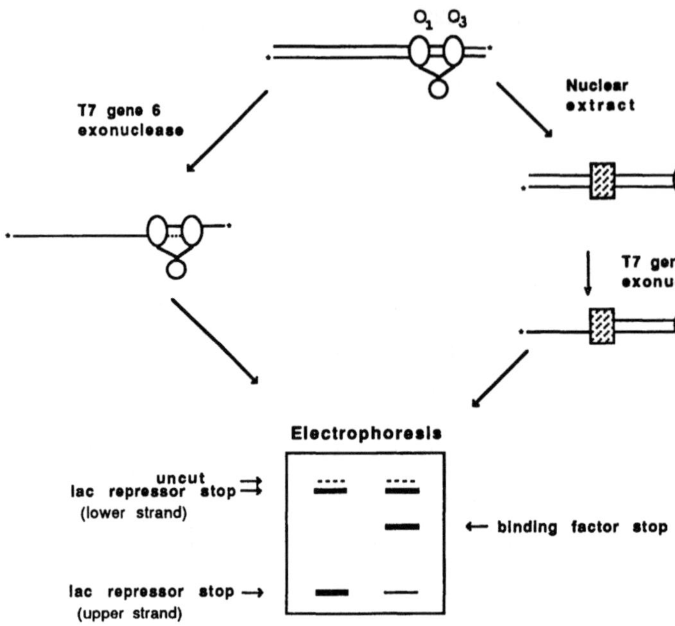

Fig. 1. Schematic summary of the long-range exonuclease assay. A fragment of DNA up to 2 kb in length is cloned adjacent to the lac operator in a plasmid vector such as pUC18. After cleavage with a restriction enzyme that leaves the operator linked to the sequence of interest, the DNA is 3' end-labeled at both ends, bound to a lac repressor-beta-galactosidase fusion protein, and immobilized to a solid support via anti-beta-galactosidase. This complex is mixed with competitor nucleic acid, incubated with or without extract, and degraded from both 5' ends. The digestion products are denatured, separated by electrophoresis, and displayed by autoradiography

A PROTEIN BINDING SITE IN INTRON I OF THE C-MYC GENE IS FREQUENTLY MUTATED IN BURKITT'S LYMPHOMAS

The Burkitt's lymphoma cell line, PA682, (Magrath et al. 1983) contains a variant 8:22 chromosomal translocation at least 16 kb downstream of the c-myc gene (Zajac-Kaye et al. unpublished data). Sequence analysis of the abnormal allele revealed point mutations scattered throughout the 3' region of exon I and the 5' region of intron I (Zajac-Kaye et al. 1988).

A 38-bp deletion removed a thymidine-rich segment at the 5' end of intron I, in a region implicated in the attenuation of c-myc transcription (Bentley and Groudine 1986). Surprisingly, a nuclear run-off analysis of PA682 cells revealed undiminished attenuation, (Zajac-Kaye et al. 1988), in contrast to several other Burkitt's lymphoma cell lines (Cesarman et al. 1987). Thus, neither the point mutations nor the deletion was sufficient to abrogate attenuation. To explore the possibility that the abnormal c-myc allele's ability to interact with proteins was altered, the corresponding regions of exon I and intron I from the wild type and mutated c-myc genes were analyzed with the exonuclease protection assay. A strong barrier to exonuclease digestion was present on the normal allele but absent on the abnormal c-myc gene (Zajac-Kaye et al. 1988). Inspection of the DNA sequence in this region revealed several point mutations present in the abnormal c-myc allele (Fig. 2). Two base changes destroyed a Mae III restriction enzyme cleavage site in the abnormal gene at the locus where exonuclease hydrolysis was halted on the wild type DNA. The sequence-specific binding of a protein to the wild type allele could be abolished by the addition of excess wild type DNA as competitor, while the mutated DNA had no effect. Cleavage of the wild type fragment with Mae III rendered it completely ineffective as a competitor. Thus the Mae III site was localized within or adjacent to the binding site of a nuclear protein. A 20 bp double-stranded oligonucleotide was synthesized with the sequence encompassing a short palindrome. This oligonucleotide abolished binding when included as a competitor in the exonuclease assay against a wild type probe; in addition, this same oligonucleotide formed several specific complexes with proteins present in the extracts from a variety of cells, as resolved by gel retardation. Importantly, a mutant double-stranded oligonucleotide with the sequence of the abnormal PA682 c-myc gene was ineffective as both probe and as competitor in exonuclease as well as in gel retardation assays. DNase I footprint analysis revealed a protection of this same 20 bp element. Comparison of the wild type c-myc gene with the homologous segment of the abnormal alleles from several Burkitt's lymphomas showed that in five of seven cases, mutations had occurred within the binding site we identified. Preliminary results suggest that this sequence is a potent negative cis-element whose activity *in vivo* requires a trans-acting factor.

MULTIPLE C-MYC UPSTREAM SEQUENCES BIND PROTEINS

The approach outlined above has allowed the mapping of protein binding sites along the c-myc gene. Although the *in vivo* significance of some of these interactions is not always apparent, four methods can be used to determine which of them may play a role in the regulation of c-myc expression. 1) Transient or stable transfection of reporter genes fused to mutated or deleted c-myc regulatory sequences may allow the correlation of alterations in gene expression with particular protein binding sites. 2) Fusion of protein binding segments to heterologous promoters may reveal activity, *in vivo*, in the absence of potentially antagonistic factors binding elsewhere along the c-myc gene.

Cell Line	Binding Region	Translocation
Normal	(MaeIII over G T A A C)	None
BL37	A G A G T A G T T A T G G T A A C T G G	(8;22)
Daudi		(8:14)
PA682	— — — — — — — — — — — — A — T — — — — —	(8;22)
BL2	— — — — — — — — — — — — — — — — T — — —	(8;22)
KK124	— — — — — — — — — — — — — — T — — — — —	(8;22)
Raji	— — — A — — C C — T — — — — — — G — — —	(8;14)
LY47	— — — A — — — — — G — T C — — — — — — —	(8;22)

Fig. 2. The protein binding sequence of the c-myc intron I negative cis-element is shown. Also shown are the sequences from several Burkitt's lymphoma cell lines. Five of seven abnormal c-myc alleles from these cases contain alterations in the binding sequence (Gazin et al. 1987; Hollis et al. 1984; Rabbitts et al. 1984; Showe et al, 1987; Denny et al. 1985; Rabbitts et al. 1983)

3) Comparison of proteins binding to the c-*myc* gene before and after treatment with agents known to influence c-*myc* expression allows the correlation of particular protein binding sites with physiological changes of c-*myc* levels. 4) Comparison between different cell lines or tissues of the presence, absence, or abundance of proteins interacting with different cis-elements may implicate particular proteins with positive or negative changes in c-*myc* transcription.

One region, 320-350 bp upstream of P1, is shown by deletion analysis to be a negative cis-element that decreases expression from both P1 and P2; this segment was originally defined by the exonuclease assay as a locus that bound to nuclear factors (Hay et al. 1987). Employing both exonuclease and gel retardation analysis, it became apparent that two distinct and separable factors interact with this region. These factors have also been shown to interact with cis-elements from several viral enhancers. The correlation of an *in vivo* negative element (defined by transfection of deletion mutants) with the binding of factors also interacting with known regulatory components, suggests a potential role for this cis-element in the control of c-*myc* expression. In a similar manner, a positive element has been proposed to reside 100-160 bp upstream of P1.

The correlation of altered binding of some factors with physiologic changes suggests a potential role for those factors *in vivo*. For example, surveying c-*myc* upstream sequences from 350 to 2329 bp upstream of P1, reveals numerous protein-DNA interactions in extracts from a variety of cells, including the promonomyelocytic leukemia cell line, HL60. Upon induction of myeloid differentiation by DMSO, HL60 cell extracts lose a single binding factor approximately 1500 bp upstream of P1 while retaining binding to numerous other loci. The kinetics of the disappearance of the factor at 1500 correspond with both a decrease in c-*myc* transcription initiation and commitment to differentiation (Siebenlist et al. 1988). Biochemical characterization of the protein interacting with this binding site and genetic analysis of the potential physiology associated with mutant binding sites are currently being conducted. Another category of protein-DNA interactions along the c-*myc* gene, as detected with the long-range exonuclease assay, is cell type-specific binding. For example, exonuclease barriers at particular loci are not ubiquitously present with extracts of all cell lines. This suggests that different sets of positive and negative transcription factors may govern c-*myc* expression in different cells, perhaps dependent upon the specific physiology of that cell. In fact, different cell lines generate specific and even unique patterns of binding when compared over 3 kb of c-*myc* sequence. The assessment of the physiological significance of these differences will require extensive genetic analysis.

WHY SO MANY FACTORS?

Figure 3 illustrates the plethora of protein-DNA interactions along the c-*myc* oncogene detected with the exonuclease protection assay. If even a fraction of these interactions represent bona fide sites for the control of c-*myc* expression, then a description of the interplay between regulatory components and an elucidation of the hierarchy of effects induced by each element will be very complex. Considering the evidence that c-*myc* expression is likely to be intimately involved in a cell's decision to replicate and/or differentiate, and considering the fact that c-*myc* levels respond to a variety of extracellular signals, regulation in response to alterations of the intracellular

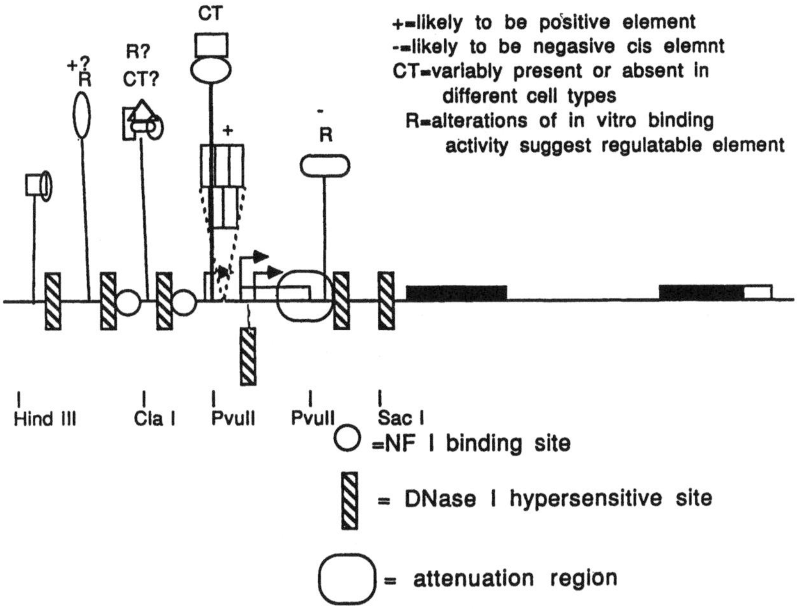

Fig. 3. Summary of the protein binding sites present in the c-myc gene as detected by the exonuclease protection assay. A map of the c-myc gene showing several restriction enzyme cleavage sites is shown. Arrows indicate the positions of P0, P1, and P2. Above the map the positions of newly detected binding sites are indicated. The loci of DNase I hypersensitive sites, NF I binding, and attenuation are indicated (Bentley and Groudine 1986; Siebenlist et al. 1984)

environment might also be expected. In view of the wide variety of differentiating agents that can alter c-myc expression, the search for a single, master switch for c-myc may prove fruitless. It is conceivable that each of a variety of signals for growth and differentiation employs a different set of transducers, all funneling into alterations of c-myc expression. In this regard, different sets of signals may be used in different cells and under different physiological circumstances. By altering the array of protein factors present in a given cell or cell type, distinctive patterns of c-myc expression may be established. In this manner certain mutations may contribute to B-cell neoplasia (perhaps including the mutated binding site in intron I) whereas other mutations may be phenotypically manifested only in other tissues.

Considering the potentially malignant consequences of c-myc deregulation, a selective advantage may arise from a certain degree of redundancy in its regulatory apparatus. Parallel utilization of more than one cis-element, and potentially more than one trans-element, in response to a physiological stress or signal may provide a fail-safe mechanism guaranteeing that the necessary change in c-myc expression occurs. Some of the interactions we detect may, therefore, represent parallel pathways responding to similar physiological signals. Redundancy presents a particularly difficult problem for the genetic dissection of c-myc cis-elements because a phenotypic alteration may require the simultaneous disruption of distinct and possibly separated sequences. An analysis of the combinatorial ablation of possible cis-elements may be required to unmask their roles *in vivo*.

The long-range exonuclease assay should facilitate the nomination of particular sequences as candidates for cis-elements involved in particular regulatory circuits and thereby allow the construction of appropriate plasmids to demonstrate their role *in vivo*.

REFERENCES

Bentley DL, Groudine M (1986) A block to elongation is largely responsible for decreased transcription of c-*myc* in differentiated HL60 cells. Nature 321:702-706

Cesarman E, Dalla-Favara R, Bentley D, Groudine M (1987) Mutations in the first exon are associated with altered transcription of c-*myc* in Burkitt lymphoma. Science 238:1272-1275

Cole M (1986) *myc* oncogene. Ann Rev Genetics 20:361-384

Denny CT, Hollis GF, Magrath I, Kirsch IR (1985) Burkitt lymphoma cell line carrying a variant translocation creates new DNA at the breakpoint and violates the hierarchy of immunoglobulin gene rearrangement. Mol Cell Biol 5:3199-3207

Gazin C, Rigolet M, Briand JP, Van Regenmortel MHV, Galibert F (1984) Nucleotide sequence of the human c-*myc* locus: provocative open reading frame within the first exon. EMBO J 3:383-387

Hay N, Bishop JM, Levens D (1987) Regulatory elements that modulate expression of human c-*myc*. Genes and Dev 1:659-671

Hollis GF, Mitchell F, Battery S, Potter H, Taub R, Lenoir GM, Leder P (1984) A variant translocation places the γ immunoglobulin genes 3' to the c-*myc* oncogene in Burkitt's lymphoma. Nature 307:752-755

Levens D, Howley P (1985) Novel method for identifying sequence-specific DNA binding proteins. Mol Cell Biol 5:2307-2315

Magrath I, Erikson J, Whang-Peng J, Sieverts H, Armstrong G, Benjamin D, Triche T, Alabaster O, Croce CM (1983) Synthesis of kappa light chains by cell lines containing an 8:22 chromosomal translocation derived from a male homosexual with Burkitt's lymphoma. Science 222:1094-1098

Quinn JP, Holbrook N, Levens D (1987) Cellular protein binding to GALV enhancer. Mol Cell Biol 5:2735-2744

Rabbitts TH, Forster A, Hamlyn P, Baer R (1984) Effect of somatic mutation within translocated c-*myc* oncogenes in Burkitt's lymphoma. Nature 309:592-597

Rabbitts TH, Hamlyn PH, Baer R (1983) Altered nucleotide sequences of a translocated c-*myc* gene in Burkitt lymphoma. Nature 306:760-765

Showe LC, Moore RCA, Erikson J, Croce CM (1987) *Myc* oncogene involved in a t(8:22) chromosome translocation is not altered in its putative regulatory regions. Proc Natl Acad Sci USA 84:2824-2828

Siebenlist U, Bressler P, Kelly K (1988) Transcriptional regulation of c-*myc*. Mol Cell Biol 8:867-873

Siebenlist U, Henninghausen L, Battey J, Leder P (1984) Chromatin structure and protein binding in the putative regulatory region of the c-*myc* gene in Burkitt lymphoma. Cell 37:381-391

Zajac-Kaye M, Gelmann E, Levens D (1988) A point mutation in the c-*myc* locus of a Burkitt lymphoma abolishes binding of a nuclear protein. Science 240: June 24.

Negative Control Elements Within and Near the Murine c-*myc* Gene

K.B. Marcu[1], C. Asselin[1], A. Nepveu[2], G. Weisinger[1], and J.Q. Yang[3]

INTRODUCTION

The c-myc proto-oncogene is highly conserved in vertebrate evolution and is expressed in most mammalian cell types. The gene consists of three exons, the first of which is a large non-coding sequence (Reviewed in Marcu, 1987). The last two exons encode a nuclear associated, DNA-binding protein with an apparent molecular mass of 62-64 Kd. The c-myc gene is regulated by two promoters with independent transcription initiation sites, P_1 and P_2. Normal c-myc RNAs are highly unstable in vivo and can be superinduced by growth factors in the presence of protein synthesis inhibitors, which has implicated labile negative factors in their regulation.

Alterations of normal c-myc expression either by gene amplification, retroviral insertion or chromosome translocation are considered important steps in the development of a variety of neoplasms (reviewed in Klein and Klein, 1985; Cory, 1986; Kelley and Siebenlist, 1986; Marcu, 1987). The c-myc gene product can immortalize cells in culture and can cooperate with other activated oncogenes in the tumorigenic process (Land et al., 1983). Transgenic mice harboring a c-myc gene linked to tissue-specific regulatory sequences develop clonal tumors in tissues expressing the c-myc transgene (Stewart et al., 1984; Adams et al., 1985). The c-myc polypeptide is believed to be a cellular mediator of growth factor action and its expression may reflect the competency of cells to enter and progress through the cell cycle (reviewed in Kelley and Siebenlist, 1986; Marcu, 1987).

The c-myc gene is regulated at the transcriptional and post-transcriptional levels in proliferating and differentiating cells (reviewed in Kelley and Siebenlist, 1986; Marcu, 1987; Nepveu et al., 1987a; Nepveu et al., 1987b), and cis-acting negative and positive regulatory elements have been identified within the c-myc locus (Remmers et al., 1986; Bentley and Groudine, 1986; Nepveu and Marcu, 1986; Eick and Bornkamm 1986; Chung et al., 1986; Yang et al., 1986; Lipp et al., 1987; Hay et al., 1987). One of these negative modulators blocks the progression of RNA polymerase II through the gene's first exon (Bentley and Groudine, 1986; Nepveu and Marcu, 1986; Eick and Bornkamm, 1986). A second class of negative elements reside upstream of the gene's promoters and appear to function in an opposite manner to transcriptional enhancers (Remmers et al., 1986). Here, we describe some of the molecular requirements and properties of these negative modulators.

---

[1]Biochemistry Department, S.U.N.Y. at Stony Brook, Stony Brook, NY..11794-5215; [2]Lugwig Institute, 687 Ave. des Pins ouest, Montreal, Canada..H3A 1A1; [3]Imclone Systems Inc., 180 Varick St., New York, NY..10014

Sequences required for c-*myc* transcriptional blockage and P_2 promoter activity

To determine the sequences required for transcriptional attenuation within the *myc* gene's first exon, we prepared populations of stable HeLa cell transformants containing *myc*-CAT gene vectors. Each of these vectors retained different portions of a 1.71 Kb BglII fragment which contains 5' flanking sequences and the entire first exon of the murine c-*myc* gene (Yang et al., 1986) (Figure 1). Nuclei were incubated *in vitro* in the presence of [α^{32}P]UTP and labeled nascent nuclear RNAs were hybridized to nitrocellulose filters bearing an excess of single-stranded probes from the c-*myc* first exon and double-stranded fragments of the CAT or β-globin genes (Figure 1). Probe A (exon 1) provides an estimation of the relative transcription initiation rates, while probe B (either CAT of β-globin) measures the amount of RNA polymerase loading downstream of the first exon transcriptional block (Nepveu and Marcu, 1986). Transcriptional attenuation occurs with the pMHHBgβGlob vector, which contains the entire first exon and 7 bp 5' of the P_1 start site (Figure 1, lanes 1-3). However, removal of the 5' 123 bp HaeIII-BamHI fragment (-7 to +116) of pMMHBgβGlob (see pMBmBgCAT) or replacement of all c-*myc* regulatory sequences 5' of the P_2 start site (including its TATA box) by SV40 promoter and enhancer sequences (see SV2M(H)$_R$CAT in Figure 1) completely relieves the elongation block. These data suggest that P_2 promoter sequences located in the 123 bp HaeIII-BamHI fragment are necessary for transcriptional blockage. S_1 nuclease mapping in Figure 2 confirms that this 123 bp region is required for P_2 usage. P_2 is detected in cells expressing pMHHBgβGlob but a read through transcript is seen in the cells harboring pMBmBgCAT.

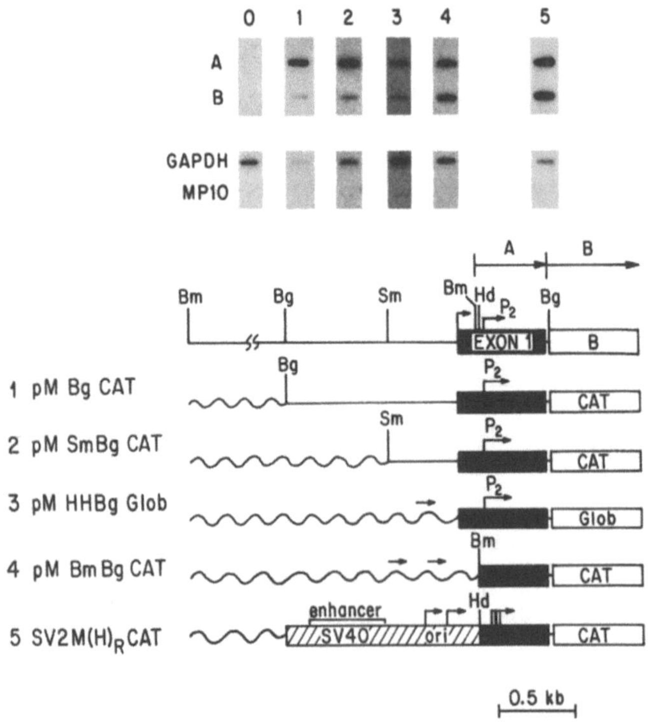

Fig. 1. Run-on transcription analysis of myc chimeric genes expressed in stably transfected populations of HeLa cells. Nascent, ^{32}P labeled, nuclear transcripts were hybridized to an excess of the indicated DNA probes on nitrocellulose filters: the c-myc M13 probe contains 423 nt of exon 1 sequences (probe A); the GAPDH probe is derived from a glyceraldehyde-3-phosphate-dehydrogenase cDNA clone, pRGAPDH13 (Piechaczyk et al., 1984); the CAT probe is a 550 bp BglII-NcoI fragment from the 5′ end of the CAT gene in pCAT(ΔEP) (Yang et al., 1986); and the βglobin probe is a 437 bp HindIII-BglII fragment containing the first and a portion of the second exon of a rabbit β globin gene. The myc-CAT vectors are shown with restriction sites designated as follows: Bg, BglII; Bm, BamHI; Hd, HindIII; Sm, SmaI.

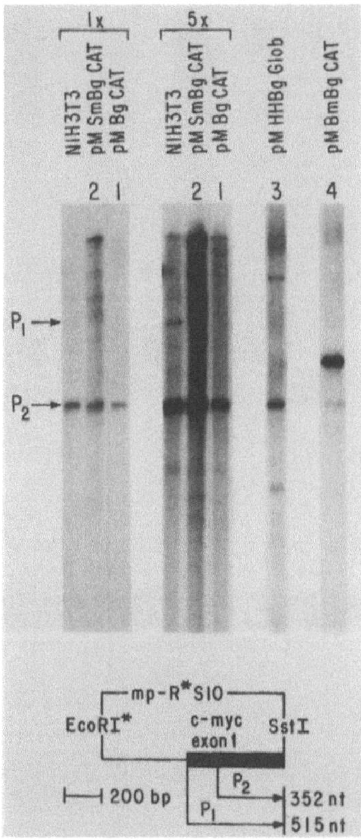

Fig. 2. S1 nuclease protection analysis of a uniformly labeled, single-stranded myc exon 1 probe, mpR*S10 (Yang et al., 1986) by CAT RNAs expressed in populations of HeLa cells stably transfected with the indicated myc-CAT vectors. 1X and 5X are 1 and 5 day exposures with intensifier screens respectively

We next determined the 3′ boundary of the exon one transcriptional block. A set of Bal 31 deletions were prepared in a 3′ direction from the exon 1 BamHI site and used as single-stranded DNA probes in a nuclear run-on experiment with nuclei from the A-MuLV transformed NIH3T3 line 54C12 (Nepveu et al., 1985) (Figure 3).

Labeled RNAs were hybridized to these exon one DNA segments (probe A series) and to a larger DNA probe containing downstream intron and exon sequences (probe C) (Figure 3). The intensities of the hybridization signals provide a measure of the distribution of RNA polymerases on the c-myc gene (Nepveu and Marcu, 1986). The probe A region contains about six times fewer adenine residues than probe C. Thus, when the signals from probe A and C are of equal intensity, the rate of transcription is about six fold higher in region A. The data in Figure 3 show a dramatic decrease in probe A signal intensity with deletion mutants A3 and A4 with the A4 probe detecting no elongation block. We conclude that the 3′ boundary of the murine c-myc gene's transcriptional block resides between +305 and +449, relative to P_1 at +1 (Figure 1), about 150 bp 5′ of the exon 1/intron 1 splice junction.

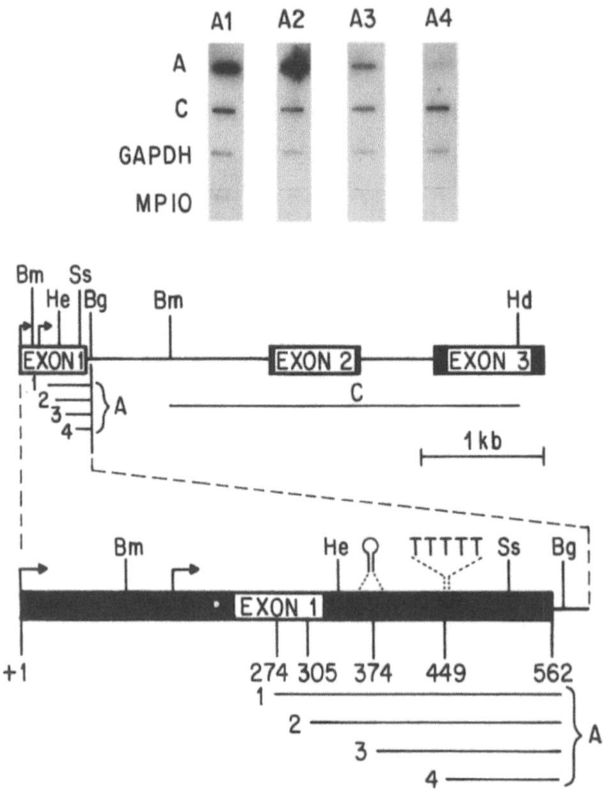

Fig. 3. Mapping of the transcriptional elongation block within the myc gene's first exon by nuclear run-on analysis with nuclei of 54C12 cells (Nepveu et al., 1985) and single-stranded exon 1 deletion mutants as probes. The relative sizes of the A1 to A4 exon 1 probes are indicated and probe C contains 3.4 Kb of downstream myc sequences. The GAPDH probe is the same as described in Fig. 1. Restriction enzyme sites are as follows: Bg, BglII; Bm, BamHI; Hd, HindIII; He, HaeIII; Ss, SstI

Nuclear factor binding nearby the myc P_2 initiation site

DNase I footprinting was employed to identify nuclear factor binding sequences nearby the *myc* P_2 start site. Two regions of protection were identified within the HaeIII-HindIII fragment encompassing sequences -7 to +140 (e.g., +57 to +77 and +96 to +119 for factors ME1a2 and ME1a1 respectively) which are just 5′ of P_2 at +163 (Figure 4). A major DNase I hypersensitive site was detected at the 5′ border of the ME1a1 binding site. The ME1a1 factor recognizes the sequence $(G)_3A(G)_4A(G)_3A$ which was also found sufficient for binding as an *in vitro* synthesized oligonucleotide (data not shown). The addition of a 100 fold molar excess of a restriction fragment carrying the SV40 early promoter sequences partially abolished complex formation at these two binding sites (data not shown). It is well known that the Sp1 transcription factor binds to multiple GC box motifs in the SV40 early promoter (Dynan and Tjian, 1983; Gidoni et al., 1984). Purified Sp1 appeared to bind weakly to the ME1a site which would be considered a low affinity Sp1 binding site (data not shown) (Asselin et al., 1988). Comparative gel retardation assays performed with crude HeLa extract and purified Sp1 revealed different gel shift patterns with the *myc* HaeIII-BamHI fragment (Asselin and Marcu, unpublished results). It would seem that ME1a is a novel transcription factor which binds to sequences required for murine c-*myc* P_2 activity. Furtheremore, our results suggest that DNA sequences needed for P_2 function and transcriptional attenuation reside between the P_1 and P_2 start sites.

Multiple negative elements upstream of the murine c-myc gene

We had previously identified a cis-acting negative modulator 424 to 1140 bp 5′ of the murine c-*myc* P_1 transcription initiation site (Remmers et al., 1986) and demonstrated that it had the properties of a transcriptional "dehancer". When placed at a distance of >1.0 Kb in either orientation from the SV40 enhancer in the pSV2CAT vector, it repressed CAT gene activity up to 50 fold (Remmers et al., 1986). We have better defined the DNA sequences required for the "dehancer effect" by inserting smaller segments of the *myc* 5′ flanking 716 bp BglII-SmaI fragment 3′ of the poly A adition site in pSV2CAT (see Figure 5 for construct maps and summary of CAT assays). These constructs were tranfected into human (HeLA, BJAB), monkey (CV1) and mouse (F9, M12.4) cell lines by either the DEAE-dextran (Mosthof et al., 1985) or calcium phosphate (Graham and Van der Eb, 1973; Chen and Okayama, 1987) techniques (see Figure 5 for summary of data). The negative elements are restricted to two regions: a 5′ 137 bp segment (-923 to -1060 bp with respect to the *myc* P_1 start site at +1) and a more 3′ 191 bp fragment (-424 to -615). The 137 and 191 bp DNA segments each down regulated CAT expression by 40 to 50 fold in all five transfected cell lines. The 5′ 137 bp segment required its 3′ terminal 33 bp (-923 to -956) for negative activity in all cell lines studied. This 33 bp fragment displayed no negative effect in HeLa, CV1 and F9 cells, but it did inhibit CAT activity by 5 fold in two lymphoid cell lines (BJAB and M12.4) (Figure 5). The 308 bp of spacer DNA sequences between the 5′ 137 bp and 3′ 191 bp "dehancer" segments displayed no negative effector activity in this assay, but was somewhat stimulatory in F9 teratocarcinoma cells (data not shown).

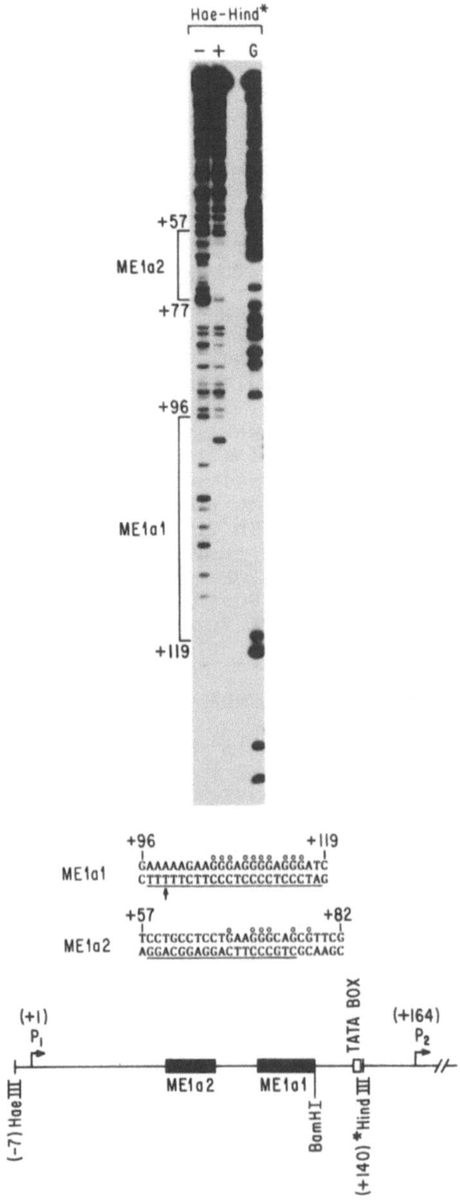

Fig. 4. DNase I footprinting of HeLa nuclear factor binding sites in a 147 bp HaeIII-HindIII fragment upstream of the P_2 initiation site. The indicated fragment was ^{32}P end-labeled at residue +140 on its non-coding strand, chemically cleaved at G residues to provide a marker and separately digested with DNase I with (+) or without (-) pre-incubation in crude HeLa nuclear extract. Lines below the sequence represent the nuclear factor binding site protected from DNase I digestion. A DNase I hypersensitive site is indicated by an arrow and open circles represent guanines which upon methylation block

nuclear factor binding in an independent DMS interference assay (data not shown). The locations of the nuclear factor binding sites with respect to the P_2 start site are indicated.

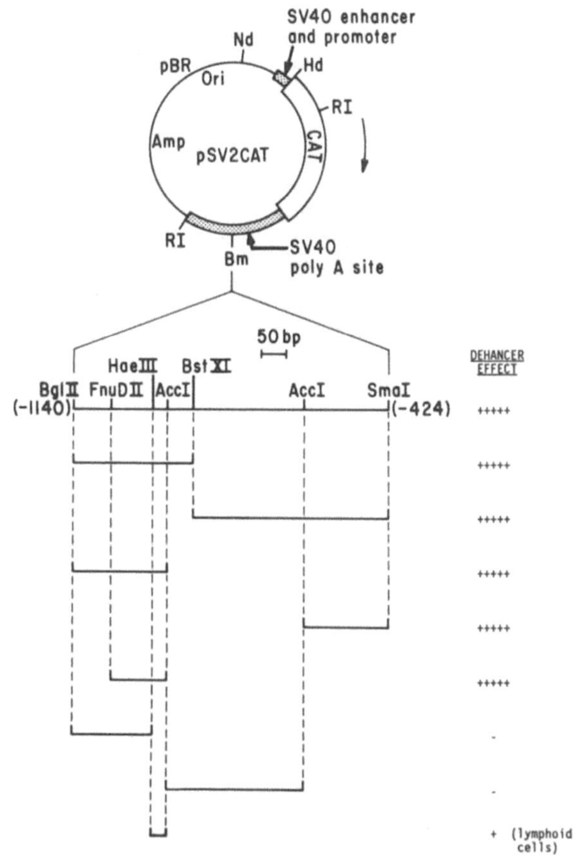

Fig. 5. Description of myc-pSV2CAT vectors. Shown are restriction maps of the pSV2CAT expression vector (Gorman et al., 1982) and the 716 bp myc "dehancer" region 424 to 1140 bp 5' of the murine c-myc gene's P_1 start site (Remmers et al., 1986). The indicated DNA fragments were inserted into the BamHI site of pSV2CAT. Relative CAT activities in transfected HeLa, CV1, F9, BJAB and M12.4 cells are indicated. CAT assays are averages of at least 3 to 6 independent experiments corrected for variations in transfection efficiency by quantitation of plasmid DNAs in Hirt lysates (Hirt, 1967). Restrition sites are as follows: Bm, BamHI; Hd, HindIII; Nd, NdeI; RI, EcoRI

Nuclear factor binding to the myc dehancer fragments

We used the gel retardation technique (Singh et al., 1986) to screen for nuclear factor binding sites within each of the myc "dehancer" segments. HeLa cell nuclear extracts were incubated with ^{32}P 5' end labeled fragments, BglII-AccI (-923 to -1140) and AccI-SmaI (-424 to -615), which respectively correspond to the 5' and 3' negative segments of the "dehancer" region. BglII-AccI gave rise to three major retained bands (5'Ma, 5'Mb and 5'Mc) and AccI-SmaI to two (5'Md, 5'Me) (see Figure 6). Proteinase K abolished all retained bands, confirming them to be protein DNA complexes (data not shown). Binding competition experiments were performed with a 150 fold molar excess of each negative effector segment (Figure 6). All retained bands were competed away by the homologous DNA fragments. AccI-SmaI (3' "dehancer" segment) competitor abolished all protein binding to the BglII-AccI (5' "dehancer" segment) probe while excess unlabeled BglII-AccI fragment only competed for the formation of AccI-SmaI complex 5'Me. We conclude that the 5' and 3' "dehancer" segments share nuclear factor binding sequences.

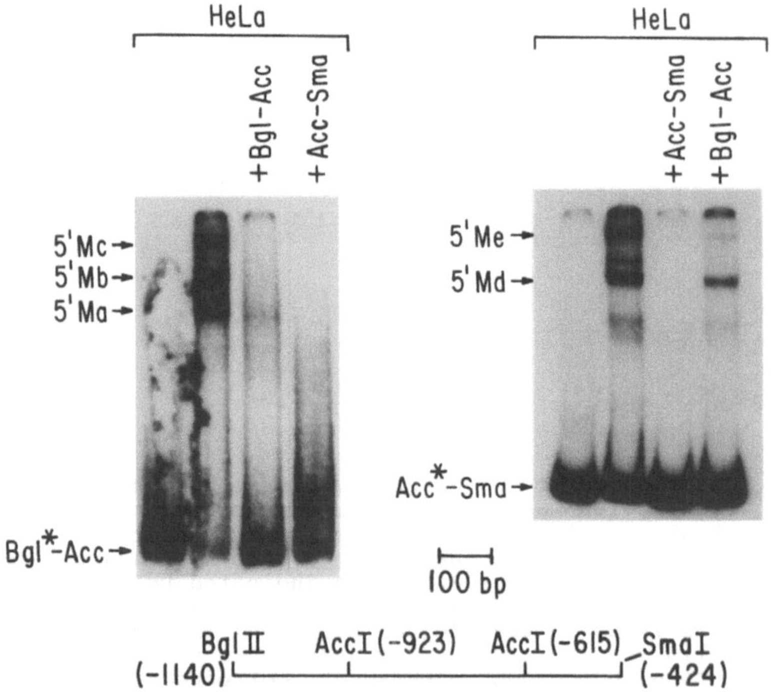

Fig. 6. HeLa cell nuclear factor binding to the myc 5' and 3' "dehancer" segments revealed by the polyacrylamide gel retention technique (Singh et al., 1986). Competitor DNAs are indicated at the top of each lane and were present in 150 fold molar excess over the labeled DNA probe. The first and second lanes of each panel are

probe alone and probe with extract respectively. The positions of migration of the major nuclear factor band shifts are indicated (e.g., 5'Ma, 5'Mb, 5'Mc, 5'Md and 5'Me)

Significance of negative control elements for normal myc regulation and activation

Chromosome translocations in murine plasma cell tumors and human Burkitt lymphomas as well as retroviral insertions in murine T cell neoplasms (reviewed in Marcu, 1987) may separate or disrupt cis-acting control element(s) from the myc gene thereby resulting in its deregulated expression. We have identified two different types of negative effector regions within and nearby the c-myc gene which are lost or disrupted in a number of these malignancies. The site of transcriptional blockage within the gene's first exon is known to modulate myc expression in proliferating and differentiating cells (Nepveu et al., 1987a; Nepveu et al., 1987b). Chromosome translocations which disrupt the myc gene are most common in murine plasma cell tumors and would remove this exon 1 transcription elongation block. The myc "dehancer" may also normally play a role in regulating c-myc transcription to varying degrees in different cell types. We have shown that independent negative elements within the myc "dehancer" region impart a 5 to 50 fold inhibitory effect (dependent on the recipient cell line) on the Sv40 promoter-enhancer in the pSV2CAT vector. In contrast, when the entire 716 bp "dehancer" was in its natural background 5' of the myc promoters, a 3 fold reduction in CAT activity was observed in the BJAB lymphoblastoid line compared to a chimeric myc CAT gene without this 716 bp sequence (Remmers et al., 1986). A similar 3-5 fold down effect was also observed in the M12.4 B lymphoma line (Remmers and Marcu, unpublished results). Therefore, these negative elements would appear to elicit a much stronger repressive effect at a distance from the SV40 promoter-enhancer than in their natural location 5' of the myc promoters. The diminished effect of the "dehancer" on the myc promoters could be due to the opposing influences of positive elements in the same DNA fragment or to intrinsic diferences in the positive control of transcription mediated by the SV40 promoter-enhancer and the myc promoters. We can not rule out the possibility that additional negative elements reside upstream of the c-myc gene which would not be revealed by our assay system. Possible discrepancies between our findings and those of other workers on the human c-myc gene (Chung et al., 1986; Lipp et al., 1987; Hay et al., 1987) may reflect different properties of murine and human c-myc regulatory sequences, the choice of recipient cell lines for DNA transfections or the precise nature of vector constructions with special attention to the spatial arrangements of negative and positive elements which are likely to be critical to their mode(s) of action.

REFERENCES

Adams JM, Harris AW, Pinkert CA, Corcoran LM, Alexander WS, Cory S, Palmiter RD, Brinster RL (1985) The c-myc oncogene driven by immunoglobulin enhancers induces lymphoid malignancy in transgenic mice. Nature 318:533-538

Asselin C, Nepveu A, Yang J-Q, Marcu KB (1988) Nuclear factor binding sites and requirements for initiation and attenuation of murine c-myc transcription. Proceedings of ICN-UCLA Symposia on Growth Factors and Receptors (in press)

Bentley DL, Groudine M (1986) Ablock to elongation is largely responsible for decreased transcription of c-myc in differentiated HL60 cells. Nature 321:702-706

Chen C, Okayama H (1987) High efficiency transformation of mammalian cells by plasmid DNA. Mol. Cell. Biol. 7:2745-2752

Chung J, Sinn E, Reed RR, Leder P (1986) Trans-acting elements modulate expression of the human c-myc gene in Burkitt lymphoma cells. Proc. Natl. Acad. Sci. USA 83:7918-7922

Cory S (1986) Activation of cellular oncogenes in hematopoietic cells by chromosome translocation. Adv. Canc. Res. 47:189-234

Dynan WS, Tjian R (1983) Isolation of transcription factors that discriminate between different promoters recognized by RNA polymerase II. Cell 32:669-680

Eick D, Bornkamm GW (1986) Transcriptional arrest within the first exon is a fast control mechanism in c-myc gene expression. Nuc. Acids Res. 14:8331-8346

Gidoni D, Dynan WS, Tjian R (1984) Multiple specific contacts between a mammalian transcription factor and its cognate promoters. Nature 312:409-413

Gorman CM, Moffat LF, Howard BH (1982) Recombinant genomes which express chloramphenicol acetyltransferase in mammalian cells. Mol. Cell. Biol. 2:1044-1051

Graham F, Van der Eb A (1973) A new technique for the assay of infectivity of human adenovirus 5 DNA. Virology 52:456-457

Hay N, Bishop M, Levens D (1987) Regulatory elements that modulate expression of human c-myc. Genes and Devel. 1:659-671

Hirt B (1967) Selective extraction of polyoma DNA from infected mouse cell cultures. J. Mol. Biol. 26:365-369

Kelley K, Siebenlist U (1986) The regulation and expression of c-myc in normal and malignant cells. Ann. Rev. Immunol. 4:317-338

Klein G, Klein E (1985) Evolution of tumors and the impact of molecular oncology. Nature 315:190-195

Lipp M, Schilling R, Wiest S, Laux G, Bornkamm GW (1987) Target sequences for cis-acting regulation within the dual promoters of the human c-myc gene. Mol. Cell. Biol. 7:1393-1400

Marcu KB (1987) Regulation of expression of the c-myc proto-oncogene. Bioessays 6:28-32

Mosthof L, Pawlita M, Gruss P (1985) A viral enhancer element specifically active in human hematopoietic cells. Nature 315:597-600

Nepveu A, Fahrlander PD, Yang J-Q, Marcu KB (1985) Amplification and altered expression of the c-myc oncogene in A-MuLV transformed fibroblasts. Nature 317:440-443

Nepveu A, Marcu KB (1986) Intragenic pausing and anti-sense transcription within the murine c-myc locus. EMBO J. 5:2859-2865

Nepveu A, Levine RA, Campisi J, Greenberg ME, Ziff EB, Marcu KB (1987a) Alternative modes of c-myc regulation in growth factor stimulated and differentiating cells. Oncogene 1:243-250
Nepveu A, Marcu KB, Skoultchi AI and Lachman HM (1987b) Contributions of transcriptional and post-transcriptional mechanisms to the regulation of c-myc expression in mouse erythroleukemia cells. Genes and Devel. 1:938-945
Piechaczyk M, Blanchard J-M, Marty L, Dani C, Panatieres F, El-Sabouty S, Fort P, Jeanteur P (1984) Post-transcriptional regulation of glyceraldehyde-3-phosphate dehydrogenase gene expression in rat tissues. Nuc. Acids Res. 12:6951-6963
Remmers EF, Yang J-Q, Marcu KB (1986) A negative transcriptional control element located upstream of the murine c-myc gene. EMBO J. 5:899-904
Singh H, Sen R, Baltimore D, Sharp PA (1986) A nuclear factor that binds to a conserved sequence motif in transcriptional control elements of immunoglobulin genes Nature 319:154-158
Stewart T, Pattengale P, Leder P (1984) Spontaneous mammary adenocarcinomas in transgenic mice that carry and express MTV/myc fusion genes. Cell 38:627-637
Yang J-Q, Remmers EF, Marcu KB (1986) The first exon of the c-myc proto-oncogene contains a novel positive control element. EMBO J. 5:3553-3562

ACKNOWLEDGEMENTS

We thank Joyce Schirmer and Chris Helmke for preparation of figures and photographs. This research was supported by NCI grant CA36246 (awarded to KBM). CA and AN were supported in part by postdoctoral fellowships from the Canadian NCI and the Quebec F.R.S.Q. respectively.

Disruption of the Putative c-*myc* Auto-Regulation Mechanism in a Human B Cell Line

S. MARTINOTTI, A. RICHMAN, and A. HAYDAY

INTRODUCTION

It has been proposed that excessive or temporally inappropriate expression of the c-myc proto-oncogene determines a genetic program for cells that directs them toward proliferation, and away from differentiation. In support of this proposal, constitutive expression of c-myc in either murine erythroleukaemia cells, or 3T3-L1 cells largely prevents the progression of their respective differentiation schedules(1,2). Additionally, when resting cells of numerous types are treated with various mitogens, proliferation is preceded by rapid induction of c-myc expression (3,4). The inductive role that the mitogens play in proliferation can be in part substituted for by c-myc (5). An issue that remains outstanding is how the induction of c-myc is controlled. It has been theorised that c-myc auto-regulates its own transcription (6). This was provoked by the observation that in Burkitts lymphoma (BL), there is de-regulated expression of c-myc derived from the translocated allele, whilst the unrearranged allele is generally silent (6). This situation was to some extent re-capitulated by Lombardi et al (7) who witnessed the down-regulation of c-myc genes in human B cell lines into which they had stably introduced de-regulated c-myc. Because the products of the c-myc gene are two nuclear phosphoproteins with DNA-binding capacity, it has been considered that the auto-regulation is direct, requiring only the c-myc product and the c-myc template. Alternatively, the effect may be indirect, operating via pathways that may be disrupted and hence rendered inoperative in certain cell types. In order to examine this, we have investigated the generality of auto-regulation. Here we report an established human B cell line that is not susceptible to auto-regulation.

THE INTRODUCTION AND EXPRESSION OF HUMAN C-MYC IN B CELLS

Previously-described procedures (8) for electro-poration were used to introduce single haploid copies of human c-myc into the genome of the human B cell line, GM4672, that is a subline of the EBNA(+) line, GM1500. The cells are positive for IgG - kappa. As a source of potentially de-regulated c-myc, we utilised a cDNA to the translocated c-myc gene of the non-Hodgkin lymphoma, Manca (9), that was cloned via XhoI linkers into a vector containing an immunoglobulin heavy chain (IgH) enhancer (to guarantee basal expression in B cells) combined with a metallothionein promoter to facilitate induction of c-myc over and above the basal level. Southern analysis revealed two clones, GM3 and GM8, to contain intact c-myc inserts (Fig. 1).

Dept. of Biology, Yale University, 06511, USA

Current Topics in Microbiology and Immunology, Vol. 141

1. A 2.2kb c-myc cDNA derived from the translocated gene of Manca cells, cloned behind a mouse metallothionein (mt) promoter, downstream of an immunoglobulin enhancer (IgH), in a vector conferring mycophenolic acid resistance on successfully transfected cells. Southern analysis of XhoI-cleaved cellular DNA reveals two clones (3,8) to have intact inserts.

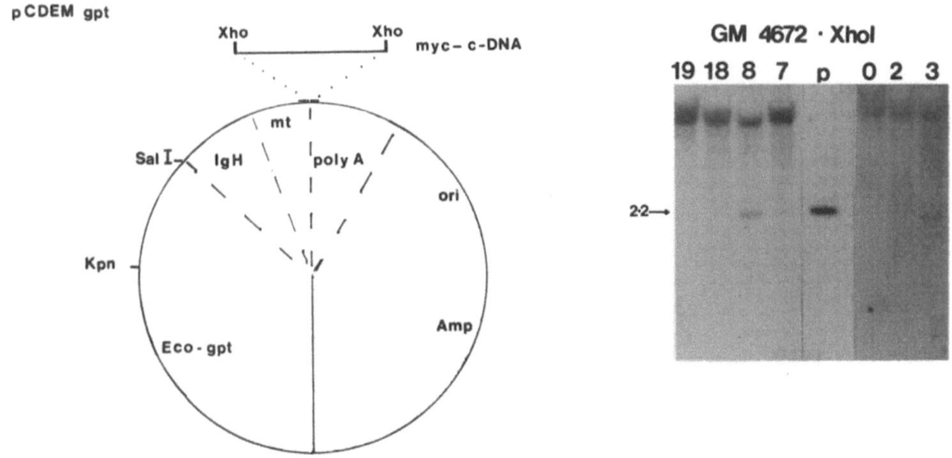

Northern analysis (Fig. 2) of RNA from clones 3 and 8, and from the parental line (GM.0), showed the presence of c-myc transcripts of orthodox size (10) in all cases (Although the c-myc cDNA lacks c-myc exon 1, the truncation is compensated for by the inclusion of intron sequences that are retained contiguous with exons 2 and 3 (9). Consistent with the contribution to the signal of at least some transcripts from the exogenously-added genes, there is clear positive modulation of the Northern signal by metal (5 micromolar cadmium for two hours) in clones 3 and 8, but not in the parental line, where if anything, the modulation is down a little. Clone 3 expresses a low basal level, that is highly induced, and clone 8 a high basal level, that is moderately induced.

2.

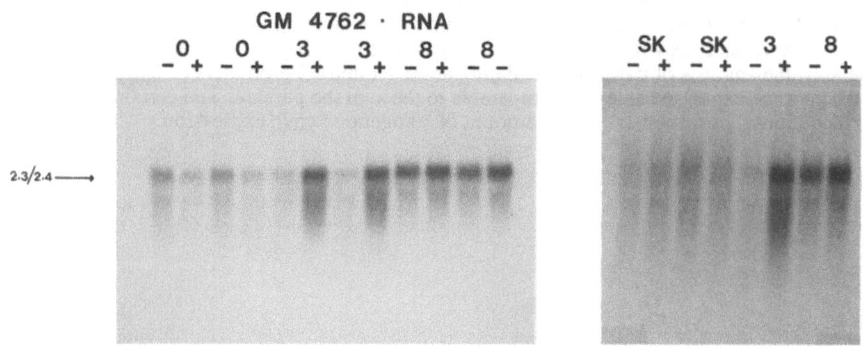

Northern analysis of total RNA from GM4672 parental cells (clone 0), and the transfected derivatives, and of SK (Manca) cells. Cells were grown with (+)/ without (-) a supplement of 5 micromolar cadmium chloride that is an effective inducer of the metallothionein promoter. Cadmium was present for 2h. The myc messenger RNA of 2.3/2.4 kb was detected with a DNA probe for exon III.

3.
S-1
assay
(see
fig.2)

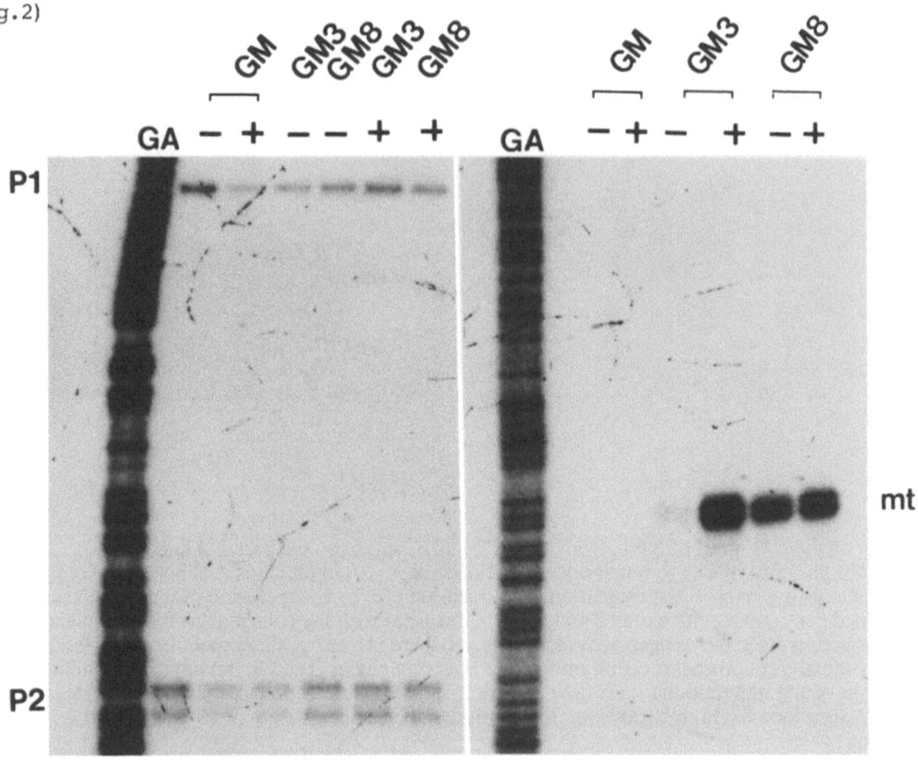

This is confirmed by S-1 analysis (Fig. 3) of transcripts using probes that span either the murine metallothionein promoter, or the dual c-myc promoters. Incubating clones 3 and 8 with and without cadmium, affords us cells with four levels ranging from low to high of exogenous c-myc expression. Despite this, the analysis of transcripts initiating at the P1 and P2 promoters of the endogenous genes shows them to be expressed at levels comparable to those in the parental, and certainly , amongst the transfected clones, unresponsive to the amount of 'exogenous' myc expression.

It may not be argued that the added c-myc genes are transcribed but not translated, since immunoprecipitation of extracts of phosphate labeled cells reveals both basal and metal-inducible myc protein expression that parallels strongly the level of transcription of the added genes (Fig. 4). Additionally, the level of protein expression is quite comparable with, that in SK (Manca) cells, from which the added myc cDNA was derived, and in which there is absolutely no activity at either the P1 or P2 promoters of the unrearranged allele (11).

4.

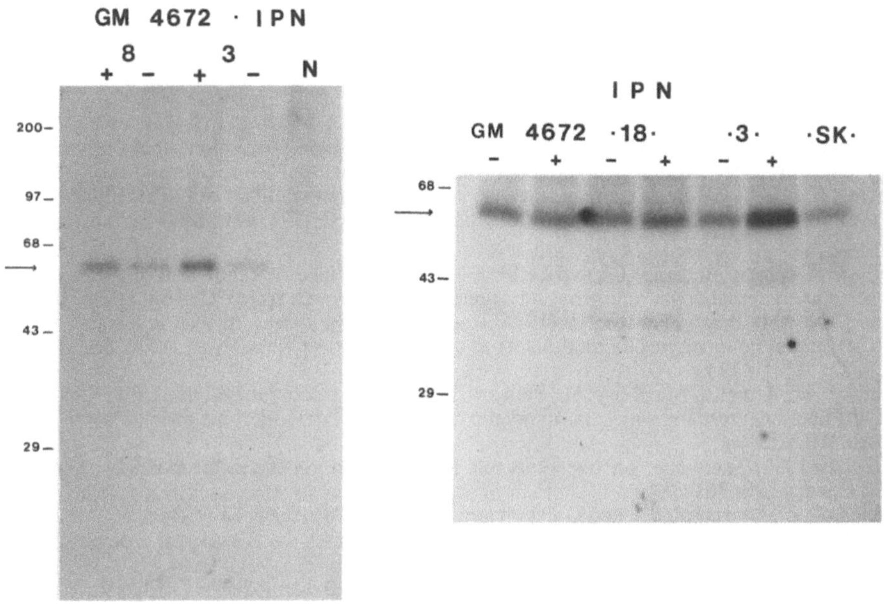

Immunoprecipitation of c-myc from the nuclei of the indicated cells, labeled for one hour with 32-P inorganic phosphate. The serum was directed against the C-terminus. N - normal serum control; SK - Manca cells.

CONCLUSIONS:

The expression of c-myc from a de-regulated gene failed to effect significantly the expression of the endogenous, unrearranged genes, in a number of independent clones of the human B cell line, GM 4672 that were transfected with a c-myc expression vector system. The de-regulated genes were present as single copy, haploid inserts, analogous to the state of the translocated genes in BL. The B cell line is a mature B cell, also somewhat analogous to the cell type in BL. The exogenously-added c-myc is expressed and translated at levels quite comparable to those of the tumor cells (Manca) from which it was cloned. However, in those cells, there is absolutely no activity at either P1 or P2 of the unrearranged alleles. This suggests that the auto-regulation is not a result of the co-existence of a de-regulated product and the target sequences per se, but that in some cells, such as GM 4672, the capacity for 'auto-regulation' to be effective is disrupted. It is not clear by what means this disruption occurs, but the fact that a number of cells of other types also show little auto-regulation (1, 2,12, 13) makes it possible that there are numerous means by which disruption can occur. It is plausible that a transformed state may render the regulation inoperable, but in that case, there must be a unique aspect to the co-existence of regulation and tumorigenicity in the B-L cells themselves. Additionally, in B-L cells, the complete loss of endogenous c-myc expression, that is common, appears not to be dose-dependent as opposed to that seen by Lombardi et al (7). Therefore, we conclude that auto-regulation, if it is a general phenomenon, is:

1. an action of one product on its gene by a route that is extremely susceptible to disruption even in human B cells, and
2. probably a product of at least two mechanisms: one, a dose-dependent one; and, two, a non-dose dependent mechanism seen in B-L. It remains plausible that the former evolves into the latter, with the target sequences being different in either case.

REFERENCES:

1. Coppola J., and Cole, M. (1986) Constitutive *c-myc* oncogene expression blocks mouse erythroleukemia cell differentiation but not commitment. Nature 320: 760-763.
2. Freytag, S. (1988) Enforced expression of the *c-myc* oncogene inhibits cell differentiation by precluding entry into a distinct predifferentiation state in G_0/G_1. Mol. Cell Biol. 8: 1614-1624
3. Kelly,K., Cochran, B., Stiles,C., and Leder, P. (1983) Cell specific regulation of the *c-myc* gene by lymphocyte mitogens and platelet-derived growth factor. Cell 35: 603-610
4. Reed,J. Nowell, P., and Hoover, R. (1985) Regulation of the *c-myc* mRNA levels in normal human lymphocytes by modulators of cell proliferation. Proc. Natl. Acad. Sci. USA 82: 4221-4224
5. Armelin, H., Armelin, M., Kelly, K., Stewart, T., Leder, P., Cochran, B., and Stiles C. (1984) Functional role for *c-myc* in mitogenic response to platelet-derived growth factor. Nature 310: 655-660
6. Cole, M. (1986) The *c-myc* oncogene: its role in transformation and differentiation. Ann. Rev. Genetics 20: 361-384
7. Lombardi, L., Newcomb, E., and Dalla Favera, R. (1987) Pathogenesis of Burkitt Lymphoma: expression of an activated *c-myc* oncogene causes the tumorigenic conversion of EBV-infected human B-lymphoblasts. Cell 49: 161-170
8. Toneguzzo,F., Hayday,A., and Keating, A. (1986) Electric field-mediated DNA transfer: transient and stable gene expression in human and mouse lymphoid cells. Mol. Cell Biol. 6: 703-706
9. Hayday, A., Gillies, S., Saito, H., Wood, C., Wiman, K., Hayward, W., and Tonegawa S. (1984) Activation of a translocated human *c-myc* gene by an enhancer in the immunoglobulin heavy-chain locus. Nature 307: 334-340
10. Saito, H., Hayday, A., Wiman, K., Hayward, W., and Tonegawa, S. (1983) Activation of the *c-myc* gene by translocation: a model for translational control. Proc. Natl. Acad. Sci. USA 80: 7476-7480
11. Wiman, K., Clarkson, B., Hayday, A., Saito, H., Tonegawa, S., and Hayward, W. (1984) Activation of a translocated *c-myc* gene: role of structural alterations in the upstream region. Proc. Natl. Acad. Sci. USA. 81: 6798-6802
12. Keath, E., Caimi, P., and Cole, M. (1984) Fibroblast lines expressing activated*c-myc* oncogenes are tumorigenic in Nude mice and syngeneic animals. Cell 39: 339- 348
13. Stewart, T., Pattengale, P., and Leder, P. (1984) Spontaneous mammary adenocarcinomas in transgenic mice that carry and express MTV/*myc* fusion genes. Cell 38: 627-63

ACKNOWLEDGEMENTS:

We thank Drs. S. Gillies (Damon Biotech), and C. Wood (University of Kansas) for important contributions to the reagents used; Dr. R. Mann for advice concerning protein analysis; and, R. Eisenman, and R. Watt for antisera.

Deregulated Expression of an Activated Allele of Human c-*myc* in Transfected Fibroblast Cultures

A. RICHMAN and A. HAYDAY

INTRODUCTION

BL tumors characteristically exhibit chromosomal rearrangements involving the *c-myc* gene and one of the Ig loci. In these cells, the translocated *myc* allele is deregulated (or "activated") in such a way that it is expressed constitutively. The normal allele, by contrast, is either transcriptionally silent or is expressed at very low levels (1). We are analyzing the mechanisms by which *c-myc* expression is altered in cells of the Ramos BL tumor. Genomic clones representing either a wild-type (germline) *c-myc* allele or the *IgH-myc* translocation locus (2) from Ramos tumor cells were cloned in the vector pSV2.gpt (3) (fig. 1) . In parallel gene transfer experiments these constructs were transfected into the rat fibroblast cell line FR3T3. Stably transfected, clonal cell lines bearing intact human *c-myc* genes were assayed for their ablity to regulate expression of the introduced alleles in response to serum in density-arrested cultures. Modulation of *c-myc* activity by serum is characteristic of the normal gene in established fibroblast cell lines (4).

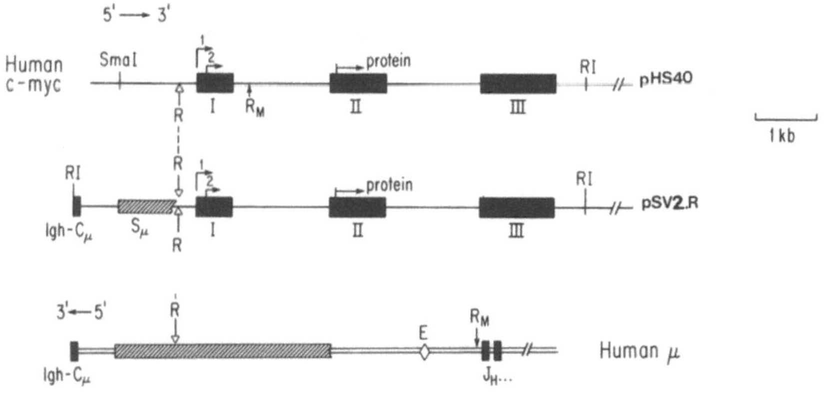

Fig.1. Genomic clones of human *c-myc* used in the transfection experiments. Also shown is the IgM locus. The rearrangement breakpoint in the Ramos translocated allele (pSV2.R) occurs approx.

Biology Department, Yale University, New Haven, Conn. 06511

Current Topics in Microbiology and Immunology, Vol. 141

340 bp upstream of exon 1. This allele is otherwise distinguished from the w.t. gene at the sequence level by only two single-base substitutions, one at 158 bp upstream of P1 and the other at a point 33 bp upstream of the 3' end of exon1. The *IgM-myc* translocation locus shown here was cloned into the EcoRI site of pSV2.gpt to create pSV2.R. The unrearranged (w.t.) *c-myc* allele used in these experiments consists of the 3-exon gene shown here and approx. 3.5 kb of additional 5' genomic sequence extending to an upstream EcoRI site. This fragment was similarly cloned in pSV2.gpt, creating the vector pHS40

REGULATION OF GENE EXPRESSION

Northern analysis of individual transfected clones indicates that both w.t. and BL-derived translocated human *c-myc* genes can be down-regulated in serum-starved cultures, and dramatically induced by addition of 15% serum for 1 hr. (fig. 2). Regulated expression is also displayed by the normal rat *c-myc* genes in the parental cells (data not shown). These observations indicate that a *c-myc* gene transcriptionally deregulated in a B cell tumor is still capable of responding appropriately to growth factor stimulation, albeit in a heterologous cell system.

It is possible that the activated phenotype is expressed differently in a transfected fibroblast cell background. We therefore analyzed steady-state levels of human *c-myc* mRNA in our transfected cell lines by more sensitive techniques. S_1 analysis using probes that detect mRNA derived from both promoters P1 and P2 of the introduced human genes indeed reveals, in most clones, a distinction in

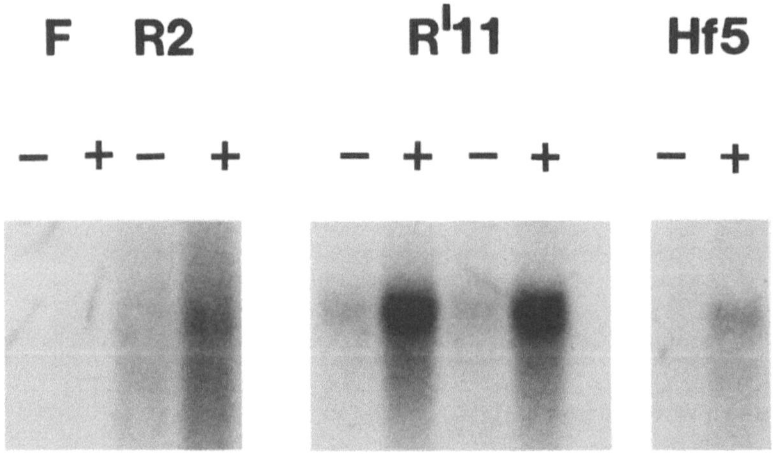

Fig. 2. Northern analysis of total RNA from untransfected parental FR3T3 cells (F) or from stable cell lines transfected with either pSV2.R (lines R2, R'11) or pHS40 (line Hf5). Cells were grown to confluence in medium containing 10% F.C.S., and subsequently maintained ("starved") in 1% serum for 4-7 days. At this time RNA was extracted from starved (-) cells and from duplicate starved cultures treated for 1 hr. with medium containing 15% serum (+). The filter was probed with the ClaI-EcoRI fragment from the 3' end of the human *c-myc* gene, and washed at high stringency

transcripts originating from the translocated human *c-myc* gene are present at the same level in serum-starved and stimulated cells. Under the same conditions, P2 activity in these cells is significantly modulated by serum. The w.t. transfected gene, by contrast, is down-regulated in most clones at <u>both</u> P1 and P2 in serum-starved cells.

Quite surprisingly, we have also found that the expression of *Eco.gpt* in these cells is regulated by serum in a manner which is dependent upon the identity of the linked human *c-myc* allele (fig. 4). In transfected clones harboring a w.t. *myc* gene, *gpt* expression is modulated by serum. In cells bearing a translocated allele it is not. Thus, *Eco.gpt* expression always mirrors the activity of the linked human promoter, P1.

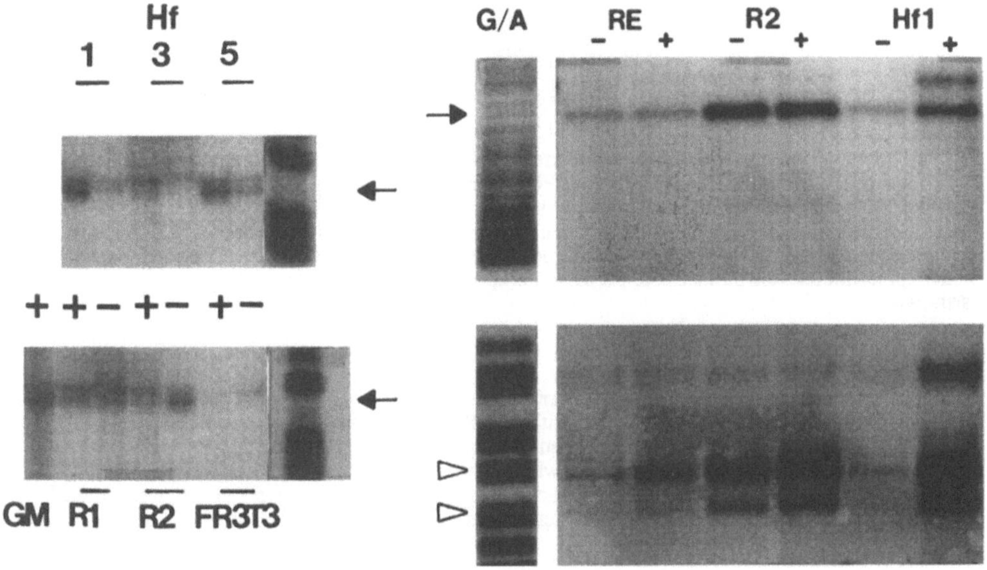

Fig. 3. S1 analysis of total RNA from stable cell lines transfected with pSV2.R (R1, R2, RE) and pHS40 (Hf1, Hf3, Hf5), or FR3T3 parental cells. Fibroblasts were serum-starved and stimulated as described in the Fig. 1 legend. GM is a human lymphoblastoid cell line bearing two normal alleles of *c-myc*, and serves as a positive control for P1 expression. Equivalence of RNA assayed in each experiment was determined by ethidium staining of samples run on agarose gels (not shown). Single-stranded, end-labelled DNA probes were specific for human *c-myc* mRNA derived from either P1 alone or from both P1 and P2. Solid arrows indicate protection at P1, open arrowhead denotes P2 protection. Precise identification of protected fragments was determined by alignment with G+A sequence ladders prepared from the end-labelled probe. Non-specific protection was determined in reactions with yeast tRNA (not shown)

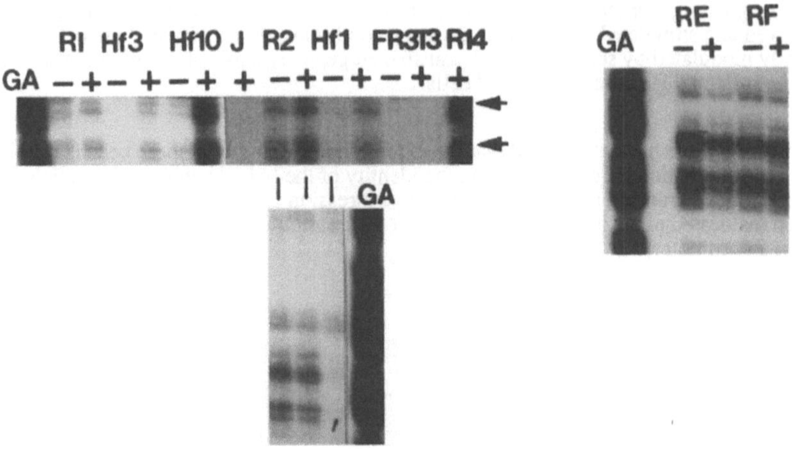

Fig. 4. S1 analysis of *gpt* transcripts in transfected clones. Cell lines are designated as in previous figure legends, except that R14 is a murine PC line transfected with pSV2.R (which serves as a positive control for protection of the *Eco.gpt*-specific, end-labelled DNA probe used here) and J is a non-transfected murine PC line (used as a negative control). Arrows indicate prominent bands specifically protected in these assays. Multiple protected fragments are generated by heterogeneous transcript initiation sites in the SV40 promoter. Duplicate analysis of clone R2 is shown

This result is particularly interesting in light of our studies of the activity of the SV40 early region promoter alone in control cells transfected with the pSV2.gpt parent vector. In the cell lines analyzed so far, modulation of *gpt* expression by serum is subject to significant clonal variation. These results suggest that alterations in *c-myc* regulation which render the translocated allele serum-independent for P1 expression operate at the level of transcription, and may exert an effect in cis upon a linked transcription unit driven by an SV40 promoter-enhancer sequence.

CONCLUSIONS

Our experiments thus far indicate that the rearranged allele of *c-myc* activated in the BL tumor Ramo expresses an altered phenotype in a biologically significant, heterologous assay system. Human *c-myc* P1 and P2 promoter activities are shown to be capable of independent modulation by factors present in whole serum. Deregulated expression of P1 may well be of biological significance in a B cell background. A shift in relative strengths of promoter utilization from P2 to P1 is associated with the activation of translocated *c-myc* in some BL tumors (1). We are currently using the serum-induction of transfected fibroblasts to analyze the sequence requirements for proper regulation of promoter activity in response to growth conditions. The extent to which attenuation (5) and post-transcriptional mechanisms (6) modulate the expression of an activated *c-myc* gene are also being investigated in this system.

REFERENCES

(1) Cole, M. D. (1986) The *myc* oncogene: its role in transformation and differentiation. *Annual Review of Genetics* 20:361-384.

(2) Wiman, K. G., Clarkson, B., Hayday, A. C., Saito, H., Tonegawa, S., and Hayward, W. S. (1984) Activation of a translocated *c-myc* gene: role of structural alterations in the upstream region. *P.N.A.S.* 81: 6798-6802.

(3) Mulligan, R. G. and Berg, P. (1980) Expression of a bacterial gene in mammalian cells. *Science* 209: 1422-1427.

(4) Kelly, K., Cochran, B. H., Stiles, C. D., and Leder, P. (1983) Cell-specific regulation of the *c-myc* gene by lymphocyte mitogens and platelet-derived growth factor. *Cell* 35: 603-610.

(5) Bentley, D. L., and Groudine, M. (1986) A block to elongation is largely responsible for decreased transcription of *c-myc* in differentiated HL60 cells. *Nature* 321: 702-706.

(6) Piechaczyk, M., Blanchard, J.-M., and Jeanteur, P. (1987) *c-myc* gene regulation still holds its secret. *Trends in Genetics* 3: 47-51.

Comparison of c-, N-, and L-*myc* Transcriptional Regulation

G. KRYSTAL, J. WAY, and J. BATTEY

ABSTRACT

The molecular mechanisms reported to regulate the expression of the c-myc gene include gene amplification, transcriptional activation, a block to transcriptional elongation, and mRNA stability. We have investigated which of these mechanisms determine the striking variation in mRNA levels of the three myc family members (c-, N-, and L-myc) observed in human small cell lung cancer cells, a system in which all three genes are expressed at widely varying levels. Myc family gene amplification is frequently observed. However there are numerous examples of elevated expression for each of these genes without amplification. In these instances, a block to transcriptional elongation appears to be a significant regulatory mechanism controlling the steady-state levels of c-myc and L-myc mRNA, but not N-myc. When this block to elongation is relieved, elevated levels of c-myc and L-myc gene expression are observed in several cell lines in the absence of gene amplification. All primary transcription detected from c-, N-, and L-myc in a nuclear run-on assay results from RNA polymerase II activity, with no detectable contribution of RNA polymerase III transcription observed.

INTRODUCTION

Alterations in the regulation of myc genes are often observed in tumors, and appears to be important in establishing and maintaining transformation. In Burkitt lymphoma, a specific chromosomal translocation joining the human c-myc gene with immunoglobulin gene sequences and regulatory elements is invariably observed. This genetic rearrangement is associated with high levels of transcription from the c-myc allele involved in the translocation, while the unrearranged allele is transcriptionally silent. It is both interesting and perplexing to note that this altered regulation occurs whether the translocation occurs within the gene, on either side of the gene, close to the gene, or over ten kilobases from the c-myc locus (for review, see Leder et. al. 1983). Nevertheless, this specific genetic change and associated c-myc deregulation are consistently observed in this B cell lymphoma.

Growth deregulation in human small cell lung cancer involves a host of diverse molecular mechanisms, including autocrine growth

NCI-Navy Medical Oncology Branch, Naval Hospital Bethesda, Bethesda, MD 20814

Current Topics in Microbiology and Immunology, Vol. 141

factor synthesis, chromosomal deletion, and alteration in the structure and regulation of proto-oncogenes (for review see Minna et. al. 1986). Increased expression of c-myc in SCLC cell lines correlates with increased cloning efficiency and decreased doubling times (Gazdar et. al. 1985, Johnson et. al. 1986), as well as poorer patient prognosis (Johnson et. al. 1987). These observations suggest that myc expression may be one important step in establishing or maintaining transformation and growth deregulation in SCLC.

The c-myc proto-oncogene is the cellular homolog of the avian myelocytomatosis virus transforming gene v-myc (Hayward et. al. 1981). The N-myc and L-myc genes were identified by their structural homology to c-myc (Kohl et. al. 1983, Schwab et. al. 1983, Nau et. al. 1985). These three genes all share a three exon structure and extensive homology in their protein coding domain, which is limited to exons two and three (for review see Alt et. al. 1986). All three genes can cooperate with ras to transform primary rat embryo fibroblasts in a co-transfection assay (Land et. al. 1983, Yancopoulous et. al. 1985, Schwab et. al. 1985, Alt et. al. 1986, Birrer et. al. 1988). They form a family of related genes both on the basis of structural and functional considerations.

The regulation of the c-myc gene occurs at several levels. Increased transcription of the c-myc gene is observed in cultured quiescent fibroblasts stimulated with growth factors (Kelly et. al. 1983, Greenberg and Ziff 1984). Variations in mRNA stability regulate the levels of c-myc mRNA in differentiating cells (Mechti et. al. 1986, Dean et. al. 1986) and in stimulated fibroblasts (Blanchard et. al. 1985). An alternative to promoter mediated transcriptional regulation of the c-myc gene is observed to modulate the decrease in c-myc mRNA levels that occurs when HL60 cells differentiate. A block to transcription elongation exists somewhere near the intron 1-exon 1 boundary. Attenuation of transcription at this block increases dramatically upon differentiation, with a consequent decline in steady state mRNA levels (Bentley and Groudine 1986). An analogous block is found in the same location in the murine c-myc gene (Nepveu and Marcu 1986).

With this variety of regulatory mechanisms described for the c-myc gene, we analyzed and compared the regulatory mechanisms determining the steady state levels of c-, N-, and L-myc mRNAs in SCLC cell lines that express one or more of the three myc family members. It has been previously shown that selected SCLC cell lines have c-, N-, and L-myc gene amplification with a consequent increase in expression. A number of cell lines express these genes without apparent amplification, providing an opportunity to explore other regulatory mechanisms important in modulating steady-state mRNA levels of these three related genes.

RESULTS AND DISCUSSION

Table 1 presents a summary of transcription regulatory studies examining the mechanism of c-, N-, and L-myc gene regulation in six representative small cell lung cancer cell lines. Among the six examples are cell lines with amplification of each of the

TABLE 1

	Gene Copies[a]			Relative mRNA Levels[b]			Initiation Levels[c]			Attenuation Ratio[d]		
	C	N	L	C	N	L	C	N	L	C	N	L
H146	1	1	1	8	-	-	+	-	+	1.0		>>5
H82	25	1	1	23	-	1	+++	-	+	5.0		4.1
H187	1	1	1	-	1	-	-	+	+		1.1	>>5
H249	1	150	1	-	10	-	+	+++	+	>>5	1.3	>>5
H209	1	1	1	-	-	17	-	-	++			1.6
H510	1	1	15	1	-	110	+	-	+++	4.8		4.6

[a] Gene copy number is determined by comparing band intensities on genomic blots of DNA from the indicated cell lines (Nau et al. 1985, Krystal et al., manuscript submitted).

[b] Relative mRNA levels are determined by Northern blot analysis of 10 μg total RNA. The autoradiographic band was scanned with a densitometer, and the area under the peak determined. The area from the lowest detectable signal was arbitrarily designated as 1, and all other values were determined as ratios of the unit area. (-) = no detectable signal above background (Krystal et al., manuscript submitted).

[c] Initiation levels are determined by qualitatively examining the ^{32}P labelled RNA hybridizing to an exon one sense target from the three genes in a nuclear run-on experiment. Comparisons between cell lines are normalized using a β-actin reference run-on signal; [(+++) = high levels of initiation; (++) = intermediate levels; (+) = detectable levels; (-) = undetectable levels]. The details of experimental procedures involved in performing the nuclear run-on analysis is as described (Krystal et al., manuscript submitted).

[d] Attenuation ratio is determined from nuclear run-on experiments performed on the six cell lines. The ^{32}P-RNA hybridizing to exon one and exon two sense target (immobilized on slot blots) is determined by autoradiography. Scanning densitometry is used to quantitate the signal as a peak integral. The areas are adjusted for differences in the relative U content of the exon one and exon two regions targets, since the ^{32}P incorporated during the nuclear run-on is exclusively ^{32}P-UTP. After adjustment, the ratio of adjusted exon 1: exon 2 signal is determined, and reported in the table. Blank spaces correspond to situations where the exon one signal was undetectable; >>5 indicates that no detectable exon two signal was present

three myc family genes, and examples which expresses high levels of c-, N-, and L-myc mRNA without amplification for comparison. We analyzed the gene copy number, relative mRNA levels, level of initiation, and transcription attenuation between exons one and two for each example (Krystal et. al., manuscript submitted). As previously described, the highest levels of mRNA from each of the three myc family genes is accompanied by gene amplification (H82, c-myc; H249, N-myc; H510, L-myc) (Little et. al. 1983, Nau et. al 1986, Nau et. al. 1985). It is interesting to note, however, that elevated levels of mRNA are frequently observed in the absence of gene amplification. A comparison of transcription regulation between these two situations reveals that differences in other regulatory mechanisms also modulate the levels of c-, N-, and L-myc mRNA in SCLC cell lines.

The c-myc gene is regulated at the level of initiation and elongation

Cell line H146 expresses abundant amounts of c-myc mRNA without gene amplification. Examination of the exon 1:exon 2 attenuation ratio shows that this is the only cell line studied here that shows no block to transcription elongation at the exon 1-intron 1 boundary, since the ratio of elongated transcripts hybridizing to exon 1 and exon 2 targets is 1.0 (Table 1). All other cell lines show an elevated exon 1:exon 2 ratio indicative of a block to transcription elongation. For example, cell line H82 with c-myc amplification has an exon 1:exon 2 ratio of 5.0, indicating that about four times in five RNA polymerase II fails to elongate a nascent chain across intron 1. Two other cell lines (H249 and H510) show detectable initiation of transcription of the c-myc gene in the absence of gene amplification. H249 does not have detectable levels of c-myc mRNA, while H510 has low levels. No run-on transcription is observed 3' to exon 1 in H249 (attenuation ratio >>5). In contrast, H510 shows some transcription reading through the entire gene (attenuation ratio 5). In each of these instances, a lower attenuation ratio correlates with elevated steady state c-myc mRNA if there is no gene amplification. The remaining two cell lines, H187 and H209, show no detectable exon 1 RNA labelled in the nuclear run-on assay indicating that promoter-mediated initiation is either very low or absent in these examples. As expected, no detectable c-myc mRNA is found in either cell line. Taken together, it is apparent that c-myc mRNA levels are modulated by alterations in transcription initiation and a block to elongation at the exon 1-intron 1 region.

N-myc mRNA levels are regulated at the level of initiation without any apparent block to elongation

Two cell lines studied express detectable steady-state levels of N-myc mRNA, H249 and H187. H249 has N-myc gene amplification, while H187 has a single copy of N-myc per haploid genome. These are the only two lines that show detectable levels of transcription initiation as detected by an exon 1 nuclear run-on target (Table 1). Neither cell line shows a significant block to transcription elongation of N-myc, in contrast to c-myc. Gene dosage is the likely explanation for the difference in mRNA levels observed between these two cell lines (Krystal et. al.,manuscript submitted). Transcription initiation appears to be the important regulatory step for N-myc in SCLC cells, in addition to gene dosage.

L-myc mRNA levels are regulated primarily by modulation of a block to transcription elongation between exon 1 and exon 2

Three cell lines among the six studied here contain detectable levels of L-myc mRNA. One of these, H510, has L-myc gene amplification while the other two, H209 and H82, have a single copy per haploid genome. In contrast to both c- and N-myc, there is detectable initiation of transcription for the L-myc gene in all cell lines studied to date. The three cell lines which do not express L-myc mRNA despite initiation (H146, H187, and H249) all have exon 1:exon 2 attenuation ratios of much greater that 5, with no detectable nuclear run-on counts on any probes 3' to exon 1. H510 has L-myc gene amplification, transcription initiation, and an attenuation ratio of 4.6 leading to high levels of mRNA. H209 and H82 have a single copy of the L-myc gene per haploid genome, and both show detectable levels of L-myc mRNA (Table 1). However, H209 has ten to twenty times more L-myc mRNA that H82. There is a small difference (roughly two-fold) in the initiation level when these two cell lines are compared. More significantly, the attenuation ratio for L-myc transcription in H82 is 4.6, while the ratio is only 1.6 for H209 resulting in increased steady-state mRNA. It appears that a loss in the transcription elongation block that is observed at varying levels in the five other lines analyzed is responsible at least in part for the elevated L-myc mRNA levels observed in H209. H82, with the second lowest attenuation ratio (4.1), is the only other cell line reported here that expresses detectable L-myc mRNA without gene amplification.

All detectable primary transcription from c-,N-, and L-myc genes is alpha-amanitin sensitive and derives from RNA polymerase II

Nuclear run-on experiments using nuclei from SCLC cell lines expressing each of the three myc gene family members were performed in the presence of two micromolar alpha-amanitin, a concentration sufficient to supress RNA polymerase II but not polymerase III transcription (Lindell et. al. 1970). Under these conditions, labelled RNA hybridizing to a beta-actin target (pol II) is completely inhibited, while transcription of 5S RNA genes (pol III) or 28S RNA (pol I) is unchanged from levels obtained in the absence of alpha-amanitin. Alpha-amanitin addition resulted in the abolition of transcription monitored from any target in the c-, N-, or L-myc genes in SCLC nuclei (Krystal et. al., mauuscript submitted). This observation indicates that the primary transcription analyzed in these experiments derives from RNA polymerase II, with no detectable contribution from either pol I or pol III.

Post-transcriptional mRNA stability is potentially an additional mechanism for regulating myc family mRNA levels. The half-lives of c-, N-, and L-myc mRNAs were determined in several cell lines expressing varying levels of the gene product, using actinomycin D to block new RNA synthesis. After actinomycin blockade, all cell lines studied showed a short half-life for histone H2B mRNA (roughly 30 min; short half-life control) and a long half-life for beta-actin mRNA (greater than four hours; long half-life control)

(Krystal et. al., manuscript submitted). No correlation was noted between longer half-lives and higher levels of steady state mRNA, suggesting that post-transcriptional stability does not play a significant role in regulating myc family mRNA levels in cultured SCLC cells studied here.

In conclusion, the molecular mechanisms which regulate expression of c-, N-, and L-myc in SCLC cells include variation in initiation levels, a block to transcription elongation, and gene copy number. Variation in initiation and elongation may be due either to cis mutations in the myc alleles, or differences in other genes expressing transcription factors critical for mediating initiation and elongation of c-, N-, and L-myc. Further studies will be needed to determine exactly where the differences between cell lines exist, and how these genetic alterations affect the regulatory mechanisms identified here. Pathologic changes in any one or a combination of these mechanisms can clearly alter steady-state mRNA levels, which in turn may result in a growth advantage important in tumorigenesis or tumor progression. Further analysis of the genetic differences responsible for the variation in regulation observed may contribute to a more complete understanding of myc family gene regulation in normal cells, and how this regulation may be altered in malignancy.

ACKNOWLEDGEMENTS

The authors wish to thank Dr. Timothy Bender and Dr. Craig Thompson for many helpful discussions and assistance with some of the nuclear run-on experiments. We also thank Dr. Jay Brooks, Dr. Michael Birrer, Dr. John Minna, Dr. Edward Sausville, and Marian Nau for providing N- and L-myc cloned DNA fragments used in some of these studies. In addition we wish to thank Herbert Oie, Ed Russell, Sylvia Stephenson, and Dr. Adi Gazdar for establishing and maintaining SCLC cell lines. Dr. Michael Kuehl and Dr. Anne-Marie Lebacq-Verheyden provided many helpful suggestions during the progress of the work. Finally, Gena Parris's help in preparation of the manuscript has as always been invaluable.

REFERENCES

Alt FW, Depinho R, Zimmerman K, Legouy E, Hatton K, Ferrier P, Tesfaye A, Yancopoulos GD, Nisen PD (1986) The human myc gene family. CSHCQB LI:931-941

Bentley DL, Groudine M (1986) A block to elongation is largely responsible for decreased transcription of c-myc in differentiated HL60 cells. Nature 321:702-706

Birrer MJ, Segal S, DeGreve JS, Kaye F, Sausville EA, Minna JD (1988) L-myc cooperates with ras to transform primary rat embryo fibroblasts. Mol.Cell.Biol., in press

Blanchard JM, Piechaczyk M, Dani C, Chambard JC, Franchi A, Pouyssegur J, Jeanteur P (1985) c-myc gene is transcribed at high rate in G_0-arrested fibroblasts and is post-transcriptionally regulated in response to growth factors. Nature 317:443-445

Dean M, Levine RA, Campisi J (1986) c-myc regulation during retinoic acid-induced differentiation of F9 cells is post-transcriptional and associated with growth arrest. Mol. Cell. Biol. 6:518-524
Gazdar AF, Carney DN, Nau MM, Minna JD (1985) Characterization of variant subclasses of cell lines derived from small cell lung cancer having distinctive biochemical, morphological, and growth properties. Cancer Res. 45:2924-2930
Greenberg ME, Ziff EB (1984) Stimulation of 3T3 cells induces transcription of the c-fos proto-oncogene. Nature 311:433-438
Hayward WS, Neel BG, Astrin SM (1981) Activation of cellular onc genes by promoter insertion in ALV-induced lymphoid leukosis. Nature 290:475-479
Johnson BE, Battey JF, Linnoila I, Becken KL, Makuch RW, Snider RH, Carney DN, Minna JD (1986) Changes in the phenotype of human small cell lung cancer cell lines after transfection and expression of the c-myc proto-oncogene. J. Clin. Invest. 78:525-532
Johnson BE, Idhe DC, Makuch RW, Gazdar AF, Carney DN, Oie H, Russell E, Nau MM, Minna JD (1987) myc family oncogene amplification in tumor cell lines established from small cell lung cancer patients and its relationship to clinical status and course. J. Clin. Invest. 79:1629-1634
Kelly K, Cochran BH, Stiles CD, Leder P (1983) Cell-specific regulation of the myc gene by lymphocyte mitogens and platelet-derived growth factor. Cell 35:603-610
Kohl NE, Kanda N, Shreck R, Bruns G, Latt S, Gilbert F (1983) Transposition and amplification of oncogene related sequence in human neuroblastomas. Cell 35:359-367
Krystal G, Birrer M, Way J, Nau MM, Sausville EA, Thompson C, Minna JD, Battey JF (1988) Multiple mechanisms for transcriptional regulation of the myc gene family in small cell lung cancer, manuscript submitted
Land H, Parada LF, Weinberg R (1983) Tumorigenic conversion of primary embryo fibroblasts requires at least two cooperating oncogenes. Nature 304:596-602
Leder P, Battey JF, Lenoir G, Moulding C, Murphy W, Potter H, Stewart T, Taub R (1983) Translocations among antibody genes in human cancer. Science 222:765-771
Lindell T, Weinberg F, Morris B, Roeder R, Rutter W (1970) Specific inhibition of RNA polymerase II by alpha-amanitin. Science 170:447-448.
Little CD, Nau MM, Carney DN, Gazdar AF, Minna JD (1983) Amplification and expression of the c-myc oncogene in human lung cancer cell lines. Nature 306:194-196
Mechti N, Piechaczyk M, Blanchard JM, Marty L, Bonnieu A, Jeanteur P, Lebleu B (1986) Transcriptional and post-transcriptional regulation of c-myc expression during the differentiation of murine erythroleukemia Friend cells. Nuc. Acids Res. 14:9653-9666
Minna JD, Battey JF, Brooks BJ, Cuttitta F, Gazdar AF, Johnson BE, Ihde DC, Lebacq-Verheyden AM, Mulshine J, Nau MM, Oie HK, Sausville EA, Seifter E, Vinocour M (1986) Molecular genetic analysis reveals chromosomal deletion, gene amplification, and autocrine growth factor production in the pathogenesis of human lung cancer. CSHSQB LI:843-853

Nau MM, Brooks BJ, Battey JF, Sausville EA, Gazdar AF, Kirsch IR, McBride OW, Bertness V, Hollis GF, Minna JD (1985) L-myc, a new myc-related gene amplified and expressed in human small cell lung cancer. Nature 318:69-73

Nau MM, Brooks BJ, Carney DN, Gazdar AF, Battey JF, Sausville EA, Minna JD (1986) Human small cell lung cancers show amplification and expression of the N-myc gene. Proc. Natl. Acad. Sci. USA 83:1092-1096

Nepveu A, Marcu KB (1986) Intragenic pausing and anti-sense transcription within the murine c-myc locus. EMBO 5:2859-2965

Schwab M, Alitalo K, Klempnauer L, Varmus H, Bishop JM, Gilbert F, Brodeur G, Goldstein M, Trent J (1983) Amplified DNA with limited homology to myc cellular oncogene is shared by human neuroblastoma cell lines and a neuroblastoma tumor. Nature 305:245-248

Schwab M, Varmus HE, and Bishop JM (1985) Human N-myc gene contributes to neoplastic transformation of mammalian cells in culture. Nature 316:160-162

Yancopoulous GD, Nisen PD, Tesfaye A, Kohl NE, Goldfarb MP, Alt F (1985) N-myc can cooperate with ras to transform normal cells in culture. Proc. Natl. Acad. Sci. USA 80:1195-1198

Delivery of c-*myc* Antisense Phosphorothioate Oligodeoxynucleotides to Hematopoietic Cells in Culture by Liposome Fusion: Specific Reduction in c-*myc* Protein Expression Correlates with Inhibition of Cell Growth and DNA Synthesis

S. L. LOKE[1], C. STEIN[2], X. ZHANG[1],
M. AVIGAN[1], J. COHEN[2], and L. M. NECKERS[1]

INTRODUCTION

To understand the role of individual genes in the regulation of growth and differentiation, their specific inhibition is necessary. When the gene products are surface proteins, such as growth factor receptors, such inhibition may be accomplished by use of monoclonal antibodies capable of interfering with the binding of ligand to receptor (Neckers and Cossman 1983). However, when the protein of interest is localized to an intracellular compartment, antibodies must be delivered by microinjection to one cell at a time. While possible, such methodology is tedious, requires specialized apparatus and techniques, and does not lend itself to the study of cells growing in suspension culture (i.e., most hematopoietic cells).

A novel approach to the problem of specifically inhibiting the production of intracellular proteins is the administration, by either plasmid transfection or viral infection, of constructs capable of producing intracellular antisense RNA complementary to the mRNA of interest (Izant and Weintraub 1984). In some instances, these constructs are linked to inducible promoters so that their activity can be exogenously regulated (Holt et al. 1986). The antisense RNA is thought to form a duplex with the host mRNA of interest, thereby preventing its efficient translation by ribosomes. While successful in some *in vitro* systems, this approach does not work well with most hematopoietic cells, which may be difficult to transfect and in which inducible promoters sometimes fail to stimulate enough production of the antisense RNA.

The past several years have seen an explosive growth in the use of antisense oligodeoxynucleotides (ODNs) as modulators of gene expression in model systems (Stein and Cohen in press). Although DNA and proteins are among the most stable of biological substances, mRNA is inherently unstable. Inhibition of one copy of mRNA will also be theoretically much more efficient than inhibiting the protein product, since one mRNA molecule gives rise to multiple protein copies. In principle, using the current technology of DNA synthesizers, a single-stranded DNA sequence of desired size can be produced which is complementary to a known region of a particular mRNA. By annealing with the mRNA in the region of complementarity, a double-stranded species is produced, resulting in at least two phenomena. First, the double-stranded DNA-RNA hybrid becomes a substrate for RNAse H and is degraded, potentially freeing up the ODN to hybridize to another copy of mRNA (Dash et al. 1987). Second, if the antisense ODN is made complementary to a region around the initiation sequence, the ribosome cannot attach efficiently and translation is arrested (Kawasaki 1985). Thus, not only is the amount of mRNA reduced (due to RNAse H digestion), but efficient translation of remaining messages is also inhibited. The net result is inhibition of specific protein synthesis. If both the protein and mRNA of interest have short half-lives (on the order of minutes), a rapid loss of total protein may occur.

Although ODNs are much more stable than RNA, there are exonucleases and endonucleases in serum which degrade them (Stein et al. in press). This in turn necessitates the exogenous addition of a large excess of ODN to be effective. Recently, the automated production of phosphorothioate ODNs (PS-ODNs) has been achieved (Stec et al. 1984). These compounds, in which a sulfur atom replaces an oxygen atom in the phosphotriester moiety of the ODN (see *Fig. 1*), are very resistant to exo- and endonuclease activity (Stein et al. in press). The utility of these compounds as inhibitors of gene expression has been suggested (Matsukura et al. 1987).

Laboratory of Pathology[1] and Medicine Branch[2], NCI, NIH, Bethesda, MD 20892

Current Topics in Microbiology and Immunology, Vol. 141

The *c-myc* protein is thought to be involved in the regulation of cell growth and differentiation and yet the precise mechanism(s) of its action is (are) unknown. This is in part true because, of all the methods known to inhibit *c-myc* transcription, none are specific. In order to study the role of *c-myc* in cell growth and differentiation, specific inhibition of its expression is essential. Because the half-life of both the *c-myc* protein and mRNA is short, it is a perfect candidate for antisense inhibition. In fact, we have already demonstrated that an antisene ODN complementary to the first 15 bases of *c-myc* mRNA (beginning with the AUG start codon) effectively blocks mitogen stimulated c-*myc* expression and proliferation in normal T cells (Heikkila et al. 1987). In this study, we describe the utility of both unsubstituted and phosphorothioate ODNs as inhibitors of *c-myc* expression in the promyelocytic leukemia cell line HL60.

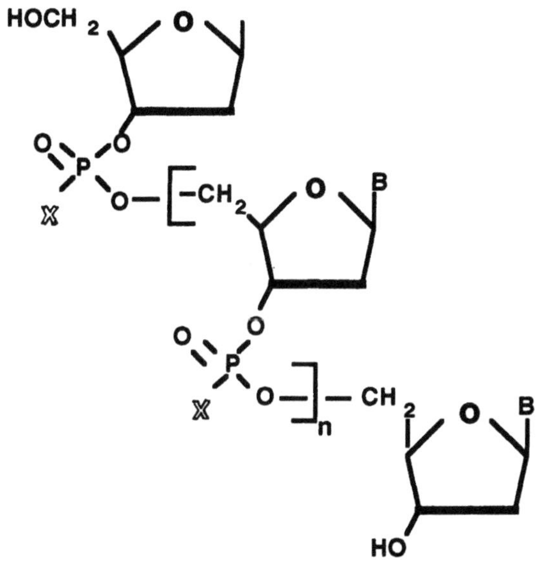

Fig. 1. Deoxyoligonucleotide structure. When the "X" is an oxygen atom, the compound is a normal (PO) oligomer. When the "X" is a sulfur atom, the compound is a phosphorothioate (PS) oligomer

MATERIALS AND METHODS

Both unsubstituted (PO-ODNs) and phosphorothioate (PS-ODNs) oligodeoxynucleotides were synthesized and purified as described previously (Heikkila et al. 1987, Matsukura et al. 1987). The *c-myc* antisense sequence used was 5' - dAACGTTGA-GGGGCAT - 3' and the corresponding nonsense sequence was 5' - dCTGAAGTGGC-ATGAG - 3'. Intracellular *c-myc* protein was quantified by flow cytometry using a murine anti-c-*myc* monoclonal antibody (Evan and Hancock 1985). Cells were washed in cold PBS and fixed in 1% paraformaldehyde (pH 7) for 30 minutes. The cells were then pelleted and post-fixed in 0.1% Triton/10% goat serum for 10 minutes. Following pelleting and resuspension in PBS, the cells were ready for staining. Cellualr uptake of both PO- and PS-ODNs was monitored by flow cytometric analysis of intracellular ODN modified with 2-methoxy, 6-chloro, 9-(5-hydroxypentyl)aminoacridine (Stein et al. in press).

Liposomes were prepared as described previously (Papahadjopoulos et al. 1975). Briefly, 5 mg of phosphatidylserine in chloroform was vaporized to a thin film under nitrogen using a rotating evaporator. The lipid was resuspended in 0.2 ml EDTA buffer with vigorous vortexing to form multilamellar vessicles, and Ca^{++} was added in excess to form cochleate bodies. Various ODNs to be entrapped were added and the mixtures incubated for 1 hour. EDTA was then added and the pH of the liposome suspension adjusted to 7. Completion of vessicle formation was monitored by light microscopy and the efficiency of entrapment was assessed by monitoring radiolabelled ODN incorporation. The preparation was centrifuged at 38,000 rpm in an SW 50 rotor in a Beckman ultracentrifuge for 30 minutes and washed thrice in PBS.

Cells to be lipofused were washed twice and incubated in PBS with 2 mM Ca^{++} and 0.1 mM Mg^{++} for 30 minutes. Liposomes were added to a final concentration of 1.25 mg/ml lipid and the mixture incubated for 30 minutes at 37°C. The cells were then exposed to polyethylene glycol (m. w. 6000) for 1 minute, washed thoroughly and resuspended in culture medium.

RESULTS

The effects of addition of 50 uM PO- and PS-ODNs to log phase HL60 cells can be seen in *Fig.* 2. While intracellular *c-myc* protein levels fall within about 4 hours of PO-ODN addition, reaching a level approximately 30% of control within 18 hours, addition of PS-ODN was ineffective in lowering intracellular *c-myc* protein. We thought that PS-ODNs might have difficulty in entering the cells, since their overall stability is much greater than that of PO-ODNs. This was confirmed by monitoring the uptake of acridine labelled PO- and PS-ODNs by flow cytometry (*Fig. 3*).

While uptake of the PO-ODN (oligo dT_7) was clearly observable within 4 hours and plateaued at 40 - 50 hours, the uptake of an identically sized PS-ODN was barely detectable, even with 50 hours of incubation. To overcome this difficulty and utilize the stability of the PS-ODNs, we loaded them in liposomes and fused these vessicles with the cells. The effects on *c-myc* protein expression can be seen in *Fig.* 4. Both PO- (top) and PS-ODNs (bottom) antisense to *c-myc* mRNA are effective in reducing c-myc protein expression within 1.5 - 2 hours following lipofusion. Empty liposomes and those containing nonsense sequences are ineffective. The inhibition of *c-myc* protein expression is transient and is nearly completely reversed within 24 hours of lipofusion. Nonetheless, both cell growth and DNA synthesis are inhibited specifically by lipofusion with PS-ODNs antisense to *c-myc* mRNA (*Fig. 5*). The inhibition of DNA synthesis is reversible and follows a similar time course to that of *c-myc* protein expression following lipofusion (see *Fig.* 4, bottom). The fact that growth is nearly completely inhibited for 24 hours (*Fig.* 5, top) suggests that the effects of PS-ODN antisense to *c-myc* persist for nearly that time period.

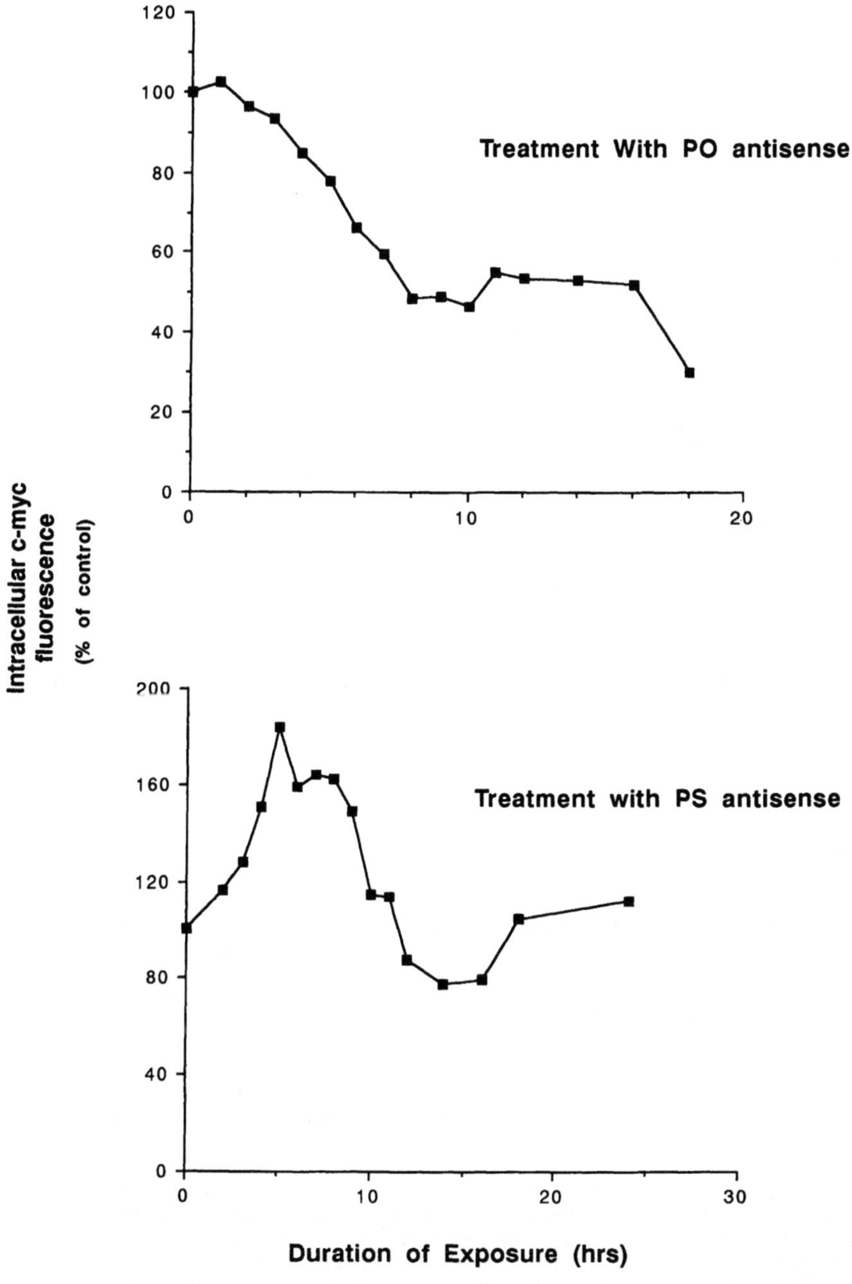

Fig. 2. Intracellular *c-myc* protein level was determined by flow cytometric analysis. Data are expressed as % of control values. Top: Exposure to 50 uM PO antisense; Bottom: exposure to 50 uM PS antisense

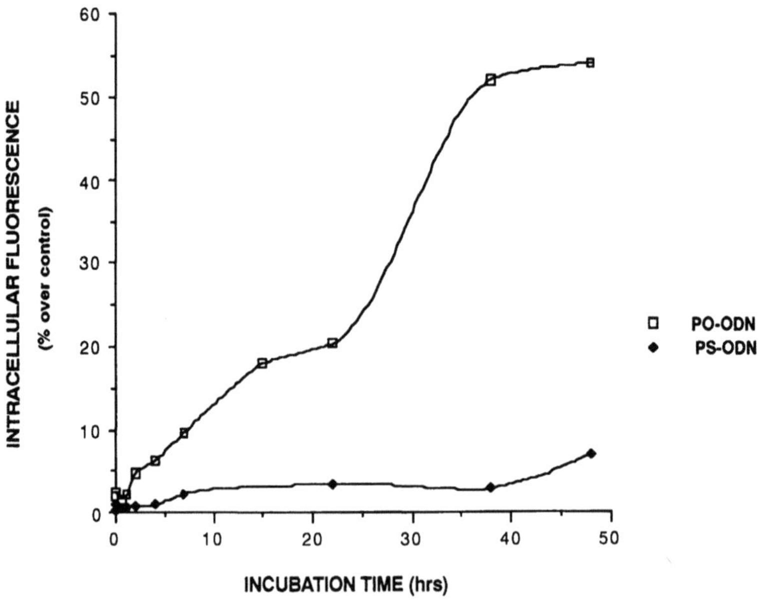

Fig. 3. Uptake of acridine labelled PO- and PS-oligo dT $_7$ by HL60 cells. Cells were incubated with 12.5 uM ODN for the times indicated and intracellular fluorescence was quantified by flow cytometry

SUMMARY AND CONCLUSIONS

Even though PS-ODNs are more resistant to breakdown, they are not taken up by cells as efficiently as PO-ODNs. In order to determine if phosphorothioate compounds can successfully serve as *myc* antisense reagents, we bypassed the cellular uptake machinery and introduced these compounds to HL60 cells by lipofusion. In this way we have demonstrated that PS-ODNs are capable of transiently reducing *c-myc* protein expression in these over-expressing cells. In addition, this effect is accompanied by an inhibition of DNA synthesis and cell growth, which persist for as long as the reduction of *myc* protein. Our data with HL60 cells and that previously reported in normal T cells (Heikkila et al. 1987) support a role for *c-myc* in DNA synthesis and proliferation.

Finally, we have demonstrated that delivery of antisense ODNs by lipofusion is a simple and successful method for introducing large quantities of these compounds intracytoplasmically. Not only does this method allow for the delivery of larger and/or modified ODNs (which would not normally penetrate the cell membrane), it also protects the ODN from extracellular degradation. The ability to antibody coat liposomes and thus target them to particular cell types raises the possibility that ODNs antisense to particular oncogenes may some day play a role in the clinical management of neoplasia.

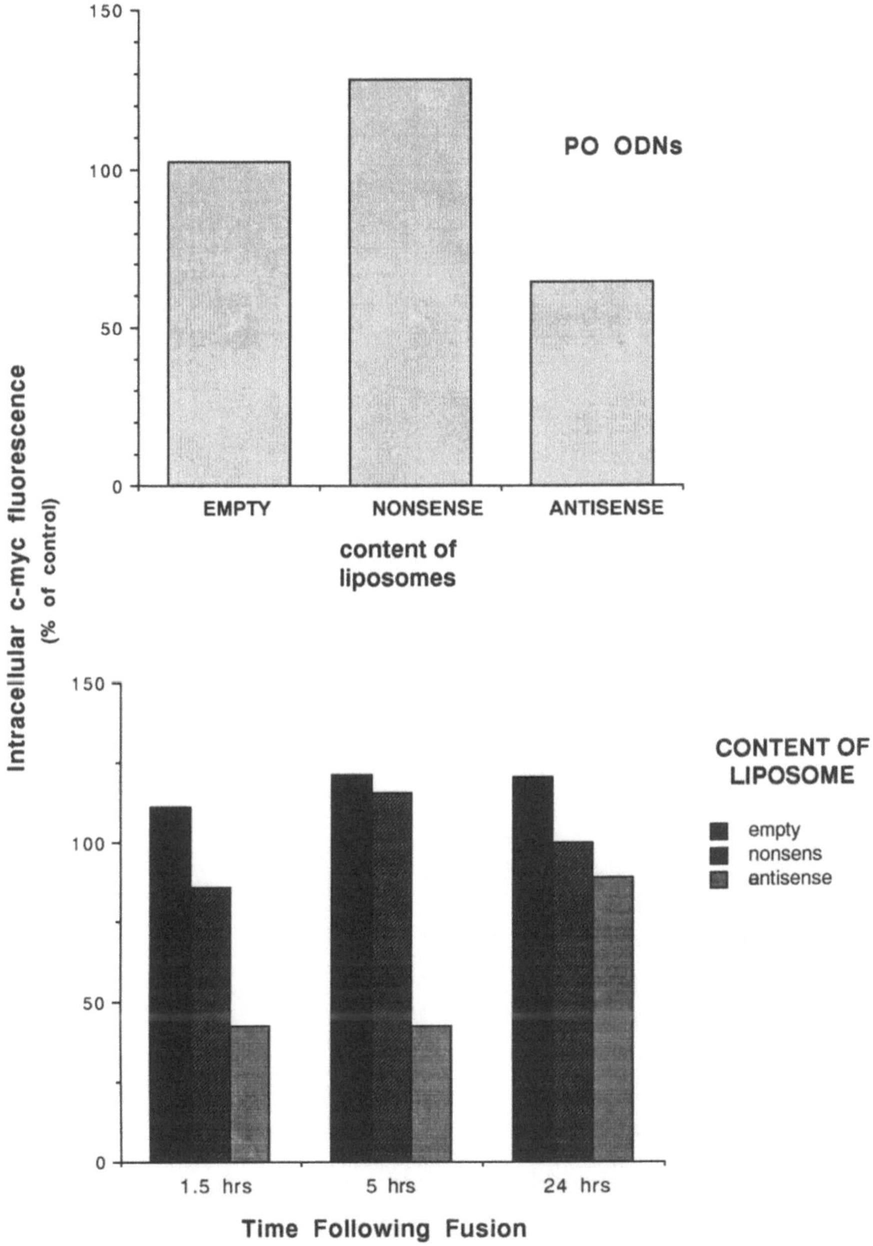

Fig. 4. Intracellular *c-myc* protein levels in HL60 cells following lipofusion with PO- (top) or PS-ODNs (bottom). Protein level is expressed as % of control and was determined by flow cytometric analysis. Cells were analyzed 2 hrs following lipofusion in the top panel

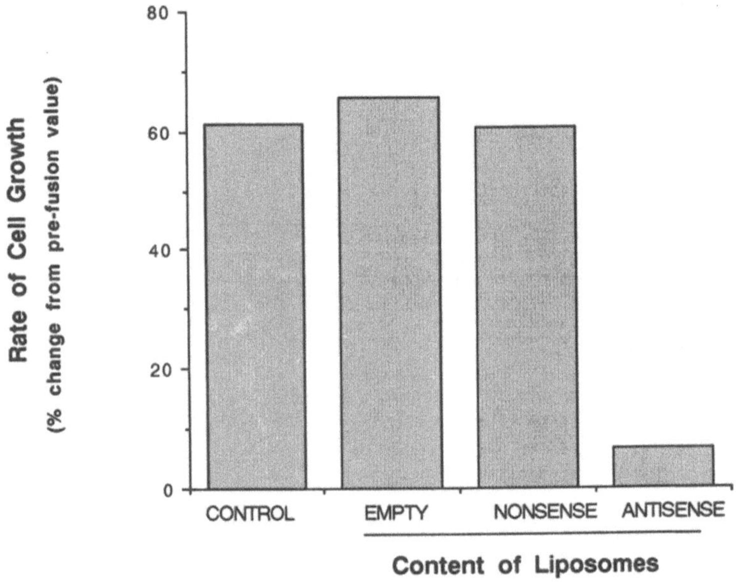

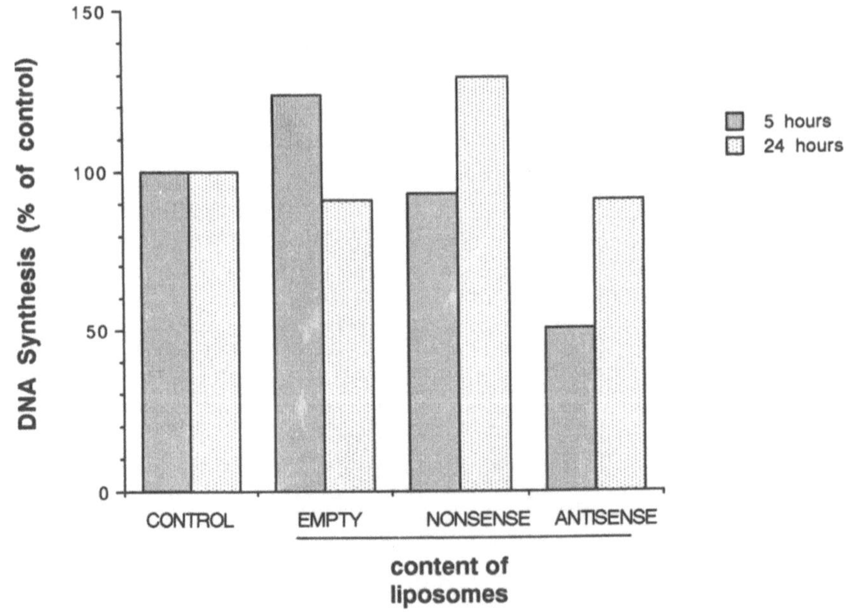

Fig. 5. Top: Following fusion with PS antisense, HL60 cell growth is markedly inhibited during the ensuing 24 hours. Bottom: DNA synthesis is transiently reduced by PS antisense ODNs following lipofusion

REFERENCES

Dash, P, Lotan, I, Knapp, M, Kandel, ER, Goelet, P (1987) Selective elimination of mRNAs *in vivo*: Complementary oligodeoxynucleotides promote RNA degradation by an RNase H - like activity. Proc. Natl. Acad. Sci. USA 84: 7896-7900 progress from G0 to G1. Nature 328: 445-449
Evan, GI, Hancock, DC (1985) Studies on the interaction of the human *c-myc* protein with cell nuclei: p62 *c-myc* as a member of a discrete subset of nuclear proteins. Cell 43: 253-261
Heikkila, R, Schwab, G, Wickstrom, E, Loke, SL, Pluznik, D, Watt, R, Neckers, LM (1987) A *c-myc* antisense oligodeoxynucleotide inhibits entry into S phase but not
Holt, JT, Venkat, GT, Moulton, AD, Nienhuis, AW (1986) Inducible production of *c-fos* antisense RNA inhibits 3T3 cell proliferation. Proc. Natl. Acad. Sci. USA 83: 4794-4798
Izant, JC, Weintraub, H (1984) Inhibition of thymidine kinase gene expression by antisense RNA: a molecular approach to genetic analysis. Cell 36: 1007-1015
Kawasaki, ES (1985) Quantitative hybridization arrest of mRNA in *Xenopus* oocytes using single stranded complementary DNA or oligonucleotide probes. Nuc. Acid Res. 13: 4991-5004
Matsukura, M, Shinozuka, K, Zon, G, Mitsuya, M, Reitz, M, Cohen, JS, Broder, S (1987) Phosphorothioate oligodeoxynucleotides : Inhibitors of replication and cytopathic effects of human immunodeficiency virus. Proc. Natl. Acad. Sci. USA 84: 7706-7710
Neckers, LM, Cossman, J (1983) Transferrin receptor induction in mitogen-stimulated human lymphocytes is required for DNA synthesis and cell division and is regulated by interleukin 2. Proc. Natl. Acad. Sci. USA 80: 3494-3498
Papahadjopoulos, D, Vail, WJ, Jacobson, K, Poste, G (1975) Cochleate lipid cylinders: formation by fusion of unilamellar lipid vesicles. Biochim. Biophys. Acta 394: 483-491
Stec, W, Zon, G, Egan, V, Stec, B (1984) Automated solid phase synthesis, separation and stereochemistry of phosphorothioate analogs of oligodeoxynucleotides. J. Amer. Chem. Soc. 106: 6077-6079
Stein, CA, Cohen, JS (in press) Oligodeoxynucleotides: Towards a new chemotherapy. Cancer Res.
Stein, CA, Mori, K, Loke, SL, Subasinghe, C, Shinozuka, K, Cohen, JS, Neckers, LM (in press) Phosphorothioate and normal oligodeoxynucleotides with 5' - linked acridine: Characterization and preliminary kinetics of of cellular uptake. Gene

Different Biological Effects of c-*myc* and H-*ras* Oncogene Expression in EBV-Infected Human Lymphoblasts

S. Seremetis[1], G. Inghirami[2], D. Ferrero[1], L. Lombardi[1], D.M. Knowlest[2], G.-P. Dotto[3], and R. Dalla-Favera[1]

INTRODUCTION

The effects of various oncogenes in the multistep transformation process have been relatively well studied in fibroblasts, while much less is known in other tissues (1,2). The hematopoietic system, and the B-lymphoid lineage in particular, affords an excellent opportunity for analysing the effects of different oncogenes. This system is extensively defined in terms of differentiation, growth and function, and a vast array of reagents is available for the precise dissection of each of these aspects.

We have initiated a series of investigations aimed at determining the biological effects of different oncogenes in human B-cells. We have previously described an in vitro transformation assay in which Epstein-Barr Virus (EBV)-immortalized human B-lymphoblastoid cells (LCL) are used as targets. We have shown that the expression of transfected c-myc oncogenes induces malignant transformation of these cells (3). Here we summarize the results of a detailed analysis of the phenotype of myc-transformed LCL, with particular focus on the observed relationship between c-myc and the expression of leukocyte-specific cell-adhesion molecules (integrins) (12). In addition, we have used the same target cells to study the effects of activated ras genes and report that expression of an activated ras is accompanied by both transformation and plasmacytoid differentiation of lymphoblastoid cells.

PHENOTYPIC ANALYSIS OF LCL TRANSFORMED BY C-MYC ONCOGENES: DOWN-REGULATION OF EXPRESSION OF THE CELL-ADHESION MOLECULE LFA-1

We have previously reported that the constitutive expression of c-myc genes under the control of heterologous enhancer/promoter elements causes the malignant transformation of transfected LCL. The main features of the transformed phenotype included the ability to grow in low serum concentrations, clonogenicity in semi-solid medium and tumorigenicity in nude mice (3). The availability of several myc-transformed LCL allowed us to analyze their phenotype in detail in order to identify changes which were consistent among different transformants and thus likely to depend upon c-myc activation. A panel of monoclonal antibodies was used to explore changes in the expression of molecules related to histocompatibility, differentation, activation, and function (4). Representative data relative to the comparative analysis of an LCL line independently transfected with three plasmids expressing c-myc genes at different levels (3) are shown in Table 1. No changes were observed in the expression of histocompatibility markers or, in general, in the expression differentiation related markers with the exception of a marked and consistently observed decrease in BA-2 expression which may reflect a slight change toward a relatively more undifferentiated phenotype. More notably, expression of activated c-myc genes consistently correlated with down-regulation of the leukocyte-specific cell-adhesion molecule LFA-1. The

[1]Department of Pathology, New York University School of Medicine, New York, NY 10016
[2]Department of Pathology, Columbia University College of Physicians and Surgeons, New York, NY 10032
[3]Department of Pathology, Yale University School of Medicine, New Haven, CT 06516

Current Topics in Microbiology and Immunology, Vol. 141

levels of down-regulation were directly proportional to the levels of exogenous c-myc expression and involved both chains (α and ß) of the heterodimeric LFA-1 molecule. These changes have been observed in 7 of 7 LCL transfected so far, ranging from down-modulation (see Table 1 and Figure 1) to complete suppression in some LCL expressing very high levels of activated c-myc genes. These changes correlate with decreased or absent intercellular adhesion which we have previously described as a decrease or loss of the clumping pattern typical of LCL growing in liquid suspension cultures (3).

Table 1. ANALYSIS OF MYC-TRANSFORMED LCL

MoAb* (specificity)	TRANSFECTED CELL LINES (MCF)x			
	SV+	Eμmyc+	SVmyc1,2,3+	SVmyc2,3+
HLA-DR (MHC)	+++	+++	+++	+++
Differentiation related				
IL2R (CD25)	-	-	-	-
BA-2 (CD9)	+++	++	+	+
PCA-1	+	+	+	+
OKT10 (CD38)	+	++	+++	+++
OKB7 (CD21)	++	++	+++	+++
OKT9 (Tfn-R)	++	++	++	+++
OKT1 (CD5)	+	++	+++	+++
Cell adhesion				
LFA-1α (CD11a)	+++	+++	++	+
LFA-1ß (CD11)	+++	+++	++	+
LeuM5 (CD11c)	-	-	-	-

(x): MCF: Mean channel fluorescence: <10:-; 10-39:+; 40-79:++; >80:+++.

(+): Cell lines are LCL (CB33) transfected with different plasmids. SV: a control plasmid lacking c-myc sequences; Eμmyc: c-myc sequences (2nd & 3rd exon) under the control of the immunoglobulin heavy chain enhancer. SVmyc1,2,3: c-myc sequences (1st, 2nd, and 3rd exons), under control of SV40 enhancer/promoter.; SVmyc2,3: c-myc sequences (2nd & 3rd exons) under control of SV40 enhancer/promoter. Levels of c-myc RNA expression are Eμmyc < SVmyc1,2,3 < SVmyc 2,3

(*): Cell lines exhibited no differences in expression of the following antigens: HLA-ABC; J5(CD10); B1(CD20); B4(CD19); BA1(CD24); LFA-3; BAC; Ki-1(CD30).

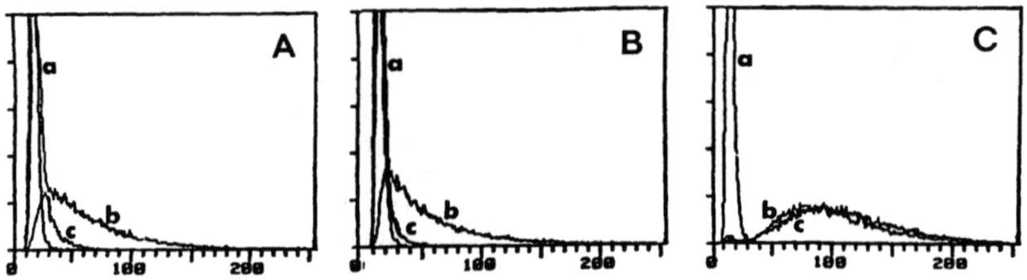

Figure 1. Expression of an activated c-myc is associated with decreased expression of LFA-1. Representative cytofluorometric analyses of LCL transfected with pHEBO SV (curves labelled b) and pHEBO SVmyc2,3. (curves c). Curves a represent reactivity of these cell lines with isotype identical control monoclonal antibodies. In Panel A are reactivities with TS-1-18 (5), whose specificity is LFA-1ß; in panel B are reactivities with TS-1-22 (5), specificity LFA-1α; and in panel C are reactivities with an anti-LFA-3 monoclonal antibody. In each case the horizontal axis reflects fluorescence intensity while the vertical axis represents cell numbers. In both panels A and B, dramatically reduced LFA-1 expression is seen in the context of high expression of activated c-myc, while in panel C, LFA-3 expression is shown to be unmodulated by myc expression.

EFFECTS OF RAS ONCOGENES EXPRESSION IN LCL: MALIGNANT TRANSFORMATION ACCOMPANIED BY PLASMACYTOID DIFFERENTIATION

Introduction Of Ras Oncogenes Into LCL

Recombinant amphotropic retroviruses containing the G418 (neomycin) resistance gene either alone (pKneo virus) or in combination with v-H-ras gene (raszip6 virus) (6) were used to infect two LCL lines, CB33 (derived from normal cord blood) and UH3 (derived from normal adult peripheral blood) (7). Antibiotic resistant lines were established from all cultures in approximately 3 weeks and an approximate 3% efficiency of infection was estimated. Southern blot hybridization analysis confirmed the presence of clonally distinct integration sites in different infected cell populations. Northern blot hybridization analysis showed the presence of viral ras genes within genomic-size viral transcripts.

Evidence for Plasmacytoid Differentiation

Phenotypic analysis of several infected LCL populations revealed changes in cell carrying activated ras genes which were consistent with plasmacytoid differentiation: i) Morphology: Raszip6 infected cells showed decreased nucleus/cytoplasm ratio, eccentric placement of the nucleus with appearance of a single prominent nucleolus and an increase in the number and size of cytoplasmic inclusions. In addition, these cells were considerably more adherent to plastic than were either the parent cell lines or the control-construct infected cell lines; ii) Immunophenotype: Data relative to the immunophenotypic analysis of infected LCLs are summarized in Table 2. While alterations are observed in expression of several antigens, the most dramatic and consistent changes were noted in the expression of the CD38 antigen, OKT10, and PCA-1. Expression of both of these antigens is considered consistent with plasma cell differentiation (8) and each of these antigens is significantly increased in v-H-ras infected cells. iii) IgM secretion: In the presence of v-H-ras expression, a marked increase in IgM secretion (Figure 2) was noted, with at least a five-fold increase noted on the sixth day of subculturing. No IgA or IgG secretion was detectable in the infected cell lines.

TABLE 2

IMMUNOPHENOTYPIC ANALYSIS OF Ras-zip6-INFECTED LYMPHOBLASTOID CELL LINES

MoAbs* (specificity)	INFECTED CELL LINES (MCF)[x]			
	CB33neo+	CB33ras°	UH3neo+	UH3ras°
HLA-DR (MHC)	+++	+++	+++	+++
Differentiation-related				
IL2R (CD25)	-	+	-	+
BA-2 (CD9)	+	++	+	+
PCA-1	+	+++	+	+++
OKT10 (CD38)	+	+++	+	+++
OKB7 (CD21)	++	+	++	+
OKT9(Tfn-R)	+++	++	+++	+++
Cell-adhesion				
LFA-1 (CD11a)	+++	+++	+++	+++
OKM1 (CD11b)	-	-	-	-
Leu M5 (CD11c)	+	++	+	+

(x): MCF: Mean channel fluorescence: <10:-; 10-39:+; 40-79:++; >80:+++.

(*): Cell lines exhibited no differences in expression of the following antigens: HLA-ABC; J5-Calla (CD10); B1 (CD20); BA1 (CD24); B4 (CD19); OKT1 (CD5); sIgM; Ki1; Ki24.

(+): Cell lines infected with pKneo
(o): Cell lines infected with raszip6

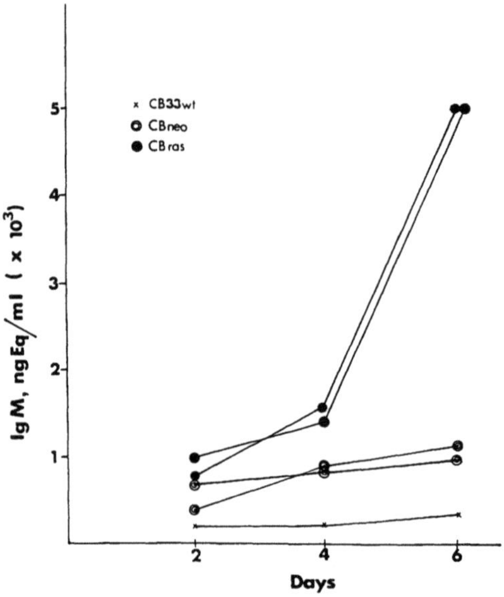

Figure 2. IgM secretion in LCL expressing activated H-ras genes. Cells were plated at a concentration of 2 x 10^5/ml and supernatants were assayed at 2, 4, and 6 days for the presence of IgM by an ELISA using an alkaline-phosphatase-coupled antibody to human IgM. CBras = CB33 infected with raszip6 and CBneo = CB33 infected with pKneo. Each curve represents data from cell lines derived from distinct infection events. Results indicate that v-H-ras expression is associated with markedly increased IgM secretion.

Evidence for malignant transformation:

As reported in Table 3, the expression of v-H-ras is associated with the acquisition of clonogenic properties in agar and tumorigenicity in nude mice (9). Based on the results obtained in both assays it can be concluded that H-ras can confer a weak, although clearly and consistently identifiable transformed phenotype. Analogous results were recently obtained using a retroviral vector carrying an activated N-ras gene.

DISCUSSION

The central observation of these studies is that although both are capable of inducing malignant transformation of LCL, c-myc and ras oncogenes induce strikingly different phenotypes. Activated c-myc expression is associated with little or no change in the stage of differentiation while its effects on LFA-expression suggest important effects on the growth pattern of transformed cells. On the other hand, the expression of an activated ras gene is associated the acquisition of malignancy but also with dramatic changes in differentiation stage. The finding of these distinct effects has implications both for the normal function of these two genes and for their role in determining the phenotype of naturally occurring tumors.

TABLE 3.

IN VITRO CLONING EFFICIENCY AND IN VIVO TUMORIGENICITY OF LYMPHOBLASTOID CELL LCL INFECTED WITH Raszip6

CELL LINE	CLONING EFFICIENCY*	TUMORS IN NUDE MICE+	
		tumors/injection	latency
BL (P3HR1)[x]	24%	4/4	2-3 wks.
CB33	0	0/12	
CB33 neo	0	0/12	
CB33 ras	0.3%	11/12	2-3 wks.
UH3	0	0/12	
UH3 neo	0	0/12	
UH3 ras	0.08%	4/12	4 wks.

(*): Colonies were examined 14 days after cells were seeded in quadruplicate at 10^3, 5×10^3, and 10^4 per culture dish in 1 ml of growth medium containing 20·/. fetal calf serum (FCS) and 0.3·/. agar (DIFCO). Data represent the average of at least three experiments.

(+): 4-week-old Swiss nude mice were injected with 5×10^6 cells in 0.2 ml of phosphate buffered saline (PBS). Data are expressed as the numbers of tumors observed after 4 weeks.

([x]): Burkitt Lymphoma cell line used as positive control.

The finding of decreased LFA-1 expression upon c-myc activation in LCL suggests a regulatory role of c-myc on LFA-1 expression which may be relevant for the control of both normal B-cell growth and for the pathogenesis of lymphomas carrying activated c-myc genes. LFA-1 is the leukocyte-specific member of the cell-adhesion receptors (integrin) superfamily of molecules which in different tissues are involved in cell-cell adhesion and in extracellular matrix-cell interactions (5,12). The LFA-1 molecule is also involved in immunorecognition of B-cells by T-cells and the genetic deficiency of LFA-1 is accompanied by impaired recognition of EBV-infected B-cells by T-cells (13). It has been reported that high grade B-cell lymphomas and Burkitt lymphomas in particular display low or undetectable levels of LFA-1 expression, and that this is accompanied by a proportionally decreased ability to stimulate T-cells in mixed lymphocyte reactions in vitro (10,11). Taken together, these results suggest that c-myc activation may contribute to Burkitt lymphomagenesis not only by directly deregulating cell growth but also by favoring escape from immunosurveillance through down regulation of LFA-1 molecules. Since expression of c-myc is ubiquitous and distinct members of the integrin superfamily are present in many tissues these observations may be generally applicable and suggest a mechanism coupling the control of cell proliferation to the extracellular environment.

The results obtained by introducing ras genes into LCLs indicate that these oncogenes can stimulate both proliferation and differentiation of LCL cells. Taken individually, these effects have been previously described (14), since ras oncogenes can contribute to transformation of a number of cell types while it causes terminal differentiation coupled with arrest of proliferation in pheochromocytoma (15) and thyroid carcinoma (16) cell lines. In LCL cells, however, proliferation and differentiation appear to be coupled, resulting in a cell population which is capable of both self renewal and stochastically determined entrance in the differentiation pathway. While our currently limited knowledge of the biochemical function of ras proteins prevents a thorough understanding of these findings it may be interesting to note that several types of naturally occurring malignancies, most notably multiple myeloma, display an analogous behaviour. In addition, we have recently determined that, distinct from lymphoma and chronic lymphocytic leukemias which do not contain activated ras genes, multiple myeloma is associated with N-ras activation in approximately 30·/. of cases (Neri, A., unpublished results). These results warrant further investigations on the role of ras activation in this type of tumors.

In general, the observations presented here further underscore the need for study on the biological effects of oncogenes in different target tissues. The B-cell transformation system described here represents a novel model for further dissection of the mechanisms through which oncogenes may contribute to B-cell lymphomagenesis.

ACKNOWLEDGEMENTS

We are grateful to Timothy Springer for a gift of anti-LFA-1 antibodies. This work was supported by N.I.H. grant CA-37165 (to R.D.-F.). R.D.-F is a Scholar of the Leukemia Society of America.

REFERENCES

1) Weinberg, R.A. (1982) Oncogenes of spontaneous and chemically induced tumors. Adv. Cancer Res. 36:149-164.
2) Bishop, J.H. (1987) The molecular genetics of cancer. Science 235:305-311.
3) Lombardi, L., Newcomb, E.W., Dalla-Favera, R. (1987) Pathogenesis of Burkitt lymphoma: Expression of an activated c-myc oncogene causes the tumorigenic conversion of EBV-infected Human B-lymphoblasts. Cell 49:161-170.
4) McMichael, A.J. ed. (1987) Leucocyte Typing III: White Cell Differention antigens. Oxford University Press, Oxford.
5) Miller, L.J., Schwarting, R., Springer, T.A. (1986). Regulated expression of the Mac-1, LFA-1, p.150,95 glycoprotein family during leukocyte differentiation. J. Immunol. 137:2891-2900.
6) Dotto, G.P., Parada, L., Weinberg, R.A. (1985) Specific growth response of ras-transformed embryo fibroblasts to tumor promoters. Nature 318: 472-475.
7) Miller, G., and Lippman, M., (1973) Release of infectious Epstein-Barr virus by transformed marmoset leukocytes. PNAS USA 70:190-194.
8) Ling, N.R., MacLennan, I.C.M., Mason, D.Y., (1987) B cell and Plasma cell antigens. In Leukocyte Typing III. Oxford University Press. Oxford: 302-335.
9) Blair, D.G., Cooper, C.S.,m Oskarsson, M.K., Eader, L.A., and Vande Woude, G.F., (1982) New method for detecting cellular transforming genes. Science 218:1122-1125.
10) Clayberger, C., Medieros, L.J., Warnke, R.A., Smith, S.A., Krensky, A.M. (1987) Absence of cell-surface LFA-1 as a mechanism of escape from immunosurveillance. Lancet 87vII:533-536.
11) Billaud, M., Calendar, A., Seigneurin, J.M., Lenoir, G.M., (1987) LFA-1, LFA-3, and ICAM-1 in Burkitt's lymphoma. Lancet 87vII:1327-1328.

12) Ruoslahti, E., Pierschbacher, M.D. (1987) New Perspectives in Cell Adhesion: RGD and Integrins. Science 238:491-497.
13) Krensky, A., Mentzer, S.J., Clayberger, C., Burakoff, S., Springer, T., (1985) Heritable Lymphocyte Function - Associated Antigen-1 deficiency. J. Immunol. 135:3102-3108.
14) Barbacid, M., (1987) Ras genes. Ann Rev. Biochem. 56:779-827.
15) Bar-Sagi, D., Feramisco, J.R., (1985) Microinjection of the ras oncogene protein into PC12 cells induces morphological differentiation. Cell 42:841-848.
16) Nakagawa, T., Mabry, M., Bustros, A., Ihle, J., Nelkin, B.D., Baylin, S.B., (1987) Introduction of v-H-ras oncogene induces differentiation of cultured human medullary thyroid carcinoma cells. PNAS USA 84: 5923-5927.

c-*abl*

Abrogation of IL-3 Dependence of Myeloid FDC-P1 Cells by Tyrosine Kinase Oncogenes Is Associated with Induction of c-*myc*

J.L. CLEVELAND[1], M. DEAN[2], J.Y. WANG[3], A.-M. HEDGE[4], J.N. IHLE[4], and U.R. RAPP[1]

INTRODUCTION

A role for the c-myc proto-oncogene in growth factor signal transduction first came from experiments showing that the gene was inducible by treating quiescent T cells and fibroblasts with experimental mitogens or growth factors (Kelly et al., 1983). Subsequently, subtractive cDNA cloning from PDGF-treated fibroblast cells identified c-myc as an early G1 or "competence" gene in these cells (Cochran et al., 1983). Significance for this induction was established from studies demonstrating that introduction of constitutively expressed source of exogenous c-myc or microinjection of purified c-myc protein relieved the requirements for PDGF in the transition from G0 into G1 (Armelin et al., 1984; Kaczmarek et al., 1985).

Concurrently we have examined the regulation and functional significance of the c-myc gene in interleukin-3 (IL-3) signal transduction in factor dependent murine myeloid FDC-P1 cells. These cells strictly require IL-3 for proliferation and viability and die rapidly when the ligand is removed. c-myc mRNA levels are tightly controlled by IL-3 in these cells and introduction of exogenous v-myc or murine c-myc either partially or fully abrogates the requirements of these cells for this growth factor (Rapp et al., 1985; Dean et al., 1987). Additionally, all partially abrogated clones containing functional exogenous c-myc gave rise to factor independent lines in a stochastic fashion. Acquisition of factor independence in these cells was not associated with alterations in IL-3 receptor levels or affinities, nor was it a consequence of inducing factors which could support the growth of these cells in an autocrine type of mechanism. However, we did observe that these clones have an activated endogenous c-myc or N-myc gene, suggesting that myc is indeed a key target in growth factor abrogation (Dean et al., 1987). Activation of myc genes in these factor independent cells was not due to gross alterations nor promoter insertion into c- or N-myc as judged by Southern blot analyses, nor was activation of endogenous c-myc due to differential promoter usage or point mutations in exon 1 (Cleveland, unpublished results), suggesting that these activations represent a new class of regulatory mutations which activate myc gene expression in trans (see Dean et al., this volume).

The next question we decided to address was how our findings of a key role for myc in IL-3 signal transduction and growth factor abrogation fit with the work of several different laboratories which have shown that introduction of expression vectors carrying tyrosine kinase oncogenes abrogates the requirement of IL-3 dependent cells for this ligand (Pierce et al., 1985; Cook et al., 1985; Mathey-Prevot et al., 1986; Wheeler et al., 1987; Pierce et al., 1988). The question we specifically asked was whether

[1]Laboratory of Viral Carcinogenesis/NCI-FCRF, Frederick, MD; [2]Program Resources Inc., Frederick, MD; [3]University of California at San Diego, La Jolla, CA; [4]Bionetics Research Institute, Frederick, MD, U.S.A.

Current Topics in Microbiology and Immunology, Vol. 141

these tyrosine kinase oncogenes abrogate factor dependence by utilizing signalling pathways which are independent of or involve c-myc.

RESULTS

A series of recombinant and natural isolates of defective murine retroviruses carrying different of tyrosine and serine/threonine kinase oncogenes and others carrying activated ras or mos oncogenes were used to infect FDC-P1 cells and virus positive cells (identified by infectious cell center assays on NIH 3T3 indicator cells) were challenged to grow in the absence of IL-3. As is illustrated in Table 1, all retroviruses carrying tyrosine kinase oncogenes were effective in giving rise to factor independent cell lines whereas the other classes of oncogenes tested were inefficient at generating factor independent clones. Again, the factor independent tyrosine kinase cell lines were not abrogated as a consequence of alterations in IL-3 receptor levels or affinities.

These tyrosine kinase oncogene containing cell lines were next examined for the status of c-myc expression by Northern blot hybridization of poly(A)+ RNA and compared with control FDC-P1 cells and cells rendered factor independent by infection with viruses carrying exogenous c-myc and v-myc (Figure 1). Several points can be made from this analysis. First, as noted above c-myc mRNA levels are tightly controlled in parental FDC-P1 cells by the presence of IL-3 since when they are depleted of the ligand for 24 hours c-myc transcripts are no longer detectable. Second, all tyrosine kinase containing independent lines express c-myc constitutively in the absence of IL-3. This constitutive expression of c-myc could reasonably be envisioned to be due to an autocrine mechanism in which these factor inependent cells produce factors which induce c-myc and hence make them factor independent. However, this does not appear likely since none of these lines express detectable IL-3 transcripts (Fig. 1) nor, from conditioned media experiments, do they produce any other factors which support their growth or that of parental FDC-P1 cells.

The constitutive expression of c-myc in the tyrosine kinase abrogated cells could be a consequence of alterations in the c-myc locus. However, Southern blot analysis did not show detectable alterations or amplification of c-myc in any of these cell lines. To examine whether constitutive c-myc expression may have been due to alternative promoter usage as has been shown in other systems, we performed an S1 nuclease experiment of c-myc RNA cap sites in FDC-P1 versus factor independent derivatives using a large probe which extends greater than 1 kb 5′ of exon 1. Promoter usage in the factor independent lines was not appreciably different than parental FDC-P1 cells and no novel RNA start sites were detected (Cleveland et al., in preparation). These factor independent lines were also checked for possible point mutations in exon 1, since it has been suggested that the point mutations which occur at the end of exon 1 in several human Burkitt lymphoma cell lines relieve c-myc from a transcription elongation block (Pelicci et al., 1986) which has been shown to occur in several cell types (Bentley and Groudine, 1986; Eick and Bornkamm, 1986; Mechti et al., 1987). Using two RNA probes which span the transcription elongation block site in mouse c-myc exon 1, RNAse A protection experiments failed to reveal any novel point mutations specific for c-myc transcripts from the factor independent cell lines (Cleveland et al., in preparation). Therefore, at least in the confines of this assay there does not appear to be any mutations which might alleviate

Table 1. Abrogation of IL-3 Dependence

Virus	Oncogene	FDC-P1 -IL-3
Svx-neo	none	-
Src-MuLV	v-src	+
Abl-MuLV	v-abl	+
Fms/neo-MuLV	v-fms	+
Trk/neo-MuLV	trk	+
3611-MSV	v-raf	-
Mo-MSV	v-mos	-
Ha-MuSV	v-Ha-ras	-

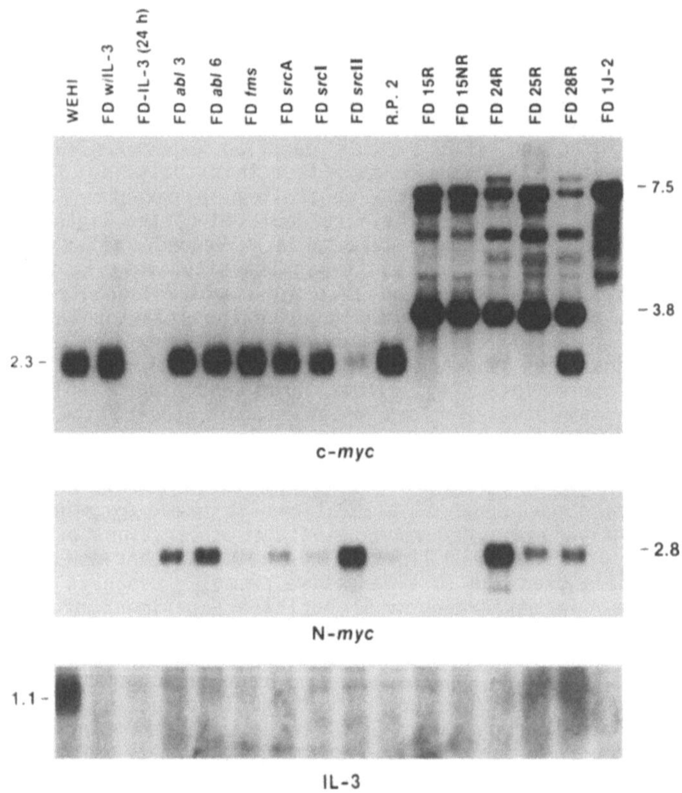

Figure 1. Northern blot analysis of myc and IL-3 expression in factor independent lines. Total poly(A)+ RNA was isolated from the independent cell lines grown in the absence of IL-3 and from control cultures (Wehi, FD) grown in the presence of IL-3. Five micrograms of RNA was analyzed per lane. Exposures for c-myc was for 20 hours; exposures for N-myc and IL-3 were for 3 days

attenuation regulation of c-myc in the tyrosine kinase containing factor independent cells.

The constitutive expression of c-myc in the factor independent derivatives could, however, still be a consequence of other secondary events which we cannot score. For example, if activation of endogenous N-myc is taken as a marker for secondary events, this would have appeared to have occurred in one of the src virus infected clones, as it expresses high levels, and perhaps also in two of the abl virus clones, which express moderate levels of N-myc (Figure 1). To try to get around this limitation we used two approaches. First, we utilized a retroviral vector which expresses the trk tyrosine kinase oncogene and the selectable neo gene so we could select for virus positive clones and assay for factor independence and c-myc expression independent of other events. The second approach involved introducing viruses containing temperature sensitive (ts) versions of the abl tyrosine kinase oncogene into FDC-P1 cells which conditionally exhibit tyrosine kinase activity depending upon temperature.

FDC-P1 cells were infected with the trk-neo virus and then immediately plated into soft agar containing IL-3 and G418. G418 resistant clones were then expanded and challenged to grow in the absence of IL-3. The trk-neo virus was very efficient in abrogating IL-3 requirements of FDC-P1 cells since frequencies of factor independent clones were approximately four orders of magnitude higher than those arising after infection with control neo virus and approximately 1000 fold higher in frequency than independent lines generated by infection with raf-neo virus (Cleveland et al., in preparation). Three trk independent clones were picked and then analyzed for expression by Northern blot hybridization (Figure 2). All clones expressed high levels of the introduced virus and again all expressed high levels of c-myc in the absence of IL-3. In contrast, none of the clones expressed detectable levels of N-myc, perhaps suggesting that secondary events were not involved in generation of factor independence or activation of c-myc.

To definitively test whether tyrosine kinase oncogenes were directly involved in generation of factor independence and induction of c-myc expression we used a series of independently derived ts abl viruses. FDC-P1 cells were infected with wild-type and ts versions of abl retroviruses at the permissive temperature (32 C) and single cell clones were screened for virus by infectious cell center assays. We then tested the virus positive clones for their ability to grow in the absence of Il-3 at 32 C and all of the clones tested were factor independent at this temperature. The clones were then expanded and examined for expression of exogenous abl and endogenous c-myc (Figure 3). All abl virus infected clones (wild-type MO-1,and ts mutants RK-4 and TS-1) expressed high levels of the expected size viral transcripts and notably all expressed c-myc in the absence of IL-3 at this temperature.

The next question was whether abrogation of factor dependence and induction of c-myc correlated with conditional abl tyrosine kinase activity. The same clones were tested for their IL-3 dependent phenotype by shifting them to 39 C, the nonpermissive temperature, in the absence of IL-3. Figure 4 shows a viability curve of these clones compared with parental FDC-P1 cells which had first been expanded at 32 C in IL-3, washed twice to remove IL-3, and then also shifted to 39 C. Parental FDC-P1 cells and clones containing either the RK or TS ts abl viruses died rapidly at 39 C, whereas the MO-1 clone containing wild type abl kinase remained viable at 39 C in the absence of IL-3. To check whether the ts phenotype in IL-3 abrogation was also

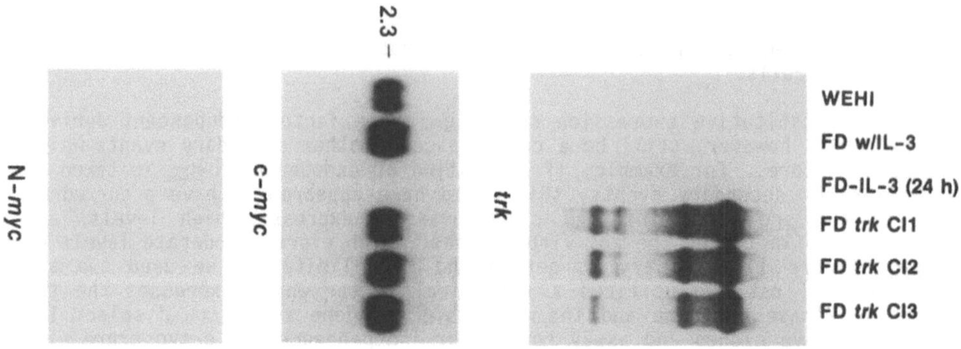

Figure 2. myc expression in FDC-P1 factor independent trk-neo clones. Total poly(A)+ RNA was isolated from the trk-neo clones grown in the absence of IL-3 and from control FDC-P1 cells grown in IL-3 and examined for expression of trk, c-myc, and N-myc. Blots were exposed for 20 hrs (trk, c-myc) to 3 days (N-myc)

Figure 3. Expression of exogenous abl and c-myc in FDC-P1 cells infected with retroviruses expressing wild-type (MO-1) and ts versions of abl tyrosine kinase (RK-4 and TS-1). abl virus clones were grown at 32 C (the permissive temperature) in the absence of IL-3 and compared to a Abl-MuLV factor independent clone and parental FDC-P1 cells grown in IL-3 at 32 C. Blots were exposed for 16 hours

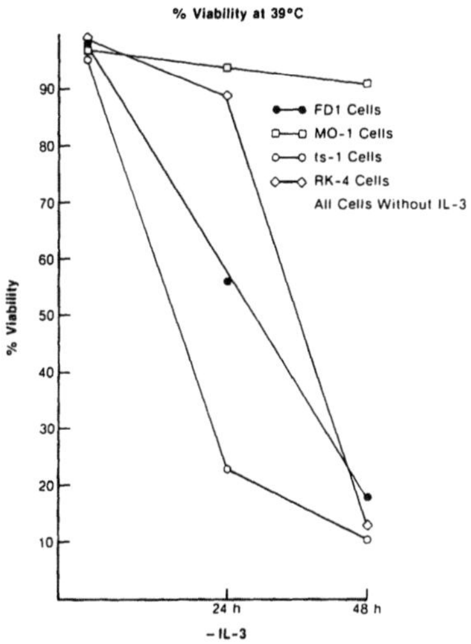

Figure 4. Loss of viability of FDC-P1 clones infected with retroviral constructs expressing wild type and ts versions of abl after shifting the cultures from the permissive temperature (32 C) to the nonpermissive temperature (39 C). At the indicated times aliquots of cells were removed from the cultures and the percentage of viable cells determined by their ability to exclude trypan blue. As a control parental FDC-P1 cells were washed to remove IL-3 and shifted to 39 C

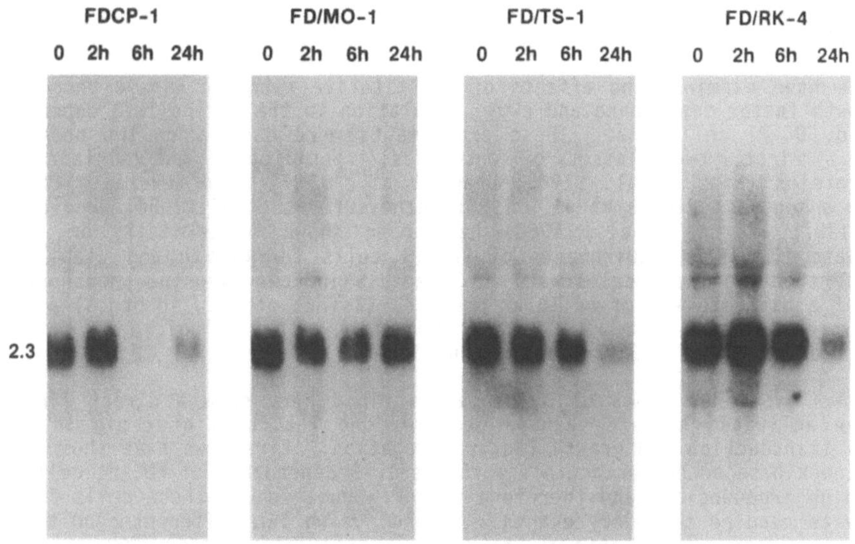

Figure 5. Effects of temperature shift upon expression of c-myc in FD/abl virus clones. Cultures of FD/abl clones expanded at 32 C in the absence of IL-3 were shifted to 39 C and at the indicated times a portion of the culture was removed and total RNA was isolated. As a control parental FDC-P1 cells were also expanded at 32 C but in the presence of IL-3, washed twice and resuspended in medium without IL-3 and shifted to 39 C. Blots were exposed for two days

reflected in a ts regulation of c-myc expression we expanded the abl virus clones at 32 C in the absence of IL-3 and RNA was isolated at specific times after shifting the cultures to 39 C. As a control, parental FDC-P1 cells were grown at 32 C in the presence of IL-3, the cells washed twice and then placed in medium without IL-3 and shifted to 39 C. Again, all abl virus clones grown at 32 C in the absence of IL-3 had levels of c-myc mRNA similar to control FDC-P1 cells grown at this temperature in the presence of IL-3 (0 time point, Figure 5). After shifting to 39 C in the absence of IL-3, levels of c-myc transcripts declined rapidly in FDC-P1 cells; a similar decrease was also seen, albeit with slightly slower kinetics, in the ts abl virus clones. In contrast, levels of c-myc in cells infected with the wild-type abl virus remained approximately equal (Figure 5). Therefore, constitutive expression of c-myc in the absence of IL-3 required functional abl kinase.

To test whether c-myc mRNA levels were directly influenced by the presence of functional abl protein we examined the status of c-myc expression after shifting FDC-P1 ts abl clones down from the nonpermissive to the permissive temperature (Figure 6). Again control FDC-P1 cells in the presence of IL-3 and abl virus clones without the ligand were expanded at 32 C, cultures were shifted to 39 C, and then after 24 hours in the absence of IL-3 at 39 C they were shifted back down to 32 C. Again in parental FDC-P1 cells, c-myc mRNA rapidly decreased after removal of IL-3 at 39 C and remained low after shifting these cells to 32 C for 4 hours. As expected, once again steady state levels of c-myc did not change appreciably in FDC-P1 cells containing wild type abl after shifting up to 39 C or back down to 32 C. Importantly, both FDC-P1 clones containing ts abl proteins behaved like parental FDC-P1 cells when shifted up to 39 C but after shifting down to 32 C for 2 hours c-myc mRNA levels were back up to "wild-type" levels. This demonstrates that induction of c-myc is directly tied to the presence of functional abl protein kinase.

DISCUSSION

We have examined the effects of constitutive tyrosine kinase oncogenes on growth factor dependence and c-myc regulation in the murine IL-3 dependent myeloid FDC-P1 cell line. These experiments were based upon two observations. First, two classes of oncogenes, specifically the nuclear myc oncoproteins (Rapp et al., 1985; Dean et al., 1987) and several tyrosine kinase oncogenes (Pierce et al., 1985; Mathey-Prevot et al.,1986; Wheeler et al., 1987; Pierce et al., 1988) have been shown to partially or fully alleviate the IL-3 requirements of FDC-P1 cells for growth and viability. Second, treating these cells with IL-3 rapidly induces tyrosine phosphorylation of a discrete set of cellular proteins (Isfort et al., in press) and c-myc expression (Dean et al., 1987), suggesting a physiologic link between these two events in IL-3 signal transduction.

Several lines of evidence described here show a direct linkage between activated tyrosine kinase proteins and induction of c-myc in IL-3 signal transduction and growth factor abrogation. First, we have shown that tyrosine kinase oncogenes acutely abrogate IL-3 dependence of FDC-P1 cells at very high frequencies, and therefore factor abrogation in these cells is not likely to require secondary events. Second, using two different conditional tyrosine kinase abl mutants we have shown that IL-3 factor abrogation is temperature sensitive and requires the presence of functional abl protein. Third, all FDC-P1 cells rendered factor independent by introduction of a

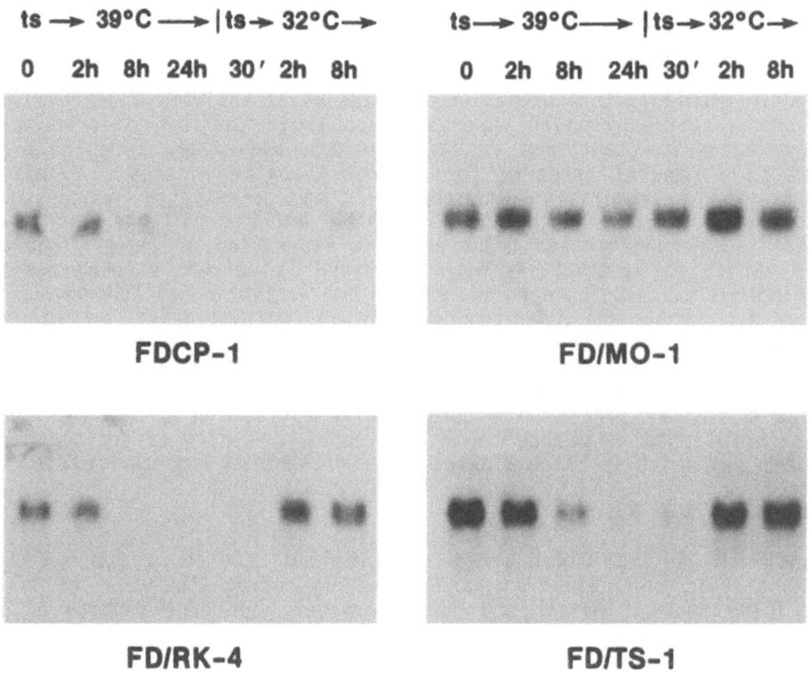

Figure 6. Trans-activation of c-*myc* RNA levels by functional *abl* protein kinase. Cultures of parental FDC-P1 cells and FD/*abl* clones were expanded and shifted to 39 C as described in Fig. 5 but after 24 hours in the absence of IL-3 at 39 C they were shifted back down to the permissive temperature and at various times total RNA was isolated and analyzed for expression of c-*myc*. Blots were exposed for 20 hours

variety of tyrosine kinase oncogenes including v-abl, v-fms, v-src, and trk constitutively express c-myc in the absence of IL-3. Therefore, although tyrosine kinase oncogenes likely perform several functions that are jointly involved in growth factor abrogation, one essential activity which they share for growth factor abrogation appears to be their ability to induce c-myc. This hypothesis was confirmed at least in the case of abl tyrosine kinase, but may be a general property for several tyrosine kinases. Experiments using ts abl retroviruses demonstrated that conditional expression of functional abl protein correlated with the ability of this oncogene to abrogate IL-3 dependence and trans-activate expression of c-myc. The exact level at which abl activates c-myc expression is currently being assessed. These findings may also have relevance for the activation of endogenous c-myc and/or N-myc which we observed concommitant with acquisition in clones partially abrogated with c-myc virus (Dean et al., 1987). Moreover, activation of myc family genes in a variety of settings where no structural alterations in these genes have been detected (see Dean et al., this volume). may have a similar basis. The constitutive expression of c-myc and N-myc could be envisioned to be due to activation of a variety of cellular trans-activators, of which tyrosine kinases may represent an important class.

ACKNOWLEDGEMENTS

We thank Naomi Rosenberg for generously providing the ts abl virus TS. The contents of this publication do not necessarily reflect the views or policies of the Department of Health and Human Services, nor does mention of trade names, commercial products, or organizations imply endorsement by the U.S. government.

REFERENCES

Armelin HA, Armelin MCS, Kelly, K, Stewart T, Leder P, Cochran BH, Stiles CD (1984) Functional role for c-myc in mitogenic response to platelet-derived growth factor. Nature (London) 310: 655-660

Bentley D, Groudine M (1986) A block to elongation is largely responsible for decreased transcription of c-myc in differentiated HL60 cells. Nature (London) 321: 702-706

Cochran BH, Reffel AC, Stiles CD (1983) Molecular cloning of gene sequences regulated by platelet-derived growth factor. Cell 33: 939-947

Cook WD, Metcalf D, Nicola NA, Burgess AW, Walker F (1985) Malignant transformation of a growth factor-dependent myeloid cell line by Abelson virus without evidence of an autocrine mechanism. Cell 41: 677-683

Dean M, Cleveland JL, Rapp UR, Ihle JN (1987) Role of myc in the abrogation of IL-3 dependence of myeloid FDC-P1 cells. Oncogene Res. 1: 279-296

Eick D, Bornkamm G (1986) Transcriptional arrest within the first exon is a fast control mechanism in c-myc gene expression. Nucleic Acids Res. 14: 8331-8346

Kaczmarek L, Hyland JK, Watt R, Rosenberg M, Baserga R (1985) Microinjected c-myc as a competence factor. Science 228: 1313-1315

Kelly K, Cochran BH, Stiles CD, Leder P (1983) Cell-specific regulation of the c-myc gene by lymphocyte mitogens and platelet-derived growth factor. Cell 35: 603-610

Mathey-Prevot B, Nabel G, Palacios R, Baltimore D (1986) Abelson virus abrogation of interleukin-3 dependence in a lymphoid cell line. Mol. Cell. Biol. 6: 4133-4135

Mechti N, Piechaczyk M, Blanchard J-M, Marty L, Jeanteur P, LeBleu B (1986) Transcriptional and post-transcriptional regulation of c-myc expression during the differentiation of murine erythroleukemia Friend cells. Nucleic Acids Res. 14: 9653-9666

Pelicci PG, Knowles DM, Magrath I, Dalla-Favera R (1986) Chromosomal breakpoints and structural alterations of the c-myc locus differ in endemic and sporadic forms of Burkitt lymphoma. Proc. Natl. Acad. Sci. U.S.A. 83: 2984-2988

Pierce JH, DeFiore PP, Aaronson SA, Potter M, Pumphrey J, Scott A, Ihle JN (1985) Neoplastic transformation of mast cells by Abelson-MuLV: abrogation of IL-3 dependence by a non-autocrine mechanism. Cell 41: 685-693

Pierce JH, Ruggiero M, Fleming TP, DeFiore PP, Greenberger JS, Varticovski L, Schlessinger J, Rovera G, Aaronson SA (1988) Signal transduction through the EGF receptor transfected in IL-3 dependent hematopoietic cells. Science 239: 628-631

Rapp UR, Cleveland JL, Brightman K, Scott A, Ihle JN (1985) Abrogation of IL-3 and IL-2 dependence by recombinant murine retroviruses expressing v-myc oncogenes. Nature (London) 317: 434-438

Wheeler EF, Askew D, May S, Ihle JN, Scherr CJ (1987) The v-fms oncogene induces factor independent growth and transformation of the IL-3 dependent cell line, FDC-P1. Mol. Cell. Biol. 7: 1673-1680

The Abelson Protein Is Required for Initiation and Maintenance for Transformation in Murine Pre-B Cells

A. ENGELMAN and N. ROSENBERG

INTRODUCTION

Abelson murine leukemia virus (Ab-MLV) is a replication defective retrovirus that carries the onc gene abl. The virus induces a rapid thymus-independent lymphoma in vivo under most conditions and transforms cells related to pre-B lymphocytes in vitro (reviewed in Rosenberg and Witte, 1988). The Abelson protein, the sole product of the virus, is a protein tyrosine kinase and expression of an enzymatically active Abelson protein is required for transformation (Rosenberg and Witte, 1988). Recent analyses of temperature sensitive (ts) mutants of Ab-MLV have indicated that expression of an active Abelson protein is also required for maintenance of NIH 3T3 cell transformation (Engelman and Rosenberg, 1987; Kipreos et al., 1987; Takemori et al., 1987). In addition, Abelson protein expression appears to play a role in maintaining the relatively undifferentiated state of some Ab-MLV-transformed lymphoid cells (Takemori et al., 1987).

We have been studying two ts mutants of Ab-MLV, Ab-MLV-P70/H590 and Ab-MLV-P70/G536. The first of these has a single point mutation at base 2,388 which results in the substitution of a histidine for a tyrosine at residue 590 of the Abelson protein; the second has a mutation resulting in the substitution of a glycine for a lysine at reside 536 (Engelman and Rosenberg, 1987). Our previous work has shown that NIH 3T3 cells transformed with these viruses revert to a normal morphology at the nonpermissive temperature (Engelman and Rosenberg 1987; unpublished data). Here, we report the effect of these viruses on initiation and maintenance of transformation in lymphoid cells.

RESULTS

Ab-MLV-P70/H590 and Ab-MLV-P70/G536 are ts for Initiation of Lymphoid Cell Transformation

To determine if Ab-MLV-P70/H590 and Ab-MLV-P70/G536 were ts for initiation of lymphoid cell transformation, mouse bone marrow cells were infected with these viruses and appropriate wild type strains. The infected cells were plated in soft agar (Rosenberg and Baltimore, 1976) and evaluated at both the permissive (34.0^{o}) and the nonpermissive (39.5^{o}) temperatures. Colonies of transformed lymphoid cells were counted 13 days after plating for cultures

---

Department of Molecular Biology and Microbiology and Department of Pathology, Tufts University School of Medicine, 136 Harrison Avenue, Boston, Massachusetts 02111

Current Topics in Microbiology and Immunology, Vol. 141

incubated at the high temperature and 22 days after plating for cultures incubated at the low temperature.

Both Ab-MLV-P160 and Ab-MLV-P70 transformed lymphoid cells at the permissive and nonpermissive temperatures (Table 1). The slight decrease in the frequency of transformants observed at 39.5° in both cases may reflect the effect of the high incubation temperature. The difference in frequency of colonies observed between Ab-MLV-P160 and Ab-MLV-P70 probably reflects the absence of C terminal sequences in the P70 protein. This portion of the molecule has been shown to be important for lymphoid cell transformation (Rosenberg et al., 1980; Prywes et al., 1983, 1985). In contrast to the results with the wild type viruses, Ab-MLV-P70/H590 and Ab-MLV-P70/G536 induced transformation only when the infected cells were plated at the permissive temperature (Table 1). Thus, the behaviour of these mutants in lymphoid cells is similar to that observed in fibroblast cells.

Table 1. Lymphoid transformation by ts Ab-MLV

Virus	Lymphoid Foci/10^6 Cells 34.0°	39.5°	# Foci 34.0° / # Foci 39.5°
P160	220	180	1.2
P70	32	17	1.8
P70/H590	110	<0.5	>220
P70/G536	35	<0.5	>70

Ab-MLV is Required for Growth of Pre-B Cells at the Nonpermissive Temperature

Lymphoid cell lines transformed by the various viruses, individual colonies of transformed lymphoid cells were isolated from agar and transferred to liquid cultures. In addition to cells transformed with the P70/H590 and P70/G536 mutants, lymphoid cells transformed with Ab-MLV-P160/H590 were also isolated. This latter virus is not ts for transformation of NIH 3T3 (Engelman and Rosenberg, 1987). Consistent with this observation, Ab-MLV-P160/H590 does transform some lymphoid cells at 39.5° (data not shown).

To determine the effect of the ts mutations on the lymphoid cells, duplicate cultures were incubated at 34.0° and 39.5° and survival was monitored by microscopic observation. Twenty-three of twenty-three independently derived cell lines infected with Ab-MLV-P70/H590 died during the first 36 hours of incubation at the high temperature. The one cell line studied that was transformed by the P70/G536 mutant also died at this temperature. In contrast, none of the twelve cell lines transformed by Ab-MLV-P160/H590 were compromised in their growth at 39.5°. As expected, cell lines transformed by wild type Ab-MLV strains also grew well at the high temperature. Therefore, mutants that are ts for initiation of transformation are also ts for maintenance of transformation. In addition, because expression of an active protein kinase is inhibited by high temperature in NIH 3T3 cells transformed by Ab-MLV-P70/H590 (Engelman and Rosenberg, 1987) these data suggest that continued expression of an active protein kinase is required to maintain a transformed phenotype.

Wild Type Ab-MLV can "Rescue" the Transformed Phenotype

To determine if expression of wild type Abelson protein was sufficient to complement the ts defect in the cells transformed with the conditional mutants, it was necessary to identify clones of lymphoid cells transformed with these viruses that could be superinfected with ecotropic retroviruses. For this purpose, nine clones transformed with Ab-MLV-P70/H590 were examined by immunofluorescent staining for the presence of surface gag determinants. Expression of these Moloney virus encoded determinants indicates the presence of helper virus and reduces the chances that the cells can be superinfected with ecotropic retroviruses. This analysis, using the monoclonal antibody 13E10 (Schiff-Maker and Rosenberg, 1986), revealed that three of nine clones examined lacked evidence of helper virus protein expression. Two of these clones were used for subsequent studies.

Cells from the helper virus-negative clones were infected with wild type strains of Ab-MLV-P160 and Ab-MLV-P70 and plated in microtiter wells at 39.5°. One to two weeks post infection, colonies of transformed cells appeared in the cultures infected with both of the wild type viruses. Analysis of the expression of Abelson protein in cells recovered from Ab-MLV-P160 superinfected cultures revealed that these populations all expressed P160 (data not shown). These cells, unlike the parent, nonsuperinfected cell line, grew equally well at 34.0° and 39.5°.

To determine if other viruses encoding related protein tyrosine kinases could also rescue the ts phenotype of the transformed lymphoid cells, cultures were infected with MRSV, a virus which expresses the Rous sarcoma virus derived src gene (Anderson and Scolnick, 1983). As in the case of Ab-MLV-P160, cells infected with this virus grew at 39.5°. Thus, other protein tyrosine kinases can substitute for Abelson protein under these conditions. In contrast, infection of the same cells with a retroviral construct containing the myc gene and a neo gene (generous gift of Dr. J. DeVilliers and Dr. K. Marcu) did not rescue the ts phenotype. Although cells infected with the myc virus could be isolated at 34.0°, these cells did not survive at 39.5° (Fig. 1).

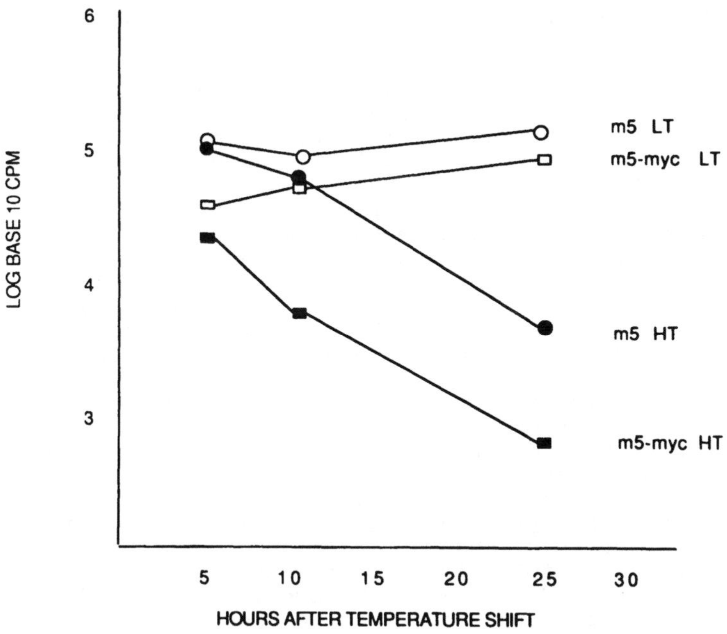

Fig. 1. Growth of m5 lymphoid cells at the permissive and nonpermissive temperatures. The m5 cell line, transformed with Ab-MLV-P70/H590, and a derivative infected with a myc virus construct were plated at 34.0° (LT) and 39.5° (HT). Replicate cultures were pulsed with ^{3}H thymidine for two hours and harvested and processed at the time points shown

DISCUSSION

Analysis of the H590 and G536 mutants in lymphoid cells suggests that viruses which are ts for NIH 3T3 transformation are also ts for initiation and maintenance of lymphoid cell transformation in vitro. The absence of ts transformation observed with the Ab-MLV-P160/H590 strain reinforces this point. This result contrasts with observations concerning the transformation potential of several nonconditional Ab-MLV mutants. In these cases, transformation of fibroblast cells occurs with high efficiency but lymphoid cell transformation is compromised (Rosenberg et al., 1980; Prywes et al., 1983, 1985). However, most of the mutants of this type encode Abelson proteins that are lacking C terminal amino acids and contain no alterations in the region of the genome encoding the catalytic domain of the protein.

The role of Abelson protein in maintenance of lymphoid cell transformation has been controversial. Several laboratories have demonstrated that lymphoid cells derived in vitro using Ab-MLV require additional signals to develop into fully malignant cells (Whitlock et al., 1983; Pillemer et al., 1984; Green et al., 1987). In addition, in some settings Ab-MLV appears to be lost from tumor cells derived from mice with Abelson disease (Grunwald et al., 1982). However, the inability of lymphoid cells transformed with the ts mutants to survive at the nonpermissive temperature demonstrates that Abelson protein is required for maintenance of the transformed state. Furthermore, any additional signals required for a fully malignant phenotype are not sufficient to maintain transformation.

The inability of the lymphoid cells to survive at the high temperature also suggests that expression of Abelson protein alone does not block differentiation of pre-B cells. Indeed, a number of lymphoid cells transformed with wild type strains of Ab-MLV undergo differentiation like changes in vitro in the absence of changes in Abelson protein expression or tyrosine kinase activity (reviewed in Alt et al., 1986). Despite this, changes in immunoglobulin expression have been observed following high temperature incubation of cells transformed by ts Ab-MLV strains (Takemori at al, 1987; our unpublished data). These apparently contrasting pieces of data highlight the complex relationship of transformation and differentiation in the Ab-MLV system.

The ability of pp60src to substitute for Abelson protein in cells transformed with the ts strains reinforces the notion that Abelson protein and pp60 interact with similar signalling pathways (Pierce et al., 1984; Mathey-Prevot et al., 1986; Hunter, 1987). However, infection of murine bone marrow cells with viruses expressing pp60 does not always lead to transformation of pre-B cells (Mathey-Prevot and Baltimore, 1985) and pp60 does not completely replace Abelson protein in IL-3 dependent cells (Mathey-Prevot et al., 1986). It is possible that changes such as those involved in complete malignant conversion co-operate with pp60 in the cells transformed by ts Ab-MLV mutants. These lymphoid cells provide a unique system in which such interactions can be studied under conditions that allow transformation to be switched on and off according to the experimental design.

REFERENCES

Alt FW, Blackwell TK, DePinho RA, Reth MG, Yancopoulos GD (1986) Regulation of genome rearrangement events during lymphocyte differentiation. Imm Rev 89:5-30.

Anderson SM, Scolnick EM (1983) Construction of a transforming murine retrovirus containing the src gene of Rous sarcoma virus. J Virol 46:594-605.

Engelman A, Rosenberg N (1987) Isolation of temperature sensitive Abelson virus mutants by site-directed mutagenesis. Proc Natl Acad Sci USA 84:8021-8025.

Green PL, Kaehler DA, Risser R (1987) Cell transformation and tumor induction by Abelson murine leukemia virus in the absence of helper virus. Proc Natl Acad Sci USA 84:5932-5936.
Grunwald DJ, Dale B, Dudley J, Lamph W, Sugden B, Ozanne B, Risser R, (1982) Loss of viral gene expression and retention of tumorigenicity by Abelson lymphoma cells. J Virol 43:92-103.
Hunter T (1987) A thousand and one protein kinases. Cell 50:823-829.
Kipreos ET, Lee GJ, Wang J-YJ, (1987) Isolation of temperature-sensitive tyrosine kinase mutants of v-abl oncogene by screening with antibodies for phosphotyrosine. Proc Natl Acad Sci USA 84:1345-1349.
Mathey-Prevot B, Nabel G, Palacois R, Baltimore D (1986) Abelson virus abrogation of interleukin 3 dependence in a lymphoid cell line. Mol Cell Biol 6:4133-4135.
Mathey-Prevot B, Baltimore D (1985) Specific transforming potential of oncogenes encoding protein-tyrosine kinases. EMBO J 4:1769-1774.
Pierce JH, Aaronson SA, Anderson SM (1984) Hematopoietic cell transformation by a murine recombinant retrovrius containing the src gene of Rous sarcoma virus. Proc Natl Acad Sci USA 81:2374-2378.
Pillemer E, Whitlock C, Weissman IL (1984) Transformation-associated proteins in murine B cell lymphomas that are distinct from Abelson virus gene products. Proc Natl Acad Sci USA 81:4434-4438.
Prywes R, Foulkes JG, Rosenberg N, Baltimore D (1983) Sequences of the A-MuLV protein needed for fibroblast and lymphoid cell transformation. Cell 34:569-579.
Prywes R, Hoag J, Rosenberg N, Baltimore D (1985) Protein stabilization explains the gag requirement for transformation of lymphoid cells by Abelson murine leukemia virus. J Virol 53:123-132.
Rosenberg N, Baltimore D (1976) A quantitative assay for transformation of bone marrow cells by Abelson murine leukemia virus. J Exp Med 143:1453-1463.
Rosenberg N, Clark DR, Witte ON (1980) Abelson murine leukemia mutants deficient in kinase activity and lymphoid cell transformation. J Virol 36:766-774.
Rosenberg N, Witte ON (1988) The viral and cellular forms of the Abelson (abl) oncogene. In: Shatkin A (ed) Advances in virus research, Academic Press, New York, in press.
Schiff-Maker L, Rosenberg N (1986) Gag-derived but not abl-derived determinants are exposed on the surface of Abelson-virus transformed cells. Virology 154:286-301.
Takemori T, Miyazoe I, Shirasawa T, Taniguchi M, Graf T (1987) A temperature-sensitive Abelson murine leukemia virus confers inducibility of IgM expression to transfomred lymphoid cells. EMBO J 6:951-956.
Whitlock CA, Ziegler SF, Witte ON (1983) Progression of the transformed phenotype in clonal lines of Abelson virus-infected lymphocytes. Mol Cell Biol 3:596-604.

ACKNOWLEDGEMENTS

This work was supported by grants from the National Cancer Institute to NR. The authors are grateful to Nancy Green for preparing the illustration.

c-*myb*

Expression and Function of the c-*myb* Oncogene During Hematopoietic Differentiation

W. M. KUEHL, T. P. BENDER, J. STAFFORD, D. MCCLINTON, S. SEGAL, and E. DMITROVSKY

The c-myb proto-oncogene encodes a 72,000 Dalton phosphoprotein which binds DNA (Klempnauer et al, 1986). Unlike the c-myc protein, which is associated with the nuclear matrix, the c-myb protein is localized in chromatin. Although c-myb mRNA is expressed predominantly in normal and tumor cells of hematopoietic origin (Gonda et al, 1982), there is increasing evidence that it is expressed at some level in many - perhaps most - nonhematopoietic tissues (e.g. Torelli et al, 1987). However, for all hematopoietic lineages examined, c-myb mRNA is at higher levels in immature than in mature cells. The function of c-myb is unknown, although it may be involved in cell proliferation since: 1) truncated forms of c-myb mRNA expressed by a retrovirus or after insertional mutagenesis are associated with enhanced cell proliferation and tumorigenesis (Ness et al, 1987; Sheng-Ong et al, 1986), and 2) c-myb mRNA expression is up-regulated in late G1 or early S phase when normal, resting T lymphocytes are stimulated by mitogens (in contrast up-regulation of c-myc mRNA occurs as an early event when T cells progress from G0 to G1) (Reed et al, 1986). Alternatively, c-myb may be involved in cell differentiation per se: 1) c-myb mRNA expression is higher in immature compared to mature hematopoietic cell lines despite indistinguishable proliferative properties of these lines (Bender and Kuehl, 1987); and 2) it has been reported that avian myelomonoblastic cells infected with a retrovirus containing a temperature sensitive v-myb gene mature to macrophages at the nonpermissive temperature but then retrodifferentiate to myeloblastic cells at the permissive temperature (Ness et al, 1987). In this report, we will briefly summarize our studies which demonstrate a role of c-myb as a marker of B cell differentiation and as a gene which must be down-regulated to permit terminal differentiation of erythroleukemia cells.

c-myb mRNA Levels as a Marker of B Cell Differentiation

We have shown that murine pre-B lymphoma cell lines express similar high levels of c-myb mRNA, whereas plasmacytoma and B lymphoma cell lines express c-myb mRNA at levels which are five-fold or more lower than the pre-B lymphoma cell lines (Bender and Kuehl, 1987). On the basis of these results, we suggested that c-myb mRNA expression is down-regulated as cells mature from pre-B to B cells. Consistent with our proposal, plasmacytoma X pre-B lymphoma somatic

NCI-Navy Medical Oncology Branch, National Cancer Institute, Naval Hospital, Bethesda, MD 20814

Current Topics in Microbiology and Immunology, Vol. 141

cell hybrids, which have a plasmacytoma phenotype, express low levels of c-myb mRNA similar to the plasmacytoma parent.

Cell lines that co-express membrane immunoglobulin (a B cell marker) and a variety of pre-B cell markers.

It had been proposed that pre-B cell tumors can spontaneously differentiate *in vitro* to immature B cells which express surface IgM (mIgM) (Reth *et al*, 1985). To further test our proposal we examined subclones of the 1881 pre-B cell line (mu^+, L^-) which had spontaneously rearranged a kappa light chain gene (Bender and Kuehl, 1987). Two of three such subclones express mIgM but continue to express the same high levels of c-*myb* mRNA as the parental 18.81 pre-B cell line. In addition, these mIgM positive subclones express 3 other pre-B cell markers (i.e. λ5 and N-*myc* mRNAs and BP-1 surface antigen) at undiminished levels. We also found that A202J B lymphoma X 70Z/3B pre-B lymphoma somatic cell hybrids, which have a phenotype similar to the pre-B lymphoma parent, express levels of c-*myb* and λ5 mRNAs comparable to the pre-B lymphoma parent, but also express both parental light chains and low levels of mIgM. These examples of cell lines which co-express mIg and pre-B cell markers led us to examine another line which might have similar properties, i.e. the Balb-1427 line which was reported as a "mature B cell" that expresses mIgM and low levels of V(D)J recombinase (Lieber *et al*, 1987). We have found that Balb-1427 cells express levels of c-*myb* and λ5 mRNAs comparable to other pre-B lymphomas.

Table I summarizes some properties of pre-B and immature B cells. Initially we viewed the cell lines which express pre-B cell markers (λ5, high c-*myb*, N-*myc* mRNAs, and BP1 surface antigen) together with mIg to be pre-B cells. However, we now propose that such cells be designated as "nascent" B cells to preserve the established and convenient definition that B cells express mIg whereas pre-B cell do not.

TABLE I. Markers to Distinguish Pre-B from Early B Cells[a]

	PRE-B	"NASCENT" B	IMMATURE B
surface Ig	-	+	+
c-*myb* mRNA	high	high	low
λ5 mRNA	+	+	-
N-*myc* mRNA[b]	+/(-)	+/(-)	-
BP_1 antigen	+	+	-
Examples of cell lines	1881 LS8 70Z/3	1881 subclones Pre-B X B hybrids Balb-1427	WEHI 231 Balentyl 17

[a] See Bender and Kuehl, 1987 for additional information and references on markers and cell lines.

[b] N-*myc* mRNA is expressed in normal murine pre-B cells and in all murine pre-B lymphoma lines examined except 70Z/3 (Zimmerman *et al*., 1986).

Down-Regulation of c-myb mRNA is Necessary for Terminal Differentiation of Mouse Erythroleukemia Cells

Friend virus infected mouse erythroleukemia (MEL) cell lines are transformed early erythroid precursors that can be induced chemically to differentiate into more mature erythroid cells (Marks and Rifkind, 1978). When inducer is added, there is a latent period of about 12 hours before cells become irreversibly committed to the terminal differentiation program, although undergoing several rounds of cell division before ceasing to replicate. Following induction there is a biphasic decline in the steady state levels of c-myc and c-myb mRNAs (Kirsch et al., 1986). For each nuclear oncogene, the initial decline of mRNA begins within hours after inducer is added but then increases to essentially basal levels during the first 24 hours of induction. During the next few days as the cells undergo terminal differentiation (including increased expression of hemoglobin), there is a secondary decline in c-myc and c-myb mRNA levels. By expressing a transfected c-myc oncogene in MEL cells, we and others have shown that the down-regulation of c-myc mRNA is a necessary event for inducer-mediated terminal differentiation (Coppola and Cole, 1986, Dmitrovsky et al., 1986, Prochownik and Kurowska, 1986,)

We have now attempted to determine whether down-regulation of c-myb mRNA is also a necessary event for inducer-mediated differentiation of MEL cells. Our c-myb expression vector contains a mouse cDNA coding region plus 33 bp of 5' untranslated and 158 bp of 3' untranslated sequences (Bender and Kuehl, 1986) (Fig. 1).

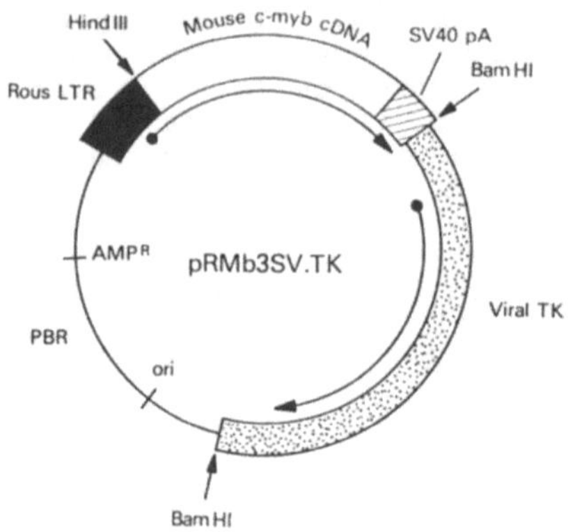

Fig. 1. Structure of mouse myb cDNA expression vector (pRMb3SVTK). The plasmid consists of four fragments: a) 3.4 Kb Bam HI-Hind III fragment from pRSVCAT vector (Gorman et al., 1982); b) mouse myb cDNA fragment from Sma I-230 (converted to Hind III site with linkers) to Bgl II-2333 (Bender and Kuehl, 1986); c) 237 bp Bcl I/Bam HI fragment containing SV40 polyadenylation sites; and d) a 3.6 Kb Bam HI fragment containing the herpes virus thymidine kinase gene. Arrows indicate direction of transcription units

The expected sizes for endogenous and exogenous c-myb mRNAs should be approximately 3.8 and 2.8 Kb, respectively. We transfected these vectors into C19 cells, a thymidine kinase (TK) negative mutant of MEL cell line 745. In our first transfection experiment, we obtained 19 transfectants. Eighteen transfectants were sibs which expressed the exogenous c-myb mRNA and one transfectant did not express exogenous c-myb mRNA (Fig. 2).

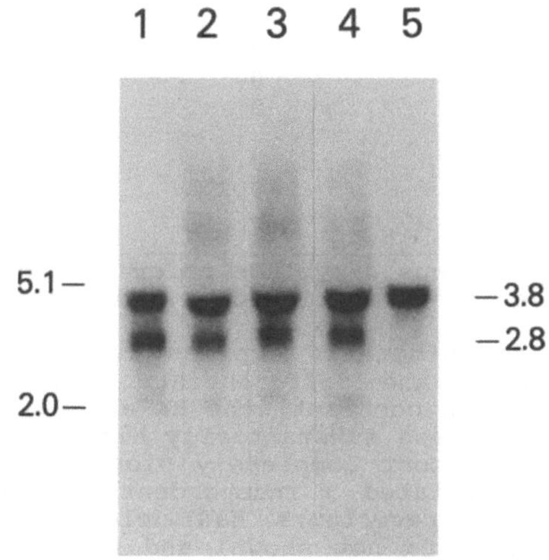

Fig. 2. RNA expression of exogenous and endogenous c-myb genes in stable MEL transfectants. The pRMb3SV.TK plasmid was transfected into C19 MEL cells as a calcium phosphate precipitate. Stable transfectants were selected in HAT medium. Total RNA was prepared, electrophoresed on a 1% agarose formaldehyde gel, blotted, and hybridized with a mouse myb cDNA probe. The blot was washed at 60°C in 0.1 X SSC and analyzed by radioautography. Each lane was loaded with 10 μg of RNA. Lanes 1-4 represent transfectant clones H4, H12, H20, H21, respectively (all of these clones are sibs - see text). Lane 5 is the C19 parent. Ribosomal RNA and myb mRNA sizes are indicated

We analyzed approximately 75 other independent transfectants in subsequent experiments. However, only 15% of these transfectants expressed detectable levels of exogenous c-myb mRNA, and none expressed high levels of exogenous c-myb mRNA (See Table II for 3 independent clones expressing highest level of exogenous c-myb mRNA). We also constructed a similar vector with a Moloney LTR replacing the Rous LTR as a promoter element. Although the Moloney vector gave rise to transfectants with slightly higher levels of exogenous c-myb mRNA, this RNA was not translated efficiently in MEL cells.

The parental cells and the 3 independent transfectants expressing the highest levels of exogenous c-myb mRNA were induced with hexamethylene bisacetamide (HMBA) to assess the ability of the cells to terminally differentiate (Table II).

TABLE II. DIFFERENTIATION OF MEL CELLS

Clone	Exogenous c-myb RNA[a]	% Benzidine Positive Cells[b]
C19	0	85
H4	1.0	4
H21	1.1	2
G6	0.4	66
25.1	0.5	58
H21TK$^-$ Revertants		
1B1	0	66
2B1	0	72
2B3	0	67
H21 Subclones		
H21E	0.3	40
H21D	0.2	66

[a]Exogenous/endogenous c-myb RNA prior to induction
[b]Scored 5 days after addition of 3.5 mM HMBA

The G6 and 25.1 transfectants express relatively low levels of exogenous c-myb mRNA, and differentiate nearly as well as the parental C19 cells. In contrast, the H4 and H21 siblings from the first transfection express substantially higher levels of exogenous c-myb mRNA, and are almost completely blocked in their ability to differentiate. We isolated 3 independent TK minus revertants of H21. Each of these revertants had deleted all traces of the transfected plasmid (data not shown) and had regained the ability to differentiate, indicating that expression of the exogenous c-myb mRNA is responsible for blocking differentiation of transfectant H21. Curiously, TK positive subclones of H21 showed decreased expression of c-myb mRNA but no evidence of gene loss (data not shown). These H21 subclones also regained the ability to differentiate, again supporting the conclusion that the level of exogenous c-myb mRNA is critical in determining whether or not the differentiation process is affected. Recently, Clark et al. (1988) published results similar to ours confirming that transfectants expressing high levels of exogenous c-myb mRNA are unable to differentiate, whereas transfectants expressing lower levels of exogenous c-myb mRNA differentiate.

In conclusion, the inducer-mediated terminal differentiation of MEL cells requires down-regulation of both the c-myc and c-myb nuclear oncogenes. In each case, it remains to be determined:
1) whether the early and/or the late phase of down-regulation is necessary and 2) the mechanism(s) by which nuclear oncogenes affect the differentiation process.

REFERENCES

Bender TP, Kuehl WM (1986) Murine myb proto-oncogene mRNA: cDNA sequence and evidence for 5' heterogeneity. Proc Natl Acad Sci USA 833204-3208
Bender TP, Kuehl WM (1987) Differential expression of c-myb proto-oncogene marks the pre-B/B cell junction in murine B lymphoid tumors. J Immunol 139:3822-3827.

Clarke MF, Kukowska-Latallo JF, Westin E, Smith M, Prochownik EV (1988) Constitutive expression of a c-myb cDNA blocks Friend murine erythroleukemia cell differentiation. Mol Cell Biol 8:884-892.
Coppola JA, Cole MD (1986) Constitutive c-myc oncogene expression blocks mouse erythroleukemia cell differentiation but not commitment. Nature 320:760-763.
Dmitrovsky E, Kuehl WM, Hollis GF, Kirsch IR, Bender TP, Segal S (1986) Expression of a transfected human c-myc oncogene inhibits differentiation of a mouse erythroleukemia cell line. Nature 322:748-750.
Gonda TJ, Sheiness DK, Bishop JM (1982) Transcripts from the cellular homologs of retroviral oncogenes: distribution among chicken tissues. Mol Cell Biol 2:617-624.
Gorman CM, Merlino GT, Willingham MC, Pastan I, Howard BH (1982) The Rous sarcoma virus long terminal repeat is a strong promoter when introduced into a variety of eukaryotyic cells by DNA-mediated transfection. Proc Natl Acad Sci USA 79:6777-6781
Kirsch IR, Bertness V, Silver J, Hollis GF (1986) Regulated expression of the c-myb and c-myc oncogenes during erythroid differentiation. J Cell Biochem 32:11-21
Klempnauer K-H, Bonifer C, Sippel AE (1986) Identification and characterization of the protein encoded by the human c-myb proto-oncogene. EMBO J 5:1903-1911.
Lieber MR, Hesse JE, Mizuuchi K, Gellert M (1987) Developmental stage specificity of the lymphoid V(D)J recombination activity. Genes and Development 1:751-761
Ness, SA, Beng H, Graf T (1987) v-myb dominance over v-myc in doubly transformed chick myelomonocytic cells. Proc Natl Acad Sci 51:41-50
Marks PA, Rifkind RA (1978) Erythroleukemic differentiation. Ann Rev Biochem 47:418-448
Prochownik EV, Kukowska JF (1986) Deregulated expression of c-myc by murine erythroleukemia cell prevents differentiation. Nature 322:848-850
Reed JC, Alpers JD, Nowell PC, Hoover RG (1986) Sequential expression of proto-oncogenes during lectin-stimulated mitogenesis of normal human lymphocytes 83:3982-3986.
Reth MG, Amirati P, Jackson S, Alt FW (1985) Regulated progression of a cultured pre-B cell line to the B-cell stage. Nature 317:353-356.
Shen-Ong GLC, Morse HC, Potter M, Mushinski JF (1986) Two modes of c-myb activation in virus-induced mouse myeloid tumors. Mol Cell Biol 6:380-392.
Torelli G, Venturelli D, Col'o A, Zanni C, Selleri L, Moretti L, Calabretta B, Tovelli V (1987) Expression of c-myb proto-oncogene and other cell cycle-related genes in normal and neoplastic human colonic mucosa. Cancer Res 47:5266-5269.
Zimmerman KA, Yancopoulos GD, Collum RG, Smith RK, Kohl NE, Denis KA, Nau MN, Witte ON, Joran-Allerand D, Gee CE, Minna JD, Alt FW (1986) Differential expression of myc family gene during murine development. Nature 319:780-783

Sense and Anti-sense Transcription in the Murine c-*myb* Attenuator Region

T. P. BENDER[1], K. M. CATRON[1], W. M. KUEHL[2], and C. B. THOMPSON[3]

INTRODUCTION

The c-*myb* proto-oncogene has been found to be expressed at high levels predominantly in normal tissues and tumor cell lines of immature hematopoietic origin (Westin 1982; Gonda 1982; Duprey 1985; Bender 1987a). In each lineage which has been examined expression of c-*myb* mRNA has been found to decrease during maturation. Recent evidence suggests, at least in the T cell lineage, that expression of c-*myb* mRNA may switch from high constitutive expression in immature T cells to a cell cycle related mode of expression at more mature stages of development (Thompson 1986; Reed 1986). These findings have suggested that the c-*myb* gene product may play a role in hematopoietic maturation and that down-regulation of c-*myb* expression is required for normal development to proceed. Indeed, transfection experiments have demonstrated that constitutive expression of *myb* may interfere with differentiation in *ts*E26 transformed monocytic cells and the inducible differentiation of murine erythroleukemia cells (Beug 1987; Clarke 1988). We have begun to examine how c-*myb* steady state mRNA levels are determined in a series of murine B lymphoid tumors which demonstrate developmentally regulated expression of c-*myb* mRNA levels. In these tumor cell lines, pre-B cell lymphomas contain 10 to greater than 100 times as much c-*myb* mRNA than B cell lymphomas or plasmacytomas (Bender 1987a). We find that these differences in steady state c-*myb* mRNA levels are regulated primarily by a block to transcription elongation which occurs in the first intron of the c-*myb* locus (Bender 1987b).

THE c-*myb* PROMOTER APPEARS TO LACK TISSUE SPECIFICITY

To initially assess the ability of c-*myb* 5' flanking/untranslated sequences (Bender 1986) to promote transcription we have constructed a chloramphicol acetyl transferase (CAT) expression vector based on pSVoCAT (provided by B. Howard). A Hind3/Ncol fragment beginning at the first base 5' of the murine c-*myb* translation start site and containing ~1.7 kb of 5' untranslated/flanking DNA was cloned into the Hind3 site adjacent to the CAT gene of pSVoCAT (Gorman 1982). This region contains all of the previously reported putative murine c-*myb* transcription start sites (Bender 1986). We have found that the sense orientation of this construct will allow expression of the bacterial CAT gene in stable Rat-1 cell transformants. The anti-sense orientation of this construct does not allow expression of the CAT gene. Thus, there does not appear to be an innate tissue specific component to the c-*myb* promoter region.

[1]Department of Microbiology, The University of Virginia, Charlottesville, VA 22908; [2]NCI-Navy Medical Oncology Branch, National Cancer Institute, National Institutes of Health, Naval Hospital, Bethesda, MD 20814; and [3]Howard Hughes Medical Institute, The University of Michigan, Ann Arbor, MI 48109.

Current Topics in Microbiology and Immunology, Vol. 141

REGULATION OF STEADY STATE c-*myb* mRNA LEVELS BY A PREMATURE BLOCK TO TRANSCRIPTION ELONGATION

To better understand the regulation of c-*myb* steady state mRNA levels in murine B lymphoid tumors we have pursued our studies primarily using the pre-B cell lymphoma 70Z/3B and the mature B cell lymphoma A20/2J (Bender 1987b). The 70Z/3B pre-B cell lymphoma contains approximately 10-20 times more c-*myb* mRNA than A20/2J. Using actinomycin D (8 μg/ml) to inhibit *de novo* RNA synthesis we have found that c-*myb* mRNA has a half life of approximately 165-190 minutes in both cell lines. We have extended these studies to several other B lymphoid tumor lines as well as the murine erythroleukemia cell line 745 with the same result. Thus, differences in message stability do not seem to account for the differences in steady state mRNA levels between these cell lines.

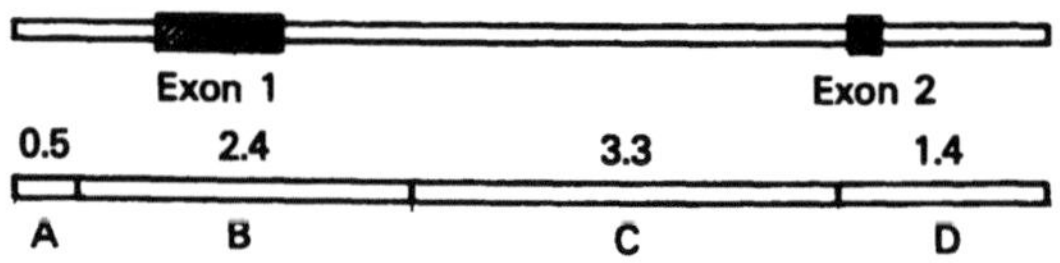

Fig. 1. Schematic representation of c-*myb* genomic targets used in nuclear run-on studies. The plasmid p*myb*13 is a 7.6 kb EcoRl murine genomic DNA fragment with c-*myb* 5′ flanking untranslated sequences, exon I, intron I and exon II cloned into the EcoRl site of pJB327. Digestion of this plasmid with EcoRl and Hind3 yields the four target fragments, A-D, used in these studies plus the vector pJB327

We have used a nuclear run-on assay (as described by Groudine 1981) to examine rates of transcription across a 7.6 kb EcoRl genomic fragment, cloned into pJB327, which contains exon I, intron I and exon II of the murine c-*myb* gene (Bender 1987b). Digestion of this plasmid with EcoRl and Hind3 conveniently divides the 7.6 kb fragment into four targets, A-D, which roughly contain 5′ flanking DNA, exon I, intron I and exon II as shown in Figure 1. When nuclear run-on transcripts from either 70Z/3B or A20/2J are hybridized to filters with these targets we do not detect transcription with target A suggesting that detectable transcription 5′ of exon I does not occur in these cell lines. Using nuclei isolated from 70Z/3B lymphoma cells transcription is detected across exon I, intron I and exon II (targets B-D respectively). However, the density of RNA polymerase is consistently 2-3 fold lower on exon II than exon I even when differences in target size are taken into account. These experiments were initially done using double stranded targets which do not differentiate between sense and anti-sense transcription. As discussed below, most of this difference is clarified by

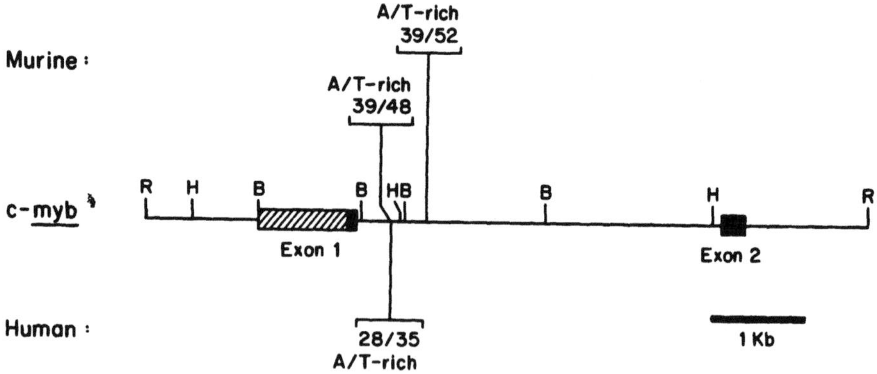

Fig. 2. Restriction map of the 7.6 kb EcoRl genomic DNA region containing the murine c-*myb* attenuator region. The positions of A/T-rich regions are shown. Exons I and II are indicated by boxes. Restriction sites are R, EcoRl; H, Hind3; and B, BamHl

Sequence I: 215-250

T T T T T T A T T T T A T T T C T C T T T T T C T T T C T T T T T T C T T T T T C T T T T T-N_{30}-$G_{22/24}$...

Sequence II: 1144 - 1198

T T T C T T T C C T T T C T T T C T T T C T T T C T T T C T T T C T T T T T T T T T T G_{11}...

Fig. 3. A/T-rich sequences found in the c-*myb* attenuator region. Positions of A/T-rich regions are numbered from the c-*myb* translation start site (Bender 1986). Sequence using single stranded targets as there is significant anti-sense transcription in the exon I region of the c-*myb* locus. Thus, in 70Z/3B most, if not all, of the sense transcription which initiates in the exon I region continues through exon II. Using nuclear run-on probes from A20/2J we found that the transcription detected with target B (exon I) was equivalent to that seen in 70Z/3B. However, transcription detected with target C was about one-fifth to one-third that seen in 70Z/3B and no transcription was detected above background with target D. Although a small amount of steady state c-*myb* mRNA is detected on total RNA blots from A20/2J, low transcription rates of moderately stable mRNA species are difficult to detect by nuclear run-on assay. This data indicates that the differences in steady state levels of c-*myb* mRNA between these cell lines is due to a premature block to transcription elongation which occurs in intron I of A20/2J rather than differences in transcription initiation.

We have recently extended the analysis of run-on transcription to the 1881 pre-B cell lymphoma, the WEH1-231 B cell lymphoma, the plasmacytoma MPC-11 and the murine erythroleukemia cell line 745. We do not detect read-through transcription (as measured using an exon II single stranded target) in either WEH1-231 or MPC-11. Interestingly, although read-through transcription is readily detected in both 1881 and MEL 745, it is consistently somewhat lower than seen in 70Z/3B suggesting that some degree of attenuation may take place in these cell lines.

All of the transcription that we detect across the 7.6 kb EcoR1 fragment, described above, is completely sensitive to α-amanitin at 2 μg/ml which has been shown to inhibit transcription by RNA polymerase II (Roeder 1976). In these experiments we control for RNA polymerase I transcription using an 18S rRNA target (provided by J. Sylvester) and RNA polymerase III transcription using a 5S rRNA target (provided by D. Bogenhagen). High levels of both polymerase I and III transcription are detected in our run-on assays, using these targets, in the presence of α-amanitin at 2 μg/ml and do not increase in its absence. Thus, the c-*myb* transcription that we are measuring appears to be the result of RNA polymerase II activity. In addition, the 7.6 kb EcoR1 fragment detects only a single band when used as a probe on genomic DNA and RNA blot analysis.

As mentioned above, we detect significant amounts of anti-sense transcription in the area of exon I in both 70Z/3B and A20/2J (Bender 1987b). The amount of anti-sense transcription detected across the exon I region using single stranded 1.1 kb BamH1 genomic targets (See Figure 2 and Bender 1987b) is approximately equivalent to the amount of sense transcription in this region. We have extended this analysis to several other pre-B cell lymphoma, B cell lymphoma and plasmacytoma cell lines and find that the ratio of sense to anti-sense transcription is quite variable ranging from roughly 1:1 to 10:1 in the exon I region. However, this variation does not appear to correlate with any specific state of differentiation and at present it is not possible to assess the biological relevance of this anti-sense transcription. Since anti-sense transcription is not detected with our exon II target nor fragment A (which covers 5′ flanking sequences), this implies that the anti-sense transcription initiates in intron I and ends near the 5′ end of exon I. Although mapping by nuclear run-on assay indicates that anti-sense transcription initiates in the 3′ half of intron I accurate mapping of this position remains to be done.

STRUCTURAL FEATURES IN THE c-*myb* INTRON I

We have recently extended our mapping studies of c-*myb* sense transcription across the 7.6 kb EcoR1 fragment shown in Figure 2. These studies indicate that most, if not all, of the sense transcription in A20/2J terminates in or slightly downstream of the ~1.5 kb BamH1 fragment just 3' of exon I. Due to the amount of anti-sense transcription in this region we cannot be certain that some sense transcription does not extend past this point but most appears to terminate here. When this region from the mouse was sequenced by the dideoxy-chain termination method we found an extraordinary AT-rich region extending from positions 215-250 3' of the translation start site (Sequence I, Figure 3). This sequence is followed, after 30 seeming random bases, by a stretch of 22/24 guanine residues. Interestingly a similar AT-rich stretch is located at approximately the same position in the human c-*myb* intron I. A second AT-rich region (Sequence 2, Figure 3) is located approximately 900 bp 3' of this site. This AT-rich sequence is immediately followed by a run of 11 guanine residues which closely resembles a polymerase III termination site (Bogenhagen 1981). We do not yet know whether this second sequence is present in the human c-*myb* intron I nor whether either AT-rich sequence is involved in attenuation. However, runs of thymidine residues followed by a run of guanine residues have been shown to be involved in premature chain termination or RNA polymerase "pausing" in prokaryotic biosynthetic operons (Platt 1986), the adenovirus major late promoter (Maderious 1984), the SV40 VP-1 gene (Hay 1985) and the c-*myc* gene (Bentley 1986; Chung 1987).

CONCLUSIONS

We have found that differences in steady state c-*myb* mRNA levels between murine B lymphoid tumors are maintained primarily by a block to transcription elongation rather than at the level of transcription initiation. Similarly, the differential expression of c-*myc* mRNA has been reported to be regulated, at least in part, by a block to transcription elongation. As expression of both c-*myc* and c-*myb* is associated with cell growth and proliferation regulation of transcription at the level of elongation may provide a more rapid and sensitive mechanism to rapidly increase or decrease steady state mRNA in response to external signals than transcription initiation. Thus, it is of interest that expression of both c-*myc* and c-*myb* mRNA increases during progression from G0 through the G1 stage of the cell cycle (Kelley 1983; Stern 1986). By keeping these genes in a transcriptionally active state simple removal or alteration of a block to elongation would allow rapid increases in steady state mRNA levels.

REFERENCES

Bender TP, Kuehl WM (1986) Murine *myb* protooncogene messenger RNA: cDNA sequence and evidence for 5' heterogeneity. Proc Natl Acad Sci 83:3204-3208.

Bender TP, Thompson CB, Kuehl WM (1987a) Differential expression of c-*myb* mRNA in murine B lymphomas by a block to transcription elongation. Science 237:1473-1476.

Bender TP, Kuehl WM (1987b) Differential expression of the c-*myb* proto-oncogene marks the pre-B cell/B cell junction in murine B lymphoid tumors. J Immunol 139:3822-3827.

Bentley DL, Groudine M (1986) A block to elongation is largely responsible for decreased transcription in differentiated HL-60 cells. Nature 321:702-706.

Beug H, Blundell PA, Graf T (1987) Reversibility of differentiation and proliferative capacity in avian myelomonocytic cells transformed by *ts*E26 leukemia virus. Genes and Development 1:277-286.
Bogenhagen DF, Brown DD (1981) Nucleotide Sequences in Xenopus 5S DNA required for transcription termination. Cell 24:261-270.
Chung J, Sussman DJ, Zeller R, Leder P (1987) The c-*myc* gene encodes superimposed RNA polymerase II and III promoters. Cell 51:1001-1008.
Clarke MF, Kukowaska-Latallo JF, Westin E, Smith M, Prochownik EV (1988) Constitutive expression of a c-*myb* cDNA blocks Friend murine erythroleukemia cell differentiation. Molec Cell Biol 8:884-892.
Gonda TJ, Sheiness DK, Bishop JM (1982) Transcripts from the cellular homologs of retroviral oncogenes: distribution among chicken tissues. Mol Cell Biol 2:617-624.
Gorman CM, Moffat LF, Howard BH (1982) Recombinant genomes which express chloramphenicol acetyltransferase in mammalian cells. Mol Cell Biol 2:1044-1051.
Groudine M, Peretz M, Wientraub H (1981) Transcriptional regulation of hemoglobin switching in chicken embryos. Molec Cell Biol 1:281-288.
Hay N, Aloni Y (1985) Attenuation of late simian virus 40 mRNA synthesis is enhanced by the agnoprotein and is temporally regulated in isolated nuclear systems. Mol Cell Biol 5:1327-1334.
Kelly K, Cochran BH, Stiles CD, Leder P (1983) Cell-specific regulation of the c-*myc* gene by lymphocyte mitogens and platelet-derived growth factor. Cell 35:603-610.
Maderious A, Chen-Kiang S (1984) Pausing and premature termination of human RNA polymerase II during transcription of adenovirus *in vivo* and *in vitro*. Proc Natl Acad Sci 81:5931-5935.
Platt, T (1986) Transcription termination and the regulation of gene expression. Ann Rev Biochem 55:339-372.
Roeder, RG (1976) RNA Polymerase. Losick, R, Chamberlin, M (eds.) Cold Spring Harbor Laboratory Press, New York, p. 285.
Stern JB, Smith KA (1986) Interleukin-2 induction of T-cell G1 progression and c-*myb* expression. Science 233:203.
Thompson CB, Challoner PB, Neiman PE, Groudine M (1986) Expression of the c-*myb* proto-oncogene during cellular proliferation. Nature 319:374-379.
Westin EH, Gallo RC, Arya SK, Eva A, Souza LM, Baluda MA, Aaronson SA, Wong-Staal F (1982) Differential Expression of the *amv* gene in human hematopoietic cells. Proc Natl Acad Sci 79:2194-2198.

bcl-2

Expression of the Murine Proto-Oncogene *bcl*-2 Is Stage Specific and Cell-Type Specific

J. F. Mushinski[1], J. D. Mountz[2], J. H. Pierce[3], J. G. Pumphrey[1], R. M. Skurla Jr.[1], F. D. Finkelman[4], D. Givol[5], and W. F. Davidson[1]

INTRODUCTION

The proto-oncogene bcl-2 is of particular interest in any consideration of neoplasia of B lymphocytes, because it has been shown to be involved in the t(14,18) (q32,q21) translocations that are found in virtually all of human follicular lymphomas, the most common B cell lymphoma in man (Bakhshi *et al.* 1985; Cleary and Sklar 1985; Tsujimoto *et al.* 1985, 1986) and in an acute pre-B lymphocytic leukemia cell line (Tsujimoto *et al.* 1984). The critical consequences of the interruption of the bcl-2 gene by this translocation appears to be a constitutive expression of bcl-2 transcripts that also contain sequences from the newly juxtaposed immunoglobulin heavy chain joining region (Ig-J_H) (Graninger *et al.* 1987; Seto *et al.* 1988). The normal function of the bcl-2 proto-oncogene protein product is unknown, but like many other proto-oncogenes it appears to be involved in control of normal growth and development of human hemopoietic cells. Support for this comes from data that show rapid induction of bcl-2 expression in mitogen-stimulated proliferation of normal human B and T lymphocytes (Graninger *et al.*, 1987; Reed *et al.*, 1987; Seto *et al.* 1988) and a down regulation of bcl-2 expression accompanying terminal B cell maturation (Graninger *et al.* 1987).

In order to gain additional information about the function of this gene, we have studied its structure and expression in normal and neoplastic murine B lymphocytes using cDNA clones derived from a mouse B cell lymphoma (Givol 1987). We reported earlier that the bcl-2 gene was located on mouse chromosome 1 (Mock *et al.* 1988) and that its expression was found only in pre B and mature B cell lymphomas but not in the very early "progenitor B cell line" HAFTL-1 nor in any plasmacytomas which represent very late, nearly "end-stage" antibody-secreting B cells (Mushinski 1987; Mushinski *et al.* 1987). These results became somewhat controversial inasmuch as 1) studies with certain human myeloma cell lines (presumably equivalent "endstage" B cells) did not show complete down regulation of bcl-2 expression (Tsujimoto 1987), and 2) LPS blasts which can differentiate into Ig-secreting cells were not followed long enough to determine if such differentiation would result in disappearance of bcl-2 RNA (Graninger *et al.* 1987; Negrini *et al.* 1987; Reed *et al.* 1987). Thus, additional experiments with normal differentiating murine B lymphocytes were undertaken as well as additional studies with the progenitor B cell lines HAFTL-1 and HAFTL-3, which are now known to be capable of differentiating along the myeloid pathway (Holmes *et al.* 1986; Davidson *et al.* 1988).

[1]Laboratory of Genetics, National Cancer Institute, Bethesda, MD 20892 USA
[2]Department of Medicine, University of Alabama, Birmingham, AL 35294 USA;
[3]Laboratory of Cellular and Molecular Biology, National Cancer Institute, Bethesda, MD 20892 USA; [4]Department of Medicine, Uniformed Services University for the Health Sciences, Bethesda, MD 20814 USA; [5] The Weizmann Institute, Rehovot, Israel

Current Topics in Microbiology and Immunology, Vol. 141

RESULTS

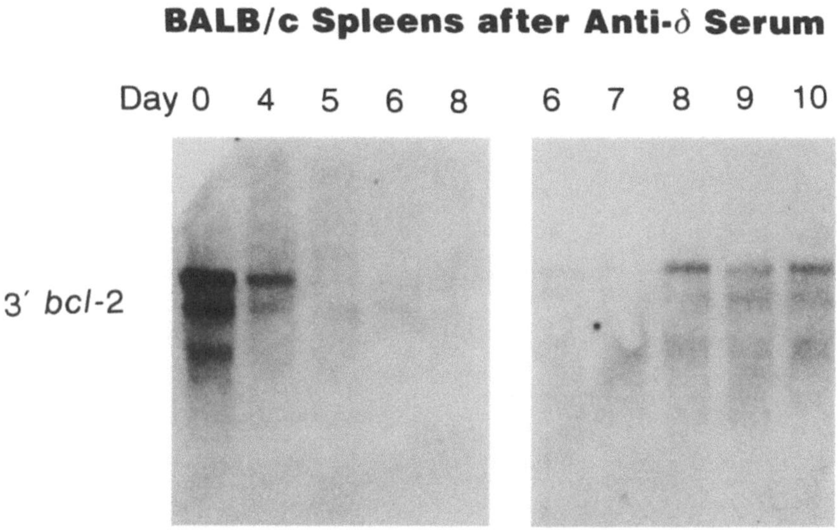

Fig. 1. Sequential analysis of bcl-2 mRNA levels in spleens following anti-IgD injection. Each lane represents a blot of 7.5 μg poly(A)$^+$ RNA extracted from spleens of BALB/c mice at various times after antiserum treatment. The two panels are data are from two independent experiments. In each case the RNA was electrophoresed in 1% agarose gel containing 2.2 M formaldehyde and transferred to a nitrocellulose membrane (Mountz et al. 1987). The membrane was hybridized to a ^{32}P-labeled 3' bcl-2 probe from the cDNA clone Ncm4 (Gurfinkel et al. 1987; Givol 1987)

In order to study the expression of bcl-2 in normal mouse B cells which are differentiating into "end-stage" antibody-secreting cells, we induced polyclonal B cell activation in intact BALB/c mice by intravenous injection of 800 μg of an affinity-purified goat antibody to IgD (Finkelman et al. 1982; Mountz et al. 1987). At various times after injection, spleens were removed and mRNA prepared from them (Mountz et al. 1987). Assays for serum immunoglobulin (Ig) levels, cytoplasmic Ig and Ig mRNA showed an increase in secretory IgM, IgG_1 and IgG_2 which peaked between 6 and 8 days after treatment. Studies of B cell proliferation have indicated that a high percentage of B cells are also synthesizing DNA during this time period. Northern blots of these RNAs were hybridized with probes for mouse bcl-2. The results of two such experiments (Fig. 1) showed that this probe detected bcl-2 transcripts which decreased to nearly undetectable levels by day 6-8 but which began to reappear at low levels thereafter. The down regulation of bcl-2 expression coincided with the maximum expression of IgM and IgG (Mountz et al. 1987), thus confirming that interpretation made from B cell tumors of varying degrees of maturation (Gurfinkel et al. 1987; Mushinski et al. 1987).

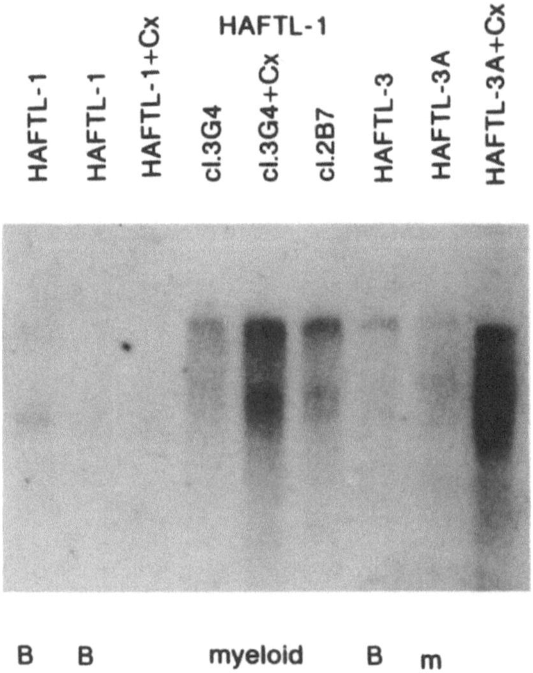

Fig. 2. Expression of *bcl*-2 RNA in progenitor B cell lines and subclones. Each lane represents a blot of 5 μg poly(A)$^+$ RNA extracted from the indicated cell lines, electrophoresed in formaldehyde/agarose gels, and blotted onto nitrocellulose (Davidson *et al.* 1988). The membranes were hybridized to a ^{32}P-labeled 3' *bcl*-2 cDNA probe. Cx indicates that the cells were treated with cyclohexamide *for* 2 hours before RNA extraction. "B" and "m" beneath the lanes indicates whether the cells belonged to the B lymphocyte lineage or the myeloid lineage, respectively

In order to study the *bcl*-2 expression of very early B cells in greater detail, RNA was prepared from HAFTL-1 and HAFTL-3 and from myeloid clones derived from these bipotential progenitor lines. When probed with the 3' *bcl*-2 probe (Fig. 2), the original HAFTL-1 cells are confirmed to have undetectable levels of *bcl*-2 mRNA, even when the cultures were exposed to 10 μg/ml cycloheximide which frequently stabilizes mRNAs with short half lives. On the other hand, clones of HAFTL-1 (1G4, 3G4, and 2B7) and of HAFTL-3 (HAFTL-3A) which had acquired the morphology, surface phenotype (Mac-1$^+$) and function (phagocytosis) characteristic of macrophages (Holmes *et al.* 1985; Davidson *et al.* 1988) did contain *bcl*-2 transcripts. Thus, the change in cell lineage from B lymphocyte to myeloid was accompanied by the activation of the *bcl*-2 gene. In contrast, certain other oncogenes which are actively expressed in HAFTL-1 and its pre-B subclones, e.g., c-*myb*, are dramatically turned off (data not shown).

DISCUSSION AND CONCLUSIONS

These studies have confirmed that expression of the putative proto-oncogene bcl-2 in B lymphocytes is confined to cells in pre-B and mature B stages. Specifically, the data show that Ig-secreting cells, whether normal spleen cells activated by anti-δ antibodies or plasma cell tumors, contain little if any bcl-2 mRNA. To this extent the ontogenic regulation of expression of bcl-2 resembles that of c-myc. In this context it is important to note that both these oncogenes are susceptible to reciprocal chromosomal translocations involving the Ig heavy chain chromosome in neoplasms of mature B lymphocytes (Pegoraro et al. 1984; Mushinski et al. 1987). It is tempting to interpret these results mechanistically and to suggest that during B cell development there are periods characterized by DNA rearrangement and switching of Ig genes which coincide with high levels of expression of bcl-2 and c-myc. This makes these proto-oncogenes accessible and potentially susceptible to accidental aberrant rearrangements by the same enzymes that are normally involved in Ig gene shuffling, leading to chromosome translocations. Some aspect of these translocations, probably the constitutive expression of bcl-2 in t(14;18) or c-myc in t(8;14), contributes to neoplastic transformation, because normal B cell differentiation is accompanied by down regulation of these genes, which is now impossible. Although the contribution of c-myc deregulation is firmly established, the direct demonstration that bcl-2 can induce neoplasia is still lacking. What is more, the normal functions of the protein products of bcl-2 and c-myc have not been discovered. It is known that their site of localization within the cell is different: c-myc is nuclear and bcl-2 is found on the inner aspect of the plasma membrane (Tsujimoto et al. 1987).

Our efforts to learn more about the physiology of bcl-2 have utilized tumor models of progenitor B cells, HAFTL-1 and HAFTL-3. These data show that bcl-2 expression is tightly regulated between the pre-B and pro-monocyte cell lineages to which these progenitor cell lines can give rise. When HAFTL-1 or HAFTL-3 develop into myeloid cells, this process is accompanied by the activation of bcl-2 expression. Whether the bcl-2 protein plays any direct role in producing the shift into the myeloid pathway remains to be demonstrated. It is interesting, however, to note that c-myc and bcl-2 expression are uncoupled in this situation, since c-myc mRNA is abundant in both HAFTL-1, its B cell and its myeloid subclones (W.F. Davidson, in preparation).

REFERENCES

Bakhshi A, Jensen JP, Goldman P, Wright JJ, McBride OW, Epstein AL, Korsmeyer SJ (1985) Cloning the chromosomal breakpoint of t(14;18) human lymphomas: clustering around J_H on chromosome 14 and near a transcriptional unit on 18. Cell 41:899-906

Cleary ML, Sklar J (1985) Nucleotide sequence of a t(14;18) chromosomal breakpoint in follicular lymphoma and demonstration of a breakpoint-cluster region near a transcriptionally active locus on chromosome 18. Proc Natl Acad Sci USA 82:7439-7443

Davidson WF, Pierce JH, Rudikoff S, Morse HC III (1988) Relationships between B cell and myeloid differentiation: studies with a B lymphocyte progenitor line, HAFTL-1. J Exp Med (in press)

Finkelman FD, Scher I, Mond JJ, Kessler S, Kung JT, Metcalf ES (1982) Polyclonal activation of the murine immune system by an antibody to IgD. II. Generation of polyclonal antibody production and cells with surface IgG. J Immunol 129:638-647
Givol D (1987) Structural studies of mouse bcl-2 genomic and cDNA clones. In: Melchers F, Potter M (eds) Mechanisms of B Cell Neoplasia. Editiones Roches, Basel, pp 257-261
Graninger WB, Seto M, Bautain B, Goldman P, Korsmeyer S (1987) Expression of Bcl-2 and Bcl-2-Ig fusion transcripts in normal and neoplastic cells. J Clin Invest 80:1512-1515
Gurfinkel N, Unger T, Givol D, Mushinski JF (1987) Expression of the bcl-2 gene in mouse B lymphocytic cell lines is differentiation stage specific. Eur J Immunol 17:567-570
Holmes KL, Pierce JH, Davidson WF, Morse HC III (1986) Murine hematopoietic cells with pre-B or pro-B myeloid characteristics are generated by in vitro transformation with retroviruses containing fos, ras, abl, and src oncogenes. J Exp Med 164:443-457
Mock BA, Givol D, D'Hoostelaere LA, Huppi K, Selden MF, Gurfinkel N, Unger T, Potter M, Mushinski JF (1988) Mapping of the bcl-2 oncogene on mouse chromosome 1. Cytogenet Cell Genet 47:11-15
Mountz JD, Smith J, Snapper CM, Mushinski JF, Finkelman FD (1987) Polyclonal activation of the murine immune system by an antibody to IgD. VIII. Stimulation of IgD secretion. J Immunol 139:2172-2178
Mushinski JF (1987) Expression of bcl-2 in mouse B lymphomas. In: Melchers F, Potter M (eds) Mechanisms of B Cell Neoplasia 1987, Editiones Roches, Basel, Switzerland, pp 266-271
Mushinski JF, Davidson WF, Morse HC III (1987) Activation of cellular oncogenes in human and mouse leukemia-lymphomas: spontaneous and induced oncogene expression in murine B lymphocytic neoplasms. Cancer Invest 5:345-368
Negrini M, Stilini E, Kozak C, Tsujimoto Y, Croce CM (1987) Molecular analysis of mbcl-2: structure and expression of the murine gene homologous to the human gene involved in follicular lymphoma. Cell 49: 455-463
Reed JC, Tsujimoto Y, Alpers JD, Croce CM, Nowell PC (1987) Regulation of bcl-2 proto-oncogene expression during normal human lymphocyte proliferation. Science 236:1295-1299
Seto M, Jaeger U, Hockett RD, Graninger W, Bennett S, Goldman P, Korsmeyer SJ (1988) Alternative promoters and exons, somatic mutation and deregulation of the Bcl-2-Ig fusion gene in lymphoma. EMBO J 7:123-131
Tsujimoto Y, Finger LR, Yunis JJ, Nowell P, Croce CM (1984) Cloning of the chromosome breakpoint of neoplastic B cells with the t(14;18) chromosome translocation. Science 226:1097-1099
Tsujimoto Y, Cossman J, Jaffe E, Croce CM (1985) Involvement of the bcl-2 gene in human follicular lymphoma. Science 228:1440-1443
Tsujimoto Y, Croce CM (1986) Analysis of the structure, transcripts, and protein products of bcl-2, the gene involved in human follicular lymphoma. Proc Natl Acad Sci USA 83:5214-5218
Tsujimoto Y (1987) Structure and expression of human bcl-2. In: Melchers F, Potter M (eds) Mechanisms of B Cell Neoplasia 1987, Editiones Roches, Basel, Switzerland, pp 251-256
Tsujimoto Y, Ikegaki N, Croce CM (1987) Characterization of the protein product of bcl-2, the gene involved in human follicular lymphoma. Oncogene 2:3-7

Recent Progress on the Human *bcl*-2 Gene Involved in Follicular Lymphoma: Characterization of the Protein Products

Y. TSUJIMOTO and C. M. CROCE

INTRODUCTION

Chromosome translocations involving the immunoglobulin heavy chain gene are common in B cell malignancies (Yunis 1983). The t(14;18) translocation has been extensively studied by us and others for the last several years because this translocation is nearly always associated with follicular lymphoma, one of the most common human B cell malignancies (Fukuhara et al. 1979; Yunis et al. 1982). By molecular cloning of the breakpoint of the t(14;18) translocation, we and others identified a gene, bcl-2, at the breakpoint region (Tsujimoto et al. 1985a; Bakshi et al. 1985; Cleary and Sklar 1985). Since the t(14;18) translocation occurs within or in the close vicinity of the bcl-2 gene (Tsujimoto et al. 1985a,b; Tsujimoto and Croce 1986; Cleary et al. 1986), and since the steady-state level of bcl-2 mRNA is elevated by the translocation (Tsujimoto et al. 1985a), the bcl-2 gene is a strong candidate for the oncogene involved in follicular lymphomagenesis, analogous to the role of the c-myc gene in Burkitt's lymphomagenesis. The bcl-2 gene consists of two exons (Tsujimoto and Croce 1986; Cleary et al. 1986) and is transcribed into several species of mRNA by splicing and differential usage of the polyA site (Tsujimoto and Croce 1986). These mRNAs encode two different bcl-2 proteins, α and β, which are identical except at the carboxyl terminus (Tsujimoto and Croce 1986; Cleary et al. 1986).

Toward the goal of understanding the function of the bcl-2 gene, we have raised polyclonal antiserum against the bcl-2 protein using the β-gal-bcl-2 fusion protein expressed in E. coli (Tsujimoto et al. 1987). This antiserum immunoprecipitates the bcl-2 α protein, prepared in vitro from RNA transcribed from bcl-2 cDNA using T7 phage RNA polymerase in vitro (Tsujimoto and Croce 1986), and also the bcl-2 β protein (Tsujimoto, unpublished). This antiserum was used to detect the bcl-2 protein in human B cells. As shown in Fig. 1, a protein with the same size (about 26 Kd) as that of the in vitro translated bcl-2 α protein is precipitated from a pre-B cell line. The identity of this 26 Kd protein as the bcl-2 α protein was confirmed by V8 protease peptide mapping (Fig. 2).

The Wistar Institute, 3601 Spruce Street, Philadelphia, PA 19104

Current Topics in Microbiology and Immunology, Vol. 141

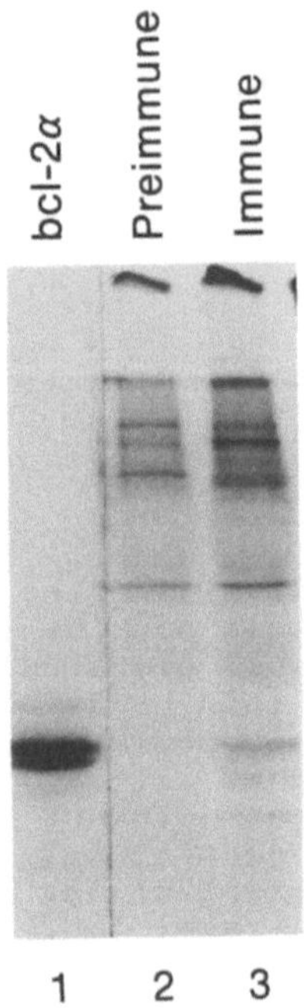

Fig. 1. Immunoprecipitation of the bcl-2 α protein from human B cells. Lane 1, [^{35}S]methionine-labeled bcl-2 α protein prepared by in vitro translation as described (Tsujimoto and Croce, 1986). Lanes 2 and 3, the proteins precipitated with pre-immune and immune serum, respectively. Cell labeling and immunoprecipitation were carried out as described (Tsujimoto et al. 1987).

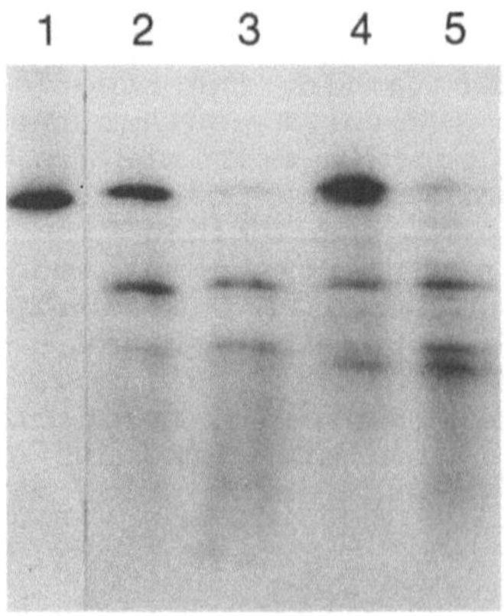

Fig. 2. Peptide mapping of bcl-2 α protein with V8 protease. Lane 1, in vitro translated bcl-2 α protein. Lanes 2 and 3, in vitro translated bcl-2 α protein digested with V8. Lanes 4 and 5, the 26 Kd protein immunoprecipitated from 697 cells and digested with V8. Fifty ng of V8 were used for lanes 2 and 4 and 500 ng for lanes 3 and 5. V8 peptide mapping was performed as described (Cleveland et al. 1977).

Subcellular fractionation of B cell proteins followed by immunoprecipitation showed that the bcl-2 α protein is recovered in the membrane fraction but not in the cytosolic or nuclear fractions (Fig. 3). Since the bcl-2 protein has no signal peptide or transmembrane domain (Tsujimoto and Croce 1986; Cleary et al. 1986), the bcl-2 protein is most probably located at the inner surface of the cytoplasmic membrane, raising the possibility that the bcl-2 protein is involved in signal transduction. This possibility, together with the observations that bcl-2 gene expression is silent in resting B and T cells and begins to be expressed 6 to 8 hr after mitogen stimulation (Reed et al. 1987), could, in turn, imply involvement of this protein in secondary activation, such as the IL2/IL2 receptor pathway, in T cells.

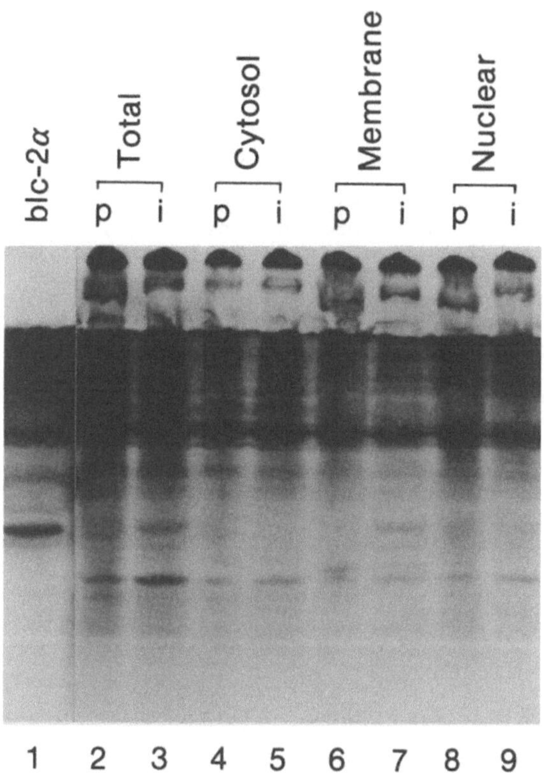

Fig. 3. Immunoprecipitation of the bcl-2 protein from subcellular fractions of human B-cells analyzed on 12.5% polyacrylamide/SDS gel. Each fraction (cytosolic, membrane and nuclear) was immunoprecipitated with pre-immune (lanes, 2, 4, 6, 8) and antiserum (lanes, 3, 5, 7, 9). The arrow indicates the bcl-2 α protein.

REFERENCES

Bakshi A, Jensen JP, Goldman P, Wright JJ, McBride OW, Epstein AL, Korsmeyer SJ (1985) Cloning the chromosomal breakpoint of t(14;18) human lymphoma: clustering around J_H on chromosome 14 and near a transcriptional unit on 18. Cell 4:899-906

Cleary ML, Sklar J (1985) Nucleotide sequence of a t(14;18) chromosomal breakpoint in follicular lymphoma and demonstration of a breakpoint cluster region near a transcriptionally active locus on chromosome 18. Proc Natl Acad Sci USA 82:7439-7443

Cleary ML, Smith SD, Sklar J (1986) Cloning and structural analysis of cDNAs for bcl-2 and hybrid bcl-2/immunoglobulin transcript resulting from the t(14;18) translocation. Cell 47:19-28

Cleveland DW, Fischer SG, Kirschner MW, Laemmli UK (1977) Peptide mapping by limited proteolysis in sodium dodecyl sulfate and analysis by gel electrophoresis. J Biol Chem 252:1102-1106

Fukuhara S, Rowley JD, Variakojis D, Golomb HM (1979) Chromosome abnormalities in poorly differentiated lymphocytic lymphoma. Cancer Res 39:3119-3128

Reed JC, Tsujimoto Y, Alpers JD, Croce CM, Nowell PC (1987) Regulation of bcl-2 proto-oncogene expression during normal human lymphocyte proliferation. Science 236:1295-1299

Tsujimoto Y, Croce CM (1986) Analysis of the structure, transcripts, and protein products of bcl-2, the gene involved in human follicular lymphoma. Proc Natl Acad Sci USA 83:5214-5218

Tsujimoto Y, Cossman J, Jaffe E, Croce CM (1985a) Involvement of the bcl-2 gene in human follicular lymphoma. Science 228:1440-1443

Tsujimoto Y, Gorham J, Cossman J, Jaffe E, Croce CM (1985b) The t(14;18) chromosome translocations involved in B-cell neoplasms result from mistakes in VDJ joining. Science 329:1390-1393

Tsujimoto Y, Ikegaki N, Croce CM (1987) Characterization of the protein product of bcl-2, the gene involved in human follicular lymphoma. Oncogene 2:3-7

Yunis JJ (1983) The chromosomal basis of human neoplasia. Science 221:227-236

Yunis JJ, Oken MM, Kaplan ME, Ensrud KM, Howe RB, Theoligides A (1982) Distinctive chromosomal abnormalities in histologic subtypes of non-Hodgkin's lymphoma. N Engl J Med 307:1231-1236

MIX
Papier aus verantwortungsvollen Quellen
Paper from responsible sources
FSC® C105338

If you have any concerns about our products,
you can contact us on
ProductSafety@springernature.com

In case Publisher is established outside the EU,
the EU authorized representative is:
Springer Nature Customer Service Center GmbH
Europaplatz 3, 69115 Heidelberg, Germany

Printed by Libri Plureos GmbH
in Hamburg, Germany